Toroidalization of Dominant Morphisms of 3-Folds

MEMOIRS
of the
American Mathematical Society

Number 890

Toroidalization of Dominant Morphisms of 3-Folds

Steven Dale Cutkosky

November 2007 • Volume 190 • Number 890 (end of volume) • ISSN 0065-9266

American Mathematical Society
Providence, Rhode Island

2000 *Mathematics Subject Classification.*
Primary 14E05, 14J30; Secondary 14B25, 14B05.

Library of Congress Cataloging-in-Publication Data

Cutkosky, Steven Dale.
Toroidalization of dominant morphisms of 3-folds / Steven Dale Cutkosky.
p. cm. — (Memoirs of the American Mathematical Society, ISSN 0065-9266 ; no. 890)
"Volume 190, number 890 (end of volume)."
Includes bibliographical references.
ISBN 978-0-8218-3998-0 (alk. paper)
1. Geometry, Algebraic. 2. Morphisms (Mathematics) 3. Algebraic varieties. 4. Commutative algebra. I. Title.

QA564.C88 2007
516′.11—dc22 2007060777

Memoirs of the American Mathematical Society

This journal is devoted entirely to research in pure and applied mathematics.

Subscription information. The 2007 subscription begins with volume 185 and consists of six mailings, each containing one or more numbers. Subscription prices for 2007 are US$649 list, US$519 institutional member. A late charge of 10% of the subscription price will be imposed on orders received from nonmembers after January 1 of the subscription year. Subscribers outside the United States and India must pay a postage surcharge of US$38; subscribers in India must pay a postage surcharge of US$43. Expedited delivery to destinations in North America US$53; elsewhere US$130. Each number may be ordered separately; *please specify number* when ordering an individual number. For prices and titles of recently released numbers, see the New Publications sections of the *Notices of the American Mathematical Society.*

Back number information. For back issues see the *AMS Catalog of Publications.*

Subscriptions and orders should be addressed to the American Mathematical Society, P. O. Box 845904, Boston, MA 02284-5904, USA. *All orders must be accompanied by payment.* Other correspondence should be addressed to 201 Charles Street, Providence, RI 02904-2294, USA.

Memoirs of the American Mathematical Society is published bimonthly (each volume consisting usually of more than one number) by the American Mathematical Society at 201 Charles Street, Providence, RI 02904-2294, USA. Periodicals postage paid at Providence, RI. Postmaster: Send address changes to Memoirs, American Mathematical Society, 201 Charles Street, Providence, RI 02904-2294, USA.

This publication is indexed in *Science Citation Index*®, *SciSearch*®, *Research Alert*®, *CompuMath Citation Index*®, *Current Contents*®/*Physical, Chemical & Earth Sciences.*
Printed in the United States of America.

♾ The paper used in this book is acid-free and falls within the guidelines established to ensure permanence and durability.
Visit the AMS home page at `http://www.ams.org/`

10 9 8 7 6 5 4 3 2 1 12 11 10 09 08 07

Contents

Abstract

This book contains a proof that a dominant morphism from a 3-fold X to a variety Y can be made toroidal by blowing up in the target and domain. We give applications to factorization of birational morphisms of 3-folds.

Received by the editor February 11, 2006.

1991 *Mathematics Subject Classification.* Primary 514E05, 14J30; Secondary 14B25, 14B05.

Key words and phrases. Toroidal, Monomial.

The first author was supported in part by NSF Grant DMS #0240819.

CHAPTER 1

Introduction

Suppose that $f : X \to Y$ is a dominant morphism of algebraic varieties, over a field $\mathbf{k}$ of characteristic zero. If X and Y are nonsingular, $f : X \to Y$ is toroidal if there are simple normal crossing (SNC) divisors D_X on X and D_Y on Y such that $f^{-1}(D_Y) = D_X$, and f is locally given by monomials in appropriate etale local parameters on X. The idea of toroidalization is developed in [**KKMS**]. A precise definition of this concept is in [**AK**], [**KKMS**] and Definition 4.3 of this paper. The problem of toroidalization is to determine, given a dominant morphism $f : X \to Y$, if there exists a commutative diagram

$$\begin{array}{ccc} X_1 & \overset{f_1}{\to} & Y_1 \\ \Phi \downarrow & & \downarrow \Psi \\ X & \overset{f}{\to} & Y \end{array} \tag{1.1}$$

such that Φ and Ψ are products of blow ups of nonsingular subvarieties, X_1 and Y_1 are nonsingular, and there exist simple normal crossing divisors D_{Y_1} on Y_1 and $D_{X_1} = f^{-1}(D_{Y_1})$ on X_1 such that f_1 is toroidal (with respect to D_{X_1} and D_{Y_1}). This is stated in Problem 6.2.1. of [**AKMW**].

In this paper we consider a stronger form of toroidalization which we will call strong toroidalization. Suppose that $f : X \to Y$ is a dominant morphism of nonsingular projective varieties, D_Y is a SNC divisor on Y and $D_X = f^{-1}(D_Y)$ is a SNC divisor on X such that the locus $\mathrm{sing}(f)$ where the morphism f is not smooth is contained in D_X. The problem of strong toroidalization is to determine if there exists a commutative diagram (1.1) such that Φ and Ψ are products of blow ups of nonsingular subvarieties which are supported in the preimages of D_X and D_Y respectively, and make SNCs with the respective preimages of D_X and D_Y, and f_1 is toroidal with respect to $D_{Y_1} = \Psi^{-1}(D_Y)$ and $D_{X_1} = \Phi^{-1}(D_X)$.

Toroidalization, and related concepts, have been considered earlier in different contexts, mostly for morphisms of surfaces. Strong torodialization is the strongest structure theorem which could be true for general morphisms. The concept of torodialization fails completely in positive characteristic. A simple counter example in characteristic $p > 0$ is obtained from the map of curves $s = t^p(1 + t)$.

In the case when Y is a curve, toroidalization follows from embedded resolution of singularities ([**H**]). When X and Y are surfaces, there are several proofs in print ([**AkK**], [**CP**], Corollary 6.2.3 [**AKMW**]). They all make use of special properties of the birational geometry of surfaces. An outline of proofs of the above cases can be found in the introduction to [**C3**].

In this paper, we prove strong toroidalization for dominant morphisms of 3-folds. Combining this with our theorem in [**C3**] of strong toroidalization in the case when X is a 3-fold and Y is a surface, we obtain the following theorem.

THEOREM 1.1. *Suppose that X is a 3-fold, and $f : X \to Y$ is a dominant morphism of nonsingular varieties over an algebraically closed field $\mathbf{k}$ of characteristic 0. Further suppose that there is a simple normal crossings (SNC) divisor D_Y on Y such that $D_X = f^{-1}(D_Y)$ is a SNC divisor which contains the non smooth locus of the map f. Then there exists a commutative diagram of morphisms*

$$\begin{array}{ccc} X_1 & \overset{f_1}{\to} & Y_1 \\ \Phi\downarrow & & \downarrow\Psi \\ X & \overset{f}{\to} & Y \end{array}$$

where Φ, Ψ are products of possible blow ups for the preimages of D_X and D_Y respectively, and f_1 is toroidal with respect to $D_{Y_1} = \Psi^{-1}(D_Y)$ and $D_{X_1} = \Phi^{-1}(D_X)$.

A variety over a field $\mathbf{k}$ is a quasi projective variety over $\mathbf{k}$. A 3-fold is a 3 dimensional variety.

A possible blow up on a nonsingular variety X with toroidal structure D_X is the blow up of a nonsingular subvariety V of D_X such that V makes simple normal crossings with D_X. That is, for all $q \in V$, local equations of the components of D_X containing q are part of a minimal set of generators of the ideal of V at q.

We also obtain the following strong toroidalization theorem for morphisms of (possibly singular) varieties.

THEOREM 1.2. *Suppose that X is a 3-fold and $f : X \to Y$ is a dominant morphism of 3-folds over an algebraically closed field $\mathbf{k}$ of characteristic 0. Further suppose that there is an equidimensional codimension 1 reduced subscheme D_Y of Y such that D_Y contains the singular locus of Y, and $D_X = f^{-1}(D_Y)$ contains the non smooth locus of the map f. Then there exists a commutative diagram of morphisms*

$$\begin{array}{ccc} X_1 & \overset{f_1}{\to} & Y_1 \\ \Phi\downarrow & & \downarrow\Psi \\ X & \overset{f}{\to} & Y \end{array}$$

where Φ, Ψ are products of blow ups of nonsingular curves and points supported above D_X and D_Y respectively, $D_{Y_1} = \Psi^{-1}(D_Y)$ is a simple normal crossings divisor on Y_1, $D_{X_1} = f_1^{-1}(D_{Y_1})$ is a simple normal crossings divisor on X_1 and f_1 is toroidal with respect to D_{Y_1} and D_{X_1}.

This paper is the synthesis of the e-prints [**C5**], [**C7**] and [**C8**]. [**C5**] proves Theorem 1.3 and proves Theorems 1.1 and 1.2 in the birational case. [**C8**] proves Theorems 1.1 and 1.2 in the case of a general morphism of 3-folds.

If we relax some of the restrictions in the definition of toroidalization, there are other constructions producing a toroidal morphism f_1, which are valid for arbitrary dimensions of X and Y. In [**AK**] it is shown that a diagram (1.1) can be constructed where Φ is weakened to being a modification (an arbitrary birational morphism).

In [**C1**], [**C2**] and [**C4**], it is shown that a diagram (1.1) can be constructed where Φ and Ψ are locally products of blow ups of nonsingular centers and f_1 is locally toroidal, but the morphisms Φ, Ψ and f_1 may not be separated. This construction is obtained by patching local solutions, at least one of which contains the center of any given valuation. The morphism f_1 is called "locally monomial".

The proof of toroidalization of morphisms of 3-folds to surfaces in [**C3**] breaks up into two parts: a reduction to prepared morphisms and then a proof of toroidalization of prepared morphisms from 3-folds to surfaces. This second step is extended to arbitrary prepared morphisms from n-folds to srufaces in [**CK**].

The first step in the proof of Theorems 1.1 and 1.2 is the following theorem, proven in Chapters 6 - 8.

THEOREM 1.3. *Suppose that $f : X \to Y$ is a dominant morphism of nonsingular projective 3-folds over an algebraically closed field $\mathbf{k}$ of characteristic zero, with toroidal structures determined by SNC divisors D_Y on Y and $D_X = f^{-1}(D_Y)$ on X such that D_X contains the singular locus of f. Then there exists a commutative diagram*

$$\begin{array}{ccc} X_1 & \stackrel{f_1}{\rightarrow} & Y_1 \\ \Phi\downarrow & & \downarrow\Psi \\ X & \stackrel{f}{\rightarrow} & Y \end{array}$$

such that Ψ and Φ are products of possible blow ups for the preimages of D_Y, D_X respectively, such that f_1 is prepared for $D_{Y_1} = \Psi^{-1}(D_Y)$ and $D_{X_1} = \Phi^{-1}(D_X)$.

A prepared morphism $f : X \to Y$ of 3-folds has the property that locally, a suitable projection of Y onto a surface S, when composed with f, is close to being a toroidal morphism from the 3-fold X to the surface S. This theorem uses the theorem of toroidalization of morphisms of 3-folds to surfaces in [**C3**] to simplify the morphism f.

It has been shown in [**AKMW**] that weak factorization of birational morphisms holds in characteristic zero, and arbitrary dimension. That is, birational morphisms of complete varieties can be factored by an alternating sequence of blow ups and blow downs of non singular subvarieties. Weak factorization of birational (toric) morphisms of toric varieties, (and of birational toroidal morphisms) has been proven by Danilov [**D1**] and Ewald [**E**] (for 3-folds), and by Wlodarczyk [**W1**], Morelli [**Mo**] and Abramovich, Matsuki and Rashid [**AMR**] in general dimensions.

Our Theorem 1.2, when combined with weak factorization for toroidal morphisms ([**AMR**]), gives a new proof of weak factorization of birational morphisms of 3-folds. We point out that our proof uses an analysis of the structure of power series of local germs of a mapping, as opposed to the entirely different proof of weak factorization, using geometric invariant theory, of [**AKMW**].

COROLLARY 1.4. Suppose that $f : X \to Y$ is a birational morphism of 3-folds which are proper over an algebraically closed field $\mathbf{k}$ of characteristic zero. Then there exists a commutative diagram of morphisms factoring f,

$$\begin{array}{ccccccccccccc} & & X_1 & & & & X_3 & & & & X_n & & \\ & \swarrow & & \searrow & & \swarrow & & \searrow & \cdots & \swarrow & & \searrow & \\ X & & & & X_2 & & & & X_4 & & & & Y \end{array}$$

where each arrow is a product of blow ups of points and nonsingular curves.

The problem of strong factorization, as proposed by Abhyankar [**Ab2**] and Hironaka [**H**], is to factor a birational morphism $f : X \to Y$ by constructing a diagram

$$\begin{array}{ccc} & Z & \\ \swarrow & & \searrow \\ X & \stackrel{f}{\rightarrow} & Y \end{array}$$

where $Z \to X$ and $Z \to Y$ factor as products of blow ups of nonsingular subvarieties. Oda [**O**] has proposed the analogous problem for (toric) morphisms of toric varieties.

A birational morphism $f : S \to Y$ of (nonsingular) surfaces can be directly factored by blowing up points (Zariski [**Z1**] and Abhyankar [**Ab1**]), but there are examples showing that a direct factorization is not possible in general for 3-folds (Shannon [**Sh**] and Sally[**S**]).

We also obtain as an immediate corollary to Theorem 1.2 the following new result, which reduces the problem of strong factorization of 3-folds to the case of toroidal morphisms.

COROLLARY 1.5. Suppose that the Oda conjecture on strong factorization of birational toroidal morphisms of 3-folds is true. Then the Abhyankar, Hironaka strong factorization conjecture of birational morphisms of complete (characteristic zero) 3-folds is true.

Abhyankar's local factorization conjecture [**Ab2**], which is "strong factorization" along a valuation, follows from local monomialization (Theorem A [**C1**]), to reduce to a locally toroidal morphism, and local factorization for toroidal morphisms along a valuation Christensen [**Ch**] (for 3-folds), and Karu [**K**] or [**CS**] in general dimensions.

CHAPTER 2

An outline of the proof

Suppose that $f : X \to Y$ is a dominant morphism of nonsingular projective 3-folds, over an algebraically closed field $\mathbf{k}$ of characteristic zero, and suppose that D_Y is a simple normal crossings divisor on Y such that $D_X = f^{-1}(D_Y)$ is also a simple normal crossings divisor, defining toroidal structures on X and Y. Further suppose that the locus of points in X where f is not smooth is contained in D_X. We will refer to points where three components of D_Y (or D_X) intersect as 3-point, points where 2-components intersect as 2-points, and the remaining points of D_Y (and D_X) as 1-points.

Permissible parameters are defined in Chapter 4. A set of regular parameters in $\mathcal{O}_{Y,q}$ (or $\hat{\mathcal{O}}_{Y,q}$) are permissible parameters at q if the product of the first of these parameters is a local equation of D_Y at q.

There is a natural stratification of Y into locally closed subsets such that f is increasingly complex in higher strata.

For $q \in Y - D_Y$, we have that f is smooth above q, so that (since f is generically finite) if $p \in f^{-1}(q)$, and u, v, w are regular parameters at q, then u, v, w are regular parameters at p.

The non finite locus of f is

$$NF_f = \{q \in Y \mid \dim f^{-1}(q) > 0\}.$$

NF_f is a closed subset of D_Y and has codimension ≥ 1 in D_Y (codimension ≥ 2 in Y). If $q \in D_Y - NF_f$, then f is finite above q, and by Abhyankar's Lemma ([**Ab4**] and XIII 5.3 [**G**]), there exist permissible parameters u, v, w at q such that if $p \in f^{-1}(q)$, there exist permissible parameters x, y, z at p such that one of the following forms hold:

q a 1-point, p a 1-point

$$u = x^a, v = y, w = z \tag{2.1}$$

q a 2-point, p a 2-point

$$u = x^a, v = y^b, w = z \tag{2.2}$$

q a 3-point, p a 3-point

$$u = x^a, v = y^b, w = z^c. \tag{2.3}$$

If q is a general point of a 1-dimensional component of NF_f, then f behaves as a dominant morphism of nonsingular surfaces (defined over a non algebraically closed field) at q. Toroidalization of maps of surfaces over non closed fields of characteristic 0 is constructed in [**CP**] and Corollary 6.2.3 [**AKMW**]. We conclude that there exists a neighborhood V of q in Y such that $NF_f \cap V$ is a nonsingular curve, and there exists a commutative diagram

$$\begin{array}{ccc} & W_1 & \xrightarrow{\overline{f}_1} & V_1 \\ & \Phi_1 \downarrow & & \downarrow \Psi_1 \\ W = f^{-1}(V) & & \xrightarrow{f} & V \end{array}$$

such that Φ_1, Ψ_1 are products of blow ups of nonsingular curves which dominate C, and for $q_1 \in D_{V_1} = \Psi_1^{-1}(D_V)$, there exist permissible parameters $u, v, w \in \mathcal{O}_{V_1,q_1}$ such that if $p \in \overline{f}_1^{-1}(q_1)$, there exist formal permissible parameters x, y, z at p such that one of the following local forms hold:

q_1 is a 1-point, p_1 is a 1-point

$$u = x^a, v = y, w = z \tag{2.4}$$

q_1 is a 2-point, p_1 is a 1-point

$$u = x^a, v = x^b(\alpha + y), w = z \tag{2.5}$$

with $0 \neq \alpha \in \mathbf{k}$.

q_1 is a 2-point, p_1 is a 2-point

$$u = x^a y^b, v = x^c y^d, w = z. \tag{2.6}$$

The final strata is a finite set of points contained in NF_f. The map f is extremely complex above these points and defies a simple analysis.

Our goal is to construct a commutative diagram

$$\begin{array}{ccc} X_1 & \xrightarrow{f_1} & Y_1 \\ \Phi_1 \downarrow & & \downarrow \Psi_1 \\ X & \xrightarrow{f} & Y \end{array}$$

such that Φ_1, Ψ_1 are products of blow ups of nonsingular curves and points which are possible blow ups (Chapter 3) such that for all $q \in D_{Y_1} = \Psi_1^{-1}(D_Y)$, there are permissible parameters at q such that for all $p \in f_1^{-1}(q)$, u, v, w are toroidal forms at p. That is, there exist permissible parameters $x, y, z \in \hat{\mathcal{O}}_{X_1,p}$ such that one of the forms 1 - 6 after Definition 4.3. hold. Such a map f_1 is called toroidal (Definition 4.3).

Observe that the forms (2.1) - (2.6) above are toroidal forms.

In this paper, we develop a series of algorithms to manipulate local germs of mappings, expressed in terms of series and polynomials. We are required to blow up in both the domain and target.

The main result of resolution of singularities [**H**] tells us that we can construct a diagram

$$\begin{array}{ccc} X_1 & & \\ \downarrow & \searrow \overline{f} & \\ X & \xrightarrow{f} & Y \end{array}$$

where for all points $p \in X_1$ and $q = \overline{f}(p) \in Y$ there are permissible parameters x, y, z at p and u, v, w at q, such that we have an expression

$$\begin{array}{ll} u & = x^a y^b z^c \gamma_1 \\ v & = x^d y^e z^f \gamma_2 \\ w & = x^g y^h z^i \gamma_3 \end{array} \tag{2.7}$$

where $\gamma_1, \gamma_2, \gamma_3$ are units at p. The algorithms of resolution give us no information about the structure of the units $\gamma_1, \gamma_2, \gamma_3$. In general, we will have that the monomials $x^a y^b z^c, x^d y^e z^f, x^g y^h z^i$ are algebraically dependent. We may then hope to blow up nonsingular subvarieties of Y (and blow up above X_1 to resolve the indeterminancy of the resulting rational map), leading to new regular parameters at a point q' above q with regular parameters u', v', w' which are obtained by dividing u, v, w by each other. We will eventually obtain an expression such as $w' = \gamma_3 - \gamma_3(p)$ by this procedure, which need not be any better than the expressions we started with for our original map $f : X \rightarrow Y$.

In attempting to obtain a sufficiently deep understanding of map germs to prove toroidalization, we must understand information about germs such as (2.7), up to the level of a series expansion of the units γ_i. In considering this problem, we are led to generalized notions of multiplicity, which can actually increase after making blow ups above X (in the proof of toroidalization of morphisms of 3-folds to surfaces [**C3**]). In summary, the problem of toroidalization requires new methods, which are not contained in any proofs of resolution of singularities.

In Definition 4.6 we define a prepared morphism $f : X \rightarrow Y$ of 3-folds. This is related to the notion of a prepared morphism from a 3-fold to a surface (Definition 6.5 [**C3**]). The basic idea of a prepared morphism $f : X \rightarrow Y$ of 3-folds is that locally an appropriate projection of Y onto a surface S, when composed with f, is (close to) a toroidal morphism from a 3-fold to a surface. If f is prepared, by a simple local calculation involving Jacobian determinants, we have expressions of the form of Definition 4.1 and Lemma 4.2 (or 2 (b), 2 (c) of Definition 4.6) at all points of X. A typical case, when p is a 1-point and q is a 2-point, is

$$u = x^a, v = x^b(\alpha + y), w = g(x, y) + x^c z$$

where $0 \neq \alpha \in \mathbf{k}$ and $g(x, y)$ is a series.

Construction of a prepared morphism

In Chapters 6 - 8 we prove Theorem 1.3. It is shown that we can construct from our given morphism $f : X \rightarrow Y$ a commutative diagram

$$\begin{array}{ccc} X_1 & \xrightarrow{f_1} & Y_1 \\ \Phi \downarrow & & \downarrow \Psi \\ X & \xrightarrow{f} & Y \end{array}$$

where Φ, Ψ are products of possible blow ups of nonsingular curves and points, and f_1 is prepared with respect to the SNC divisors $D_{X_1} = \Phi^{-1}(D_X)$ and $D_{Y_1} = \Psi^{-1}(D_Y)$.

The essential ingredient of the proof is our theorem on toroidalization of dominant morphisms from 3-folds to surfaces (Theorem 19.11 [**C3**]).

Let NF_f be the non finite locus of f. Above points of $D_Y - NF_f$, we see by (2.1) - (2.3) that f is prepared.

Above general points of one dimensional components of NF_f, we can also easily show that f is prepared (but in general is not toroidal).

Let $\Lambda \subset D_Y$ be the finite set of points above which f is not prepared. By blowing up 2-curves and 3-point above Y, we can assume that Λ is a finite set of 1-points in Y (Lemmas 7.1 and 7.2).

Suppose that $q \in \Lambda$. We can choose permissible parameters u, v, w at q. Let V be a neighborhood of q in Y such that u, v, w are uniformizing parameters ($\lambda : V \rightarrow \text{spec}(\mathbf{k}[u, v, w])$ is etale).

Let $\pi : \text{spec}(\mathbf{k}[u, v, w]) \rightarrow S = \text{spec}(\mathbf{k}[u, v])$ be the projection. We apply Theorem 6.1 to the map $\pi \circ f$ to construct a commutative diagram

$$\begin{array}{ccc} W_1 & \rightarrow & S_1 \\ \Phi \downarrow & & \downarrow \Psi \\ f^{-1}(V) & \rightarrow & S \end{array} \tag{2.8}$$

such that $W_1 \rightarrow S_1$ is toroidal.

Now Ψ is a sequence of blow ups of points which induce a sequence of blow ups of curves $Y_1 \rightarrow Y$ such that there is a factorization $W_1 \rightarrow Y_1$ which is prepared. The curves blown up above Y dominate the curve C with local equations $u = v = 0$ at q. We can choose u, v so that C is a general curve through q on D_Y. With a bit of care, we can construct the diagrams (2.8) so that they patch to a projective morphism

$$\begin{array}{ccc} Y_1 & \stackrel{f_1}{\rightarrow} & Y_1 \\ \downarrow & & \downarrow \\ X & \stackrel{f}{\rightarrow} & Y \end{array}$$

such that f_1 is prepared.

The τ-invariant

It may appear that the local forms of a prepared morphism are very simple, and we can easily modify them to obtain a toroidal form. However, a little exploration with these local forms will reveal that the notion of being prepared is stable under blow ups of 2-curves, but is in general not stable under blow ups of points and curves which are not 2-curves. It also will quickly become apparent that it is absolutely necessary to blow up subvarieties of Y other than 2-curves to toroidalize. This leads to the notion of super parameters (Definition 10.1), which is necessary for all the blow ups which we consider to preserve the notion of being prepared, and for a global invariant τ to behave well under blow ups.

To prove Theorem 1.2, we may assume that f is prepared. This condition is preserved throughout the proof.

We define the τ-invariant of a point $p \in D_X$ in Definition 9.1. $\tau_f(p) = -\infty$ if f is toroidal at p, and $\tau_f(p)$ is a positive integer otherwise.

The simplest case is when p is a 1-point and $q = f(p)$ is a 1-point. We have permissible parameters u, v, w at q and x, y, z at p such that

$$u = x^a, v = y, w = \sum_{i<c} \alpha_i(y)x^i + x^c(z + \beta)$$

with $\beta \in \mathbf{k}$. Here $u = 0$ is a local equation of D_Y and $x = 0$ is a local equation of D_X. f is toroidal at p if and only if $c = 0$. In this case we define $\tau_f(p) = -\infty$.

If $c > 0$, we define (as follows (9.11))

$$\tau_f(p) = |(a\mathbf{Z} + \sum_{\alpha_i \neq 0} i\mathbf{Z})/a\mathbf{Z}|.$$

The most complicated case is when p is a 3-point. Since f is prepared, $f(p) = q$ is a 2-point or a 3-point. There are permissible parameters u, v, w in $\mathcal{O}_{Y,q}$ and x, y, z

in $\hat{\mathcal{O}}_{X,p}$ ($xyz = 0$ is a local equation of D_X, $uv = 0$ or $uvw = 0$ is a local equation of D_Y) and there is an expression (9.1) of Definition 9.1

$$(2.9)\qquad \begin{aligned} u &= x^a y^b z^c \\ v &= x^d y^e z^f \\ w &= \textstyle\sum_{i \geq 0} \alpha_i M_i + N \end{aligned}$$

with $0 \neq \alpha_i \in \mathbf{k}$, $M_i = x^{a_i} y^{b_i} z^{c_i}$, $N = x^g y^h z^i$, $N \not| M_i$ for all i,

$$\text{rank}\begin{pmatrix} a & b & c \\ d & e & f \end{pmatrix} = 2, \det\begin{pmatrix} a & b & c \\ d & e & f \\ a_i & b_i & c_i \end{pmatrix} = 0 \text{ for all } i,$$

$$\det\begin{pmatrix} a & b & c \\ d & e & f \\ g & h & i \end{pmatrix} \neq 0.$$

f is toroidal at p (and $\tau_f(p) = -\infty$) if and only if q is a 3-point and

$$w = \gamma N$$

where γ is a unit in $\hat{\mathcal{O}}_{X,p}$.

Otherwise, define (as follows (9.1) of Definition 9.1)

$$H_p = (a, b, c)\mathbf{Z} + (d, e, f)\mathbf{Z} + \sum_i (a_i, b_i, c_i)\mathbf{Z},$$

$$A_p = \begin{cases} (a, b, c)\mathbf{Z} + (d, e, f)\mathbf{Z} + (a_0, b_0, c_0)\mathbf{Z} & \text{if } q \text{ is a 3-point (we have } w = \gamma M_0 \\ & \quad \text{where } \gamma \text{ is a unit in } \hat{\mathcal{O}}_{X,p}) \\ \mathbf{Z}(a, b, c)\mathbf{Z} + (d, e, f)\mathbf{Z} & \text{if } q \text{ is a 2-point.} \end{cases}$$

Now define

$$(2.10)\qquad \tau_f(p) = |H_p/A_p|.$$

τ_f is well defined (Lemma 9.2) and upper semi-continuous (Lemma 9.3).

We define (after Definition 9.1)

$$\tau_f(X) = \max\{\tau_f(p) \mid p \in D_X\}.$$

We have that $\tau_f(X) \geq 1$ or $\tau_f(X) = -\infty$. f is toroidal if and only if $\tau_f(X) = -\infty$.

The proof of Theorem 1.2 is by descending induction on $\tau_f(X)$. In our proof of Theorem 1.2 we may thus assume that $\tau = \tau_f(X) \neq -\infty$.

Let (Definition 9.4))

$$G_X(f, \tau) = \{p \in D_X \mid \tau_f(p) = \tau\}.$$

$G_X(f, \tau)$ is Zariski closed in X (since τ_f is upper semi continuous). Let (Definition 9.4))

$$G_Y(f, \tau) = f(G_X(f, \tau)).$$

$G_Y(f, \tau)$ is Zariski closed in Y since f is proper.

By Theorem 14.4, there exist sequences of blow ups of 2-curves

$$\begin{array}{ccc} X_1 & \xrightarrow{f_1} & Y_1 \\ \downarrow & & \downarrow \\ X & \xrightarrow{f} & Y \end{array}$$

such that f_1 is τ-prepared (Definition 9.4). That is, f_1 is prepared, and $G_{Y_1}(f_1, \tau)$ contains no 2-curves or 3-points.

From now on, we assume that $f: X \to Y$ is τ-prepared and $\tau > 1$. The case $\tau = 1$ is actually the simplest, but is a little different from the general case $\tau > 1$.

Relations, and τ-quasi-well prepared morphisms

A τ-quasi-well prepared morphism $f: X \to Y$ is a τ-prepared morphism with a relation R for f which has certain nice properties.

A Relation R (Definition 12.5) consists of a finite number of primitive relations R_i for f. To each primitive relation R_i is associated a pre-relation $\overline{R}_i$ on Y.

A pre-relation $\overline{R}_i$ (Definitions 12.1, 12.3) consists of a locally closed subset $U(\overline{R}_i)$ of D_Y consisting of 1-points and a finite number of 2-points. For each $q \in U(\overline{R}_i)$ we have permissible parameters $u = u_{\overline{R}_i(q)}, v = v_{\overline{R}_i(q)}, w = w_{\overline{R}_i(q)}$ at q such that $u_{\overline{R}_i(q)}, v_{\overline{R}_i(q)} \in \mathcal{O}_{Y,q}$.

If q is a 2-point, we have $a = a_{\overline{R}_i}(q), b = b_{\overline{R}_i}(q), e = e_{\overline{R}_i}(q) \in \mathbf{Z}$ and $0 \neq \lambda = \lambda_{\overline{R}_i}(q) \in \mathbf{k}$ such that $e > 1$, and $\gcd(a, b, e) = 1$.

If q is a 1-point, we have $a = a_{\overline{R}_i}(q), e = e_{\overline{R}_i}(q) \in \mathbf{Z}$ and $0 \neq \lambda = \lambda_{\overline{R}_i}(q) \in \mathbf{k}$ such that $e > 1$, and $\gcd(a, e) = 1$.

We can formally represent the relation $\overline{R}_i$ by the (meromorphic) form

$$w^e - \lambda u^a v^b = 0 \tag{2.11}$$

if q is a 2-point,

$$w^e - \lambda u^a = 0 \tag{2.12}$$

if q is a 1-point.

We will only allow morphisms $\Psi : Y_1 \to Y$ in our construction which are the blow up of an admissible center V for $\overline{R}_i$ (Definition 12.2). V must be a possible center, and if $q \in U(\overline{R}_i)$, then some subset of our permissible parameters $u = u_{\overline{R}_i(q)}, v = v_{\overline{R}_i(q)}, w = w_{\overline{R}_i(q)}$ are local equations of V at q.

For instance, if q is a 1-point, the possible local equations for V at q are $u = v = w = 0$ (the point q) and $u = w = 0$ (the curve which is the intersection of the (possibly formal) surface $w = 0$ with the component E of D_Y containing q). Of course this curve germ must be algebraic for V to be an admissible center.

Suppose that $\Psi : Y_1 \to Y$ is the blow up of an admissible center. Then we define (after Definition 12.2) the transform $\overline{R}_i^1$ of the pre-relation $\overline{R}_i$ on Y_1. For $q \in U(\overline{R}_i)$, we define $U(\overline{R}_i^1) \cap \Psi^{-1}(q)$ to be the points of $\Psi^{-1}(q)$ which lie on the strict transform of the surface germ $w_{\overline{R}_i(q)} = 0$.

For instance, suppose that Ψ_1 is the blow up of a curve V, and $q \in V \cap U(\overline{R}_i)$ is a 1-point. Then we must have that $u_{\overline{R}_i(q)} = w_{\overline{R}_i(q)} = 0$ are local equations of V at q. The only point $q_1 \in \Psi^{-1}(q)$ which is on the strict transform of $w_{\overline{R}_i(q)} = 0$ has permissible parameters defined by

$$u_{\overline{R}_i(q)} = u_{\overline{R}_i^1(q_1)}, v_{\overline{R}_i(q)} = v_{\overline{R}_i^1(q_1)}, w_{\overline{R}_i(q)} = u_{\overline{R}_i^1(q_1)} w_{\overline{R}_i^1(q_1)}.$$

The strict transform of the form $w^e_{\overline{R}_i(q)} - \lambda u^a_{\overline{R}_i(q)} = 0$ is the meromorphic form

$$w^e_{\overline{R}_i^1(q_1)} - \lambda u^{a-e}_{\overline{R}_i^1(q_1)} = 0.$$

We have $e_{\overline{R}_i^1(q_1)} = e$, $a_{\overline{R}_i^1(q_1)} = a - e$.

A primitive relation R_i for f (Definition 12.5) associated to the pre-relation $\overline{R}_i$ on Y consists of a locally closed subset $T(R_i)$ of $f^{-1}(U(\overline{R}_i)) \cap G_X(f, \tau)$.

Suppose that $p \in T(R_i)$, and $q = f(p)$. We must have that $u = u_{\overline{R}_i(q)}, v = v_{\overline{R}_i(q)}$ are toroidal forms at p, and there exists an expression

$$w^e_{\overline{R}_i(q)} = u^a v^b \overline{\Lambda}$$

if q is a 2-point,

$$w^e_{\overline{R}_i(q)} = u^a \overline{\Lambda}$$

if q is a 1-point, where $\overline{\Lambda}$ is a unit in $\hat{\mathcal{O}}_{X,p}$, with $\overline{\Lambda}(p) = \lambda$.

When we construct a τ-quasi-well prepared morphism, we must further insist that $u_{\overline{R}_i(q)}, v_{\overline{R}_i(q)}, w_{\overline{R}_i(q)}$ do not satisfy a relation at p with a smaller value of e (3 of Definition 13.1).

Now suppose that we have a commutative diagram

$$\begin{array}{ccc} X_1 & \stackrel{f_1}{\to} & Y_1 \\ \Phi\downarrow & & \downarrow\Psi \\ X & \stackrel{f}{\to} & Y \end{array}$$

where Ψ is an admissible blow up, f_1 is τ-prepared, and for $p \in D_{X_1}$, $\tau_{f_1}(p) \le \tau_f(\Phi(p))$. Let $\overline{R}^1_i$ be the transform of $\overline{R}_i$ on Y_1.

Suppose that for

$$p \in f_1^{-1}(U(\overline{R}^1_i)) \cap \Phi^{-1}(T(R_i)) \cap G_{X_1}(f_1, \tau),$$

we have that $u_{\overline{R}^1_i}, v_{\overline{R}^1_i}$ are toroidal forms at p. Then we can define a primitive relation R^1_i on X_1 such that

$$T(\overline{R}^1_i) = f_1^{-1}(U(\overline{R}^1_i)) \cap \Phi^{-1}(T(R_i)) \cap G_{X_1}(f_1, \tau).$$

The resulting relation R^1 is called the transform of R (Definition 12.6).

A τ-prepared morphism $f : X \to Y$ is τ-quasi-well prepared if there is a relation R for f such that the sets $T(R_i)$ are a partition of $G_X(f, \tau)$ (Definition 13.1).

The basic building block for the construction of a toroidalization will be a τ-quasi-well prepared diagram

$$\begin{array}{ccc} X_1 & \stackrel{f_1}{\to} & Y_1 \\ \Phi\downarrow & & \downarrow\Psi \\ X & \stackrel{f}{\to} & Y \end{array}$$

consisting of a commutative diagram of morphisms such that Ψ is an admissible blow up, f_1 is τ-prepared, for $p \in D_{X_1}$, $\tau_{f_1}(p) \le \tau_f(\Phi(p))$, the transform R^1 of R is defined for f_1, and f_1 is τ-quasi-well prepared for this relation. The necessary diagrams of this type are constructed in Chapter 14.

To get an idea of how this works in the simplest case, we will construct such a diagram above a germ of a 1-point $p \in T(R_i)$ which maps to a 1-point $q \in U(\overline{R}_i)$. Let

$$u = u_{\overline{R}_i(q)}, v = v_{\overline{R}_i(q)}, w = w_{\overline{R}_i(q)},$$

and suppose that the admissible center V with local equations $u = w = 0$ is blown up by Ψ. This case is worked out in detail in the proof of Lemma 14.11.

Points q_1 of $\Psi^{-1}(q)$ have local equations u_1, v_1, w_1 which are defined by one of the following cases:

q_1 is a 1-point, with

$$u = u_1, v = v_1, w = u_1(w_1 + \alpha) \tag{2.13}$$

for some $\alpha \in \mathbf{k}$, or q_1 is a 2-point,

$$u = u_1 w_1, v = v_1, w = w_1. \tag{2.14}$$

Now we must have permissible parameters x, y, z at p for u, v, w defined by

$$u = x^a, v = y, w = \sum_{j=0}^{m} a_{i_j}(y) x^{i_j} + x^n(z + \beta).$$

where $i_m < n$, $a_{i_j}(y) \neq 0$ for all j and $\beta \in \mathbf{k}$. We have that $w^{e_{\overline{R}_i}(q)} = u^{a_{\overline{R}_i}(q)}\Lambda$, where Λ is a unit in $\hat{\mathcal{O}}_{X,p}$ such that $\Lambda(p) = \lambda_{\overline{R}_i}(q)$. Thus we have that $a_{i_0}(0) \neq 0$, and $a \not| i_0$ (since $e_{\overline{R}_i}(q) > 1$ and $\gcd(a_{\overline{R}_i}(q), e_{\overline{R}_i}(q)) = 1$). Further, the rational map $X \to Y_1$ is defined at p, so we take $X_1 = X$ in a neighborhood of p.

First suppose that $a \leq i_0$. Since $a \not| i_0$, we have $a < i_0$. $q_1 = f_1(p)$ thus has permissible parameters u_1, v_1, w_1 satisfying (2.13) with $\alpha = 0$. Thus $q_1 \in U(\overline{R}_i^1)$. We have

$$u_1 = x^a, v_1 = y, w_1 = \sum_{j=0}^{m} a_{i_j}(y) x^{i_j - a} + x^{n-a}(z + \beta).$$

We have

$$w_1^{e_{\overline{R}_i}(q)} = u_1^{a_{\overline{R}_i}(q) - e_{\overline{R}_i}(q)}\Lambda$$

where $\Lambda(p) = \lambda_{\overline{R}_i}(q)$. We compute

$$\begin{aligned} \tau_{f_1}(p) &= |a\mathbf{Z} + \textstyle\sum_{j=0}^{m}(i_j - a)\mathbf{Z}/a\mathbf{Z}| \\ &= |a\mathbf{Z} + \textstyle\sum_{j=0}^{m} i_j\mathbf{Z}/a\mathbf{Z}| \\ &= \tau_f(p) = \tau. \end{aligned}$$

Thus $p \in T(R_i^1)$.

Now suppose that $a > i_0$. We have an expression $w = x^{i_0}\gamma$, where γ is the unit series

$$\gamma = a_{i_0}(y) + a_{i_1}(y)x^{i_1 - i_0} + \cdots + a_{i_m}(y)x^{i_m - i_0} + x^{n-i_0}(\beta + z).$$

Let $\overline{x} = x\gamma^{\frac{1}{i_0}}$. We have expressions (by the type of calculation worked out in Chapter 9)

$$\begin{aligned} u &= \overline{x}^a \gamma^{-\frac{a}{i_0}} = \overline{x}^a(P(y, \overline{x}^{i_1 - i_0}, \ldots, \overline{x}^{i_m - i_0}) + \overline{x}^{n-i_0}(\overline{\alpha} + \overline{z})) \\ v &= y \\ w &= \overline{x}^{i_0} \end{aligned}$$

where $\overline{x}, y, \overline{z}$ are regular parameters in $\hat{\mathcal{O}}_{X,p}$, P is a unit series, and $\overline{\alpha} \in \mathbf{k}$.

$q_1 = f_1(p)$ has permissible parameters u_1, v_1, w_1 satisfying (2.14). Thus $q \notin U(\overline{R}_i^1)$. w_1, u_1, v_1 are permissible parameters at the 2-point q_1, and we have

$$\begin{aligned} w_1 &= \overline{x}^{i_0} \\ u_1 &= \overline{x}^{a-i_0}(P(y, \overline{x}^{i_1 - i_0}, \ldots, \overline{x}^{i_m - i_0}) + \overline{x}^{n-i_0}(\overline{\alpha} + \overline{z})) \\ v_1 &= y. \end{aligned} \tag{2.15}$$

This is case 2 (b) of Definition 4.6 of a local form of a prepared morphism.

We compute $\tau_{f_1}(p)$ from (9.10) of Definition 9.1. We have

$$\begin{aligned}\tau_{f_1}(p) &\le |(i_0\mathbf{Z}+\textstyle\sum_{j=0}^m(i_j-i_0+a-i_0)\mathbf{Z})/(i_0\mathbf{Z}+(a-i_0)\mathbf{Z})|\\ &= |(a\mathbf{Z}+\textstyle\sum_{j=0}^m i_j\mathbf{Z})/(i_0\mathbf{Z}+a\mathbf{Z})|\\ &< |(a\mathbf{Z}+\textstyle\sum_{j=0}^m i_j\mathbf{Z})/a\mathbf{Z}| = \tau_f(p)=\tau\end{aligned}$$

The strict inequality in line 3 above follows from the condition $e_{\overline{R}_i}(q)>1$, which implies that $a \not| i_0$.

The construction of a τ-quasi-well prepared morphism is rather complicated. This is accomplished in Theorem 15.4. One case where the basic idea is fairly transparent is when a 1 point p with $\tau_f(p)=\tau$ maps to a 1 point q. We start with permissible parameters $u,v,w\in\mathcal{O}_{Y,q}$ such that there are permissible parameters x,y,z in $\hat{\mathcal{O}}_{X,p}$ such that there is an expression

$$u=x^a, v=y, w=\sum_{j=0}^m a_{i_j}(y)x^{i_j}+x^n(\alpha+z),$$

where $a_{i_j}(y)\neq 0$ for all j, $i_m<n$ and $\alpha\in\mathbf{k}$. Let $\phi(u)=\sum_{a|i_j}a_{i_j}(v)u^{\frac{i_j}{a}}$. We define a pre-relation $\overline{R}$ at q by setting

$$u_{\overline{R}_i(q)}=u, v_{\overline{R}_i(q)}=v, w_{\overline{R}_i(q)}=w-\phi(u).$$

Since we are assuming $\tau>1$, there exists j_1 such that $a \not| j_1$ and

$$w_{\overline{R}_i(q)}=x^{j_1}\Lambda$$

where $\Lambda\in\hat{\mathcal{O}}_{X,p}$ is such that $x \not| \Lambda$. Part of the difficulty is to arrange things so that Λ is actually a unit in $\hat{\mathcal{O}}_{X,p}$. Assume that this is the case here. Let $d=\gcd(a,j_1)$. Let $e=\frac{a}{d}>1$, $\overline{a}=\frac{j_1}{d}$. We have an expression

$$w^e_{\overline{R}_i(q)}=u^{\overline{a}}\Lambda^e.$$

Setting

$$e_{\overline{R}_i}(q)=e, a_{\overline{R}_i}(q)=\overline{a}, \lambda_{\overline{R}_i}(q)=\Lambda(p)^e,$$

we define the primitive relation R_i at p.

τ-very well prepared morphisms

A morphism $f:X\to Y$ is τ-very well prepared with respect to a relation R if it is τ-quasi-well prepared, and satisfies some other properties (Definition 13.9). One of these properties is that R has a pre-algebraic structure (Definition 13.4). That is, if $q\in U(\overline{R}_i)$ for some pre-relation $\overline{R}_i$ associated to R, we have that $w_{\overline{R}_i(q)}\in\mathcal{O}_{Y,q}$.

If f is τ-very-well prepared, then there exists a finite, distinguished set of nonsingular algebraic surfaces $\Omega(\overline{R}_i)$ in Y such that $U(R)\cap\Omega(\overline{R}_i)=U(\overline{R}_i)$, and for $q\in U(\overline{R}_i)$, $w_{\overline{R}_i(q)}=0$ is a local equation of $\Omega(\overline{R}_i)$. There is a SNC divisor F_i on $\Omega(\overline{R}_i)$ such that the intersection graph of F_i is a tree. The irreducible components of F_i are the nonsingular curves $V=E\cdot\Omega(\overline{R}_i)$, which are the intersection of a component E of D_Y containing a point of $U(\overline{R}_i)$ with $\Omega(\overline{R}_i)$. Such V are closed in Y. Further, V is a *-permissible center (Definition 13.12). That is, V is a possible center for D_Y and there exists a commutative diagram of morphisms

$$\begin{array}{ccc} X_1 & \overset{f_1}{\rightarrow} & Y_1 \\ \Phi_1 \downarrow & & \downarrow \Psi_1 \\ X & \overset{f}{\rightarrow} & Y \end{array} \tag{2.16}$$

where Ψ_1 is the blow up of V (possibly followed by blow ups of some special 2-points q not mapping to $G_Y(f,\tau)$), such that f_1 is τ-quasi-well prepared, $\tau_{f_1}(p) \le \tau_f(p)$ for $p \in D_{X_1}$, and f_1 is τ-very-well prepared.

A τ-very-well prepared morphism is constructed by constructing an appropriate sequence of τ-quasi-well prepared diagrams.

Making $\tau_f(X)$ decrease

Suppose that $f : X \to Y$ is τ-very-well prepared. In Theorem 16.1, we construct a commutative diagram

$$\begin{array}{ccc} X_n & \overset{f_n}{\rightarrow} & Y_n \\ \downarrow & & \downarrow \\ X & \overset{f}{\rightarrow} & Y \end{array}$$

such that f_n is prepared and $\tau_{f_n}(X_n) < \tau$. By induction on τ, we then obtain the proof of Theorem 1.2.

We fix an index i of the relations R_i, with corresponding surface $\Omega(\overline{R}_i)$. A curve C on Y is good if it is a component of the divisor F_i on $\Omega(\overline{R}_i)$, and if j is such that $C \cap \Omega(\overline{R}_j) \neq \emptyset$, then C is a component of F_j.

In our construction we begin with $i = 1$, and blow up a good curve V on Y, by a morphism (2.16). Let R^1 be the transform of R on X_n. Since the intersection graph of F_i is a tree, there exists a good curve. Suppose that $p \in X$ is a point with $\tau_f(p) = \tau$ and $q = f(p) \in V$. Suppose that $p_1 \in \Phi_1^{-1}(p)$. We assume that p and q are 1-points. this is the simplest but most illustrative case. Set $q_1 = f_1(p_1)$. Let

$$u = u_{\overline{R}_1(q)}, v = v_{\overline{R}_1(q)}, w = w_{\overline{R}_1(q)}.$$

V has local equations $u = w = 0$ at q. Thus q_1 has regular parameters u_1, v, w_1 with

$$u = u_1 w_1, w = w_1 \tag{2.17}$$

or

$$u = u_1, w = u_1(w_1 + \alpha) \tag{2.18}$$

with $\alpha \in \mathbf{k}$. Let

$$\overline{a} = a_{\overline{R}_1}(q), e = e_{\overline{R}_1}(q), \lambda = \lambda_{\overline{R}_1}(q).$$

The relation R_1 is determined at q by a form

$$w^e - \lambda u^{\overline{a}}. \tag{2.19}$$

We computed earlier (in the discussion on relations) that in this case $p_1 = p$, and $\tau_{f_1}(p_1) < \tau_f(p)$ unless (2.18) holds. In this case, we saw that $\alpha = 0$ in(2.18), and we have

$$a_{\overline{R}_1^1}(q_1) = \overline{a} - e, e_{\overline{R}_1^1}(q_1) = e, \lambda_{\overline{R}_1^1}(q_1) = \lambda,$$

with $a_{\overline{R}_1^1}(q_1) > 0$.

If

$$T(R_1^1) = \Phi_1^{-1}(T(R_1)) \cap f_1^{-1}(U(R_1^1)) \cap G_{X_1}(f_1, \tau) = \emptyset,$$

then we increase i to 2.

Suppose otherwise. We have SNC divisors $F_i^1 = \Psi_1^{-1}(F_i)$ on the surfaces $\Omega(\overline{R}_i^1) = \Psi_1^{-1}(\Omega(\overline{R}_i)))$. There exists a good curve on Y_1 for the SNC divisor F_i^1 on the surface $\Omega(\overline{R}_1^1))$. We continue to iterate, blowing up good curves. After a finite number of blow ups, we obtain a reduction in τ. The easiest case to see this in is above the 1-point q considered above. If $\tau_{f_1}(p_1) = \tau_f(p)$, we blow up the curve C_1 with local equations $u_1 = w_1 = 0$ at q_1. If there is a point in $T(R_1^2)$ above p_1, we must have a reduction $a_{\overline{R}_1^2}(q_2) = \overline{a} - 2e$. Since this number must be positive, after a finite number of iterations, we must have a decrease in τ.

We then continue this algorithm for the preimages of all of the surfaces $\Omega(\overline{R}_i)$. The algorithm terminates in the construction of a morphism with a drop in τ as desired.

The final proof of Theorem 1.2 is given in Chapter 17.

The relationship of the birational case to the generically finite case

If $f : X \to Y$ is a birational morphism of projective, nonsingular 3-folds, and $q \in Y$ is such that $\dim f^{-1}(q) < 2$, then the morphism f factors as a product of blow ups of nonsingular curves above a neighborhood of q [**D2**]. If $f : X \to Y$ is also prepared, the situation is even simpler. In this case the germ of the nonfinite locus NF_f of f at q is a nonsingular curve C, and $f : X \to Y$ factors through the blow up of C in a neighborhood of q (Lemma 4.8).

This statement fails completely if $f : X \to Y$ is not birational.

We give a simple example of a projective, prepared, dominant morphism of 3-folds $f : X \to Y$, with toroidal structures D_Y and $D_X = f^{-1}(D_Y)$ such that f is prepared, and there exists a 1-point $q \in D_Y$ such that the germ of F_f at q is singular.

Let

$$Y = \mathrm{spec}(\mathbf{k}[u, v, w]) = \mathbf{k}^3,$$

and $q \in Y$ be the origin. Let

$$X_0 = \mathrm{spec}(\mathbf{k}[\overline{x}, \overline{y}, \overline{z}]) = \mathbf{k}^3.$$

Let D_Y be the plane $u = 0$.

Define $f_0 : X_0 \to Y$ by the inclusion of $\mathbf{k}$-algebras $\mathbf{k}[u, v, w] \to \mathbf{k}[\overline{x}, \overline{y}, \overline{z}]$ defined by $u = \overline{x}(\overline{x} - 2)$, $v = \overline{y}$, $w = \overline{z}$. $D_{X_0} = f_0^{-1}(D_Y)$ is the (disjoint) union of the two planes $\overline{x} = 0$ and $\overline{x} = 2$.

Now define $X \to X_0$ to be the blow up of the two disjoint, nonsingular curves C_1 and C_2, where C_1 is defined by $\overline{x} = \overline{z} = 0$, and C_2 is defined by $\overline{x} - 2 = \overline{z} - \overline{y}^2 = 0$. Let $f : X \to Y$ be the composite map. X has a toroidal structure defined by $D_X = f^{-1}(D_Y)$. It can be shown after a little calculation that f is prepared. In fact, at the 1-point q, the permissible parameters u, v, w realize a form (4.5) and (4.11), or (4.6) and (4.12) of Definition 4.1 at all points of $f^{-1}(q)$, so that the conclusions of Lemma 4.9 holds for u, v, w at q.

However, we see that the nonfinite locus NF_f of f is the union of two curves γ_1 and γ_2 where γ_1 is defined by $u = w = 0$, and γ_2 is defined by $u = w - v^2 = 0$. F_f thus has a singularity at q. Observe that both components of F_f are in fact nonsingular at q, which is consistent with Lemma 4.10, which shows that all formal components of the nonfinite locus of a prepared morphism of 3-folds are nonsingular at a 1-point.

We further see that there is no point or curve V in Y such that $f : X \to Y$ factors through the blow up of V.

This contrast, between the extremely simple structure of birational morphisms of 3-folds above 1-points, and the much more complicated structure above 1-points in the general case of morphisms of 3-folds, is the major obstruction to extending the proofs of the e-print [**C5**] to the general case.

CHAPTER 3

Notation

Throughout this paper, $\mathbf{k}$ will be an algebraically closed field of characteristic zero. A variety is a quasi projective variety over $\mathbf{k}$. A curve, surface or 3-fold is a quasi projective variety over $\mathbf{k}$ of respective dimension 1, 2 or 3. If X is a variety, and $p \in X$ is a nonsingular point, then regular parameters at p are regular parameters in $\mathcal{O}_{X,p}$. Formal regular parameters at p are regular parameters in $\hat{\mathcal{O}}_{X,p}$. If X is a variety and $V \subset X$ is a subvariety, then $\mathcal{I}_V \subset \mathcal{O}_X$ will denote the ideal sheaf of V. If V and W are subvarieties of a variety X, we denote the scheme theoretic intersection $Y = \text{spec}(\mathcal{O}_X/\mathcal{I}_V + \mathcal{I}_W)$ by $Y = V \cdot W$.

Let $f : X \to Y$ be a morphism of varieties. We will denote the singular locus of f by sing(f). sing(f) is the closed set of points $p \in X$ such that f is not smooth at p. If D is a (reduced) Cartier divisor on Y, then $f^{-1}(D)$ will denote the reduced divisor $f^*(D)_{red}$.

Suppose that $a, b, c, d \in \mathbf{Q}$. Then we will write $(a, b) \leq (c, d)$ if $a \leq b$ and $c \leq d$.

Suppose that V is an affine variety over $\mathbf{k}$. $x_1, \ldots, x_n \in \Gamma(V, \mathcal{O}_V)$ are uniformizing parameters on V if the natural morphism $V \to \text{Spec}(\mathbf{k}[x_1, \ldots, x_n]) \cong \mathbf{A}^3$ is etale. Suppose that $\Lambda : V \to X$ is an etale morphism. Define $D_X \cap V = \Lambda^{-1}(D_X)$.

A toroidal structure on a nonsingular variety X is a simple normal crossing divisor (SNC divisor) D_X on X. We will say that an ideal sheaf $\mathcal{I} \subset \mathcal{O}_X$ is toroidal if $\mathcal{I}$ is locally generated by monomials in local equations of components of D_X.

We will say that a nonsingular curve C which is a subvariety of a nonsingular 3-fold X with toroidal structure D_X makes simple normal crossings (SNCs) with D_X if for all $p \in C$, there exist regular parameters x, y, z at p such that $x = y = 0$ are local equations of C, and $xyz = 0$ contains the support of D_X at p.

Suppose that X is a nonsingular 3-fold with toroidal structure D_X. If $p \in D_X$ is on the intersection of three components of D_X then p is called a 3-point. If $p \in D_X$ is on the intersection of two components of D_X (and is not a 3-point) then p is called a 2-point. If $p \in D_X$ is not a 2-point or a 3-point, then p is called a 1-point. If C is an irreducible component of the intersection of two components of D_X, then C is called a 2-curve. $\Sigma(X)$ will denote the closed locus of 2-curves on X.

A possible center on a nonsingular 3-fold X with toroidal structure defined by a SNC divisor D_X, is a point on D_X or a nonsingular curve in D_X which makes SNCs with D_X. A possible center on a nonsingular surface S with toroidal structure defined by a SNC divisor D_S is a point on D_S. We will also call the blow up of a possible center a possible blow up.

Observe that if $\Phi : X_1 \to X$ is the blow up of a possible center, then $D_{X_1} = \Phi^{-1}(D_X)$ is a SNC divisor on X_1. Thus D_{X_1} defines a toroidal structure on X_1. All blow ups $\Phi : X_1 \to X$ considered in this paper will be of possible centers, and we will impose the toroidal structure on X_1 defined by $D_{X_1} = \Phi^{-1}(D_X)$.

We define the nonfinite locus NF_f of a proper morphism $f : X \to Y$ to be

$$\{q \in Y \mid \dim f^{-1}(q) > 0\}.$$

The fundamental locus F_g of a birational map $g : X - - - \mapsto Y$ is the complement of the maximal open subset of X on which g is a morphism.

By a general point q of a variety V, we will mean a point q which satisfies conditions which hold on some nontrivial open subset of V. The exact open condition which we require will usually be clear from context. By a general section of a coherent sheaf $\mathcal{F}$ on a projective variety X, we mean the section corresponding to a general point of the $\mathbf{k}$-linear space $\Gamma(X, \mathcal{F})$.

If X is a variety, $\mathbf{k}(X)$ will denote the function field of X. A 0-dimensional valuation ν of $\mathbf{k}(X)$ is a valuation of $\mathbf{k}(X)$ such that $\mathbf{k}$ is contained in the valuation ring V_ν of ν and the residue field of V_ν is $\mathbf{k}$. If X is a projective variety which is birationally equivalent to X, then there exists a unique (closed) point $p_1 \in X_1$ such that V_ν dominates $\mathcal{O}_{X_1,p_1}$. p_1 is called the center of ν on X_1. If $p \in X$ is a (closed) point, then there exists a 0-dimensional valuation ν of $\mathbf{k}(X)$ such that V_ν dominates $\mathcal{O}_{X,p}$ (Theorem 37, Section 16, Chapter VI [**ZS**]). For $a_1, \dots, a_n \in \mathbf{k}(X)$, $\nu(a_1), \dots, \nu(a_n)$ are rationally dependent if there exist $\alpha_1, \dots, \alpha_n \in \mathbf{Z}$ which are not all zero, such that $\alpha_1\nu(a_1) + \cdots + \alpha_n\nu(a_n) = 0$ (in the value group of ν). Otherwise, $\nu(a_1), \dots, \nu(a_n)$ are rationally independent. The valuations of $\mathbf{k}(X)$ form a quasi compact topological space called the Zariski Riemann manifold of X. This is discussed in [**Z**], [**ZS**] and the Appendix [**Li**] to [**Z2**].

If $x_1, \dots, x_n$ are indeterminates, and $M_i = \prod_{j=1}^n x_j^{a_{ij}}$ are Laurent monomials for $1 \le i \le m$, then we will denote $\mathrm{rank}(M_1, \dots, M_m) = \mathrm{rank}(a_{ij})$.

Suppose that S is a finite set. Then $|S|$ will denote the cardinality of S.

CHAPTER 4

Toroidal morphisms and prepared morphisms

Suppose that X is a nonsingular variety with toroidal structure D_X.

Suppose that $q \in X$. We say that u, v, w are (formal) permissible parameters at q (for D_X) if u, v, w are regular parameters in $\hat{\mathcal{O}}_{X,q}$ such that $u = 0$ is a local equation of D_X at q if q is a 1-point, $uv = 0$ is a local equation of D_X at q if q is a 2-point and $uvw = 0$ is a local equation of D_X at q if q is a 3-point. u, v, w are algebraic permissible parameters if we further have that $u, v, w \in \mathcal{O}_{X,q}$.

DEFINITION 4.1. Let $f : X \to Y$ be a dominant morphism of nonsingular 3-folds with toroidal structures D_Y on Y and $D_X = f^{-1}(D_Y)$ on X such that $\mathrm{sing}(f) \subset D_X$. Suppose that u, v, w are (possibly formal) permissible parameters at $q \in D_Y$. Then u, v are **toroidal forms** at $p \in f^{-1}(q)$ if there exist permissible parameters x, y, z in $\hat{\mathcal{O}}_{X,p}$ such that one of the following cases hold.

1. q is a 2-point or a 3-point, p is a 1-point and

$$u = x^a, v = x^b(\alpha + y) \tag{4.1}$$

   where $0 \neq \alpha \in \mathbf{k}$.

2. q is 2-point or a 3-point, p is a 2-point and

$$u = x^a y^b, v = x^c y^d \tag{4.2}$$

   with $ad - bc \neq 0$.

3. q is a 2-point or a 3-point, p is a 2-point and

$$u = (x^a y^b)^k, v = (x^a y^b)^t(\alpha + z) \tag{4.3}$$

   where $0 \neq \alpha \in \mathbf{k}$, $a, b, k, t > 0$ and $\gcd(a, b) = 1$.

4. q is a 2-point or a 3-point, p is a 3-point and

$$u = x^a y^b z^c, v = x^d y^e z^f \tag{4.4}$$

   where

$$\mathrm{rank}\begin{pmatrix} a & b & c \\ d & e & f \end{pmatrix} = 2.$$

5. q is a 1-point, p is a 1-point and

$$u = x^a, v = y \tag{4.5}$$

6. q is a 1-point, p is a 2-point and

$$u = (x^a y^b)^k, v = z \tag{4.6}$$

   with $a, b, k > 0$ and $\gcd(a, b) = 1$

Regular parameters x, y, z as in Definition 4.1 will be called permissible parameters for u, v, w at p.

LEMMA 4.2. *Let notation be as in Definition 4.1. Suppose that $q \in D_Y$, $p \in f^{-1}(q)$ and u, v, w are permissible parameters at q such that u, v are toroidal forms at p. Then there exist permissible parameters x, y, z for u, v, w at p such that an expression of Definition 4.1 holds for u and v at p, and one of the following respective forms for w holds at p.*

1. *q is a 2-point or a 3-point, p is a 1-point, u, v satisfy (4.1) and*
$$w = g(x, y) + x^c z \tag{4.7}$$
where g is a series.
2. *q is 2-point or a 3-point, p is a 2-point, u, v satisfy (4.2) and*
$$w = g(x, y) + x^e y^f z \tag{4.8}$$
where g is a series.
3. *q is a 2-point or a 3-point, p is a 2-point, u, v satisfy (4.3) and*
$$w = g(x^a y^b, z) + x^c y^d \tag{4.9}$$
where g is a series and rank$(u, x^c y^d) = 2$.
4. *q is a 2-point or a 3-point, p is a 3-point, u, v satisfy (4.4) and*
$$w = g(x, y, z) + N \tag{4.10}$$
where g is a series in monomials M in x, y, z such that rank$(u, v, M) = 2$, and N is a monomial in x, y, z such that rank$(u, v, N) = 3$.
5. *q is a 1-point, p is a 1-point, u, v satisfy (4.5) and*
$$w = g(x, y) + x^c z \tag{4.11}$$
where g is a series.
6. *q is a 1-point, p is a 2-point, u, v satisfy (4.6) and*
$$w = g(x^a y^b, z) + x^c y^d \tag{4.12}$$
where g is a series and rank$(u, x^c y^d) = 2$.

PROOF. Choose permissible parameters x, y, z for u, v, w at p. The lemma follows from an explicit calculation of the Jacobian determinant
$$J = \text{Det}\begin{pmatrix} \frac{\partial u}{\partial x} & \frac{\partial u}{\partial y} & \frac{\partial u}{\partial z} \\ \frac{\partial v}{\partial x} & \frac{\partial v}{\partial y} & \frac{\partial v}{\partial z} \\ \frac{\partial w}{\partial x} & \frac{\partial w}{\partial y} & \frac{\partial w}{\partial z} \end{pmatrix},$$
and a change of variables of x, y, z. Observe that $J = 0$ is supported on D_X, since the singular locus of f is contained in D_X.

We indicate the proof if (4.4) holds at p. There exists a unit series γ in x, y, z and $l, m, n \in \mathbf{N}$ such that $J = \gamma x^l y^m z^n$. Write w as a series
$$w = \sum c_{ijk} x^i y^j z^k$$
with $c_{ijk} \in \mathbf{k}$. We compute
$$J = \sum c_{ijk} \text{Det}\begin{pmatrix} a & b & c \\ d & e & f \\ i & j & k \end{pmatrix} x^{a+d+i-1} y^{b+e+j-1} z^{c+f+k-1} = \gamma x^l y^m z^n$$
from which we obtain forms (4.4) and (4.10), after making a change of variables in x, y, z, multiplying x, y, z by appropriate roots of γ. □

DEFINITION 4.3. ([**KKMS**], [**AK**]) A normal variety $\overline{X}$ with a SNC divisor $D_{\overline{X}}$ on $\overline{X}$ is called toroidal if for every point $p \in \overline{X}$ there exists an affine toric variety X_σ, a point $p' \in X_\sigma$ and an isomorphism of $\mathbf{k}$-algebras

$$\hat{\mathcal{O}}_{\overline{X},p} \cong \hat{\mathcal{O}}_{X_\sigma,p'}$$

such that the ideal of $D_{\overline{X}}$ corresponds to the ideal of $X_\sigma - T$ (where T is the torus in X_σ). Such a pair (X_σ, p') is called a local model at $p \in \overline{X}$. $D_{\overline{X}}$ is called a toroidal structure on $\overline{X}$.

A dominant morphism $\Phi : \overline{X} \to \overline{Y}$ of toroidal varieties with SNC divisors $D_{\overline{Y}}$ on $\overline{Y}$ and $D_{\overline{X}} = \Phi^{-1}(D_{\overline{Y}})$ on $\overline{X}$, is called toroidal at $p \in \overline{X}$, and we will say that p is a toroidal point of Φ if with $q = \Phi(p)$, there exist local models (X_σ, p') at p, (Y_τ, q') at q and a toric morphism $\Psi : X_\sigma \to Y_\tau$ such that the following diagram commutes:

$$\begin{array}{ccc} \hat{\mathcal{O}}_{\overline{X},p} & \leftarrow & \hat{\mathcal{O}}_{X_\sigma,p'} \\ \hat{\Phi}^* \uparrow & & \hat{\Psi}^* \uparrow \\ \hat{\mathcal{O}}_{\overline{Y},q} & \leftarrow & \hat{\mathcal{O}}_{Y_\tau,q'}. \end{array}$$

$\Phi : \overline{X} \to \overline{Y}$ is called toroidal (with respect to $D_{\overline{Y}}$ and $D_{\overline{X}}$) if Φ is toroidal at all $p \in \overline{X}$.

The following is the list of toroidal forms for a dominant morphism $f : X \to Y$ of nonsingular 3-folds with toroidal structures D_Y and $D_X = f^{-1}(D_X)$. Suppose that $p \in D_X$, $q = f(p) \in D_Y$, and f is toroidal at p. Then there exist permissible parameters u, v, w at q and permissible parameters x, y, z for u, v, w at p such that one of the following forms hold:

1. p is a 3-point and q is a 3-point,

$$\begin{aligned} u &= x^a y^b z^c \\ v &= x^d y^e z^f \\ w &= x^g y^h z^i, \end{aligned}$$

where $a, b, d, e, f, g, h, i \in \mathbf{N}$ and

$$\text{Det} \begin{pmatrix} a & b & c \\ d & e & f \\ g & h & i \end{pmatrix} \neq 0.$$

2. p is a 2-point and q is a 3-point,

$$\begin{aligned} u &= x^a y^b \\ v &= x^d y^e \\ w &= x^g y^h (z + \alpha) \end{aligned}$$

with $0 \neq \alpha \in \mathbf{k}$ and $a, b, d, e, g, h \in \mathbf{N}$ satisfy $ae - bd \neq 0$.

3. p is a 1-point and q is a 3-point,

$$\begin{aligned} u &= x^a \\ v &= x^d (y + \alpha) \\ w &= x^g (z + \beta) \end{aligned}$$

with $0 \neq \alpha, 0 \neq \beta \in \mathbf{k}$, $a, d, g > 0$.

4. p is a 2-point and q is a 2-point,

$$\begin{aligned} u &= x^a y^b \\ v &= x^d y^e \\ w &= z \end{aligned}$$

with $ae - bd \neq 0$

5. p is a 1-point and q is a 2-point,

$$\begin{aligned} u &= x^a \\ v &= x^d (y + \alpha) \\ w &= z \end{aligned}$$

with $0 \neq \alpha \in \mathbf{k}$, $a, d > 0$.

6. p is a 1-point and q is a 1-point,

$$\begin{aligned} u &= x^a \\ v &= y \\ w &= z \end{aligned}$$

with $a > 0$.

We obtain the above list by observing that the germ of f at p is formally isomorphic to a localization of a monomal mapping $\mathbf{k}^3 \rightarrow \mathbf{k}^3$ at a point on the coordinate axis.

DEFINITION 4.4. Let notation be as in Definition 4.1. u, v, w have a (non toroidal) **monomial form** at $p \in f^{-1}(q)$ if

1. u, v have a form (4.1) at p, q is a 2-point and $w = x^c(z + \beta)$ for some $c \in \mathbf{N}$ with $c > 0$, $\beta \in \mathbf{k}$.
2. u, v have a form (4.2) at p, q is a 2-point and $w = x^e y^f (z + \beta)$ for some $e, f \in \mathbf{N}$, $e + f > 0$, $\beta \in \mathbf{k}$.
3. u, v have a form (4.3) at p, q is a 2-point and $w = x^c y^d$ with $ad - bc \neq 0$.
4. u, v have a form (4.4) at p, q is a 2-point and $w = x^g y^h z^i$ with

$$\det \begin{pmatrix} a & b & c \\ d & e & f \\ g & h & i \end{pmatrix} \neq 0.$$

5. u, v have a form (4.5) at p and $w = x^b(z + \beta)$ with $b \in \mathbf{N}$, $b > 0$, $\beta \in \mathbf{k}$.
6. u, v have a form (4.6) at p and $w = x^c y^d$ with $ad - bc \neq 0$.

A prepared morphism $\Phi_X : X \rightarrow S$ from a nonsingular 3-fold X to a nonsingular surface S (with respect to toroidal structures D_S and $D_X = \Phi_X^{-1}(D_S)$) is defined in Definition 6.5 [**C3**].

REMARK 4.5. *Suppose that $f : X \rightarrow Y$ is a dominant proper morphism of nonsingular 3-folds with toroidal structures determined by SNC divisors D_Y on Y and $D_X = f^{-1}(D_Y)$ on X, and D_X contains the singular locus of the morphism f. With our assumptions on f, f is generically finite. The nonfinite locus NF_f is a closed set of codimension ≥ 2 in Y. Let $\overline{X}$ be the normalization of Y in the function field of X, with induced finite morphism $\lambda : \overline{X} \rightarrow Y$. The branch locus of λ (which is a divisor by the purity of the branch locus* [**N**]*) is contained in the SNC divisor D_Y. Let E be an irreducible component of D_Y. By Abhyankar's lemma (*[**Ab4**], *XIII 5.3* [**G**]*), the irreducible components of $\lambda^{-1}(E)$ are disjoint. Thus the irreducible components of D_X which dominate E are disjoint.*

DEFINITION 4.6. A dominant morphism $f : X \to Y$ of nonsingular 3-folds with toroidal structures determined by SNC divisors D_Y on Y, and $D_X = f^{-1}(D_Y)$ on X such that the singular locus of f is contained in D_X is **prepared** for D_Y and D_X if:

1. If $q \in Y$ is a 3-point, u, v, w are permissible parameters at q and $p \in f^{-1}(q)$, then u, v and w are each a unit (in $\hat{\mathcal{O}}_{X,p}$) times a monomial in local equations of the toroidal structure D_X at p . Furthermore, there exists a permutation of u, v, w such that u, v are toroidal forms at p.
2. If $q \in Y$ is a 2-point, u, v, w are permissible parameters at q and $p \in f^{-1}(q)$, then either
   (a) u, v are toroidal forms at p or
   (b) p is a 1-point and there exist regular parameters $x, y, z \in \hat{\mathcal{O}}_{X,p}$ such that there is an expression
   $$\begin{aligned} u &= x^a \\ v &= x^c(\gamma(x,y) + x^d z) \\ w &= y \end{aligned}$$
   where γ is a unit series and $x = 0$ is a (formal) local equation of D_X, or
   (c) p is a 2-point and there exist regular parameters x, y, z in $\hat{\mathcal{O}}_{X,p}$ such that there is an expression
   $$\begin{aligned} u &= (x^a y^b)^k \\ v &= (x^a y^b)^l(\gamma(x^a y^b, z) + x^c y^d) \\ w &= z \end{aligned}$$
   where $a, b > 0$, $\gcd(a,b) = 1$, $ad - bc \neq 0$, γ is a unit series and $xy = 0$ is a (formal) local equation of D_X.
3. If $q \in Y$ is a 1-point, and $p \in f^{-1}(q)$, then there exist permissible parameters u, v, w at q such that u, v are toroidal forms at p.

REMARK 4.7. *Suppose that $f : X \to Y$ is a dominant morphism of nonsingular 3-folds, with toroidal structures determined by SNC divisors D_Y on Y and $D_X = f^{-1}(D_Y)$ on X such that the singular locus of f is contained in D_X. Suppose that $q \in Y$ is a 2-point, and u, v, w are permissible parameters at q. it follows from a computation of the Jacobian determinant that the following holds.*

1. *Suppose that $p \in f^{-1}(q)$ is a 1-point, and there exist permissible parameters x, y, z at p such that*
   $$u = x^a, v = x^c\gamma, w = y$$
   where $\gamma \in \hat{\mathcal{O}}_{X,p}$ is a unit series. Then f is prepared at p of type 2 (b),
2. *Suppose that $p \in f^{-1}(q)$ is a 2-point, and there exist permissible parameters x, y, z at p such that*
   $$u = x^e y^f, v = x^g y^h \gamma, w = z$$
   where $\gamma \in \hat{\mathcal{O}}_{X,p}$ is a unit series.

   Then f is either torodial at p (so that it is prepared of type 2 (a) at p) or f is prepared of type 2 (c) at p.

LEMMA 4.8. *Suppose that X, Y are projective, $f : X \to Y$ is birational, prepared and $q \in Y$ is a 1-point such that f is not an isomorphism over q. Then the* *fundamental locus* $F_{f^{-1}} = NF_f$ *of* f *contains a unique curve* C *passing through* q *and* C *is nonsingular at* q.

Furthermore, there exist algebraic permissible parameters u, v, w *at* q *such that a form (4.5) or (4.6) of Definition 4.1 (for this fixed choice of* u, v, w*) holds at* p *for all* $p \in f^{-1}(q)$. $u = w = 0$ *are local equations of* C *at* q.

PROOF. As f is birational and Y is nonsingular, the exceptional locus of f is a divisor on X, which is necessarily supported on D_X.

Since for all $p \in f^{-1}(q)$, there exist permissible parameters at q such that a form (4.5) or (4.6) of Definition 4.1 holds at p, we have that dim $f^{-1}(q) = 1$. Thus if E' is an exceptional component of f such that $q \in f(E')$, then $f(E')$ is a curve.

Let D be the component of D_Y containing q. Let F be the strict transform of D on X and let $p \in f^{-1}(q) \cap F$. Then p must be a 2-point, and since f is birational, there exist permissible parameters $\overline{u}, \overline{v}, \overline{w}$ at q such that there is an expression in $\hat{\mathcal{O}}_{X,p}$

$$\begin{aligned}\overline{u} &= x^a y\\ \overline{v} &= z\\ \overline{w} &= \phi(x^a y, z) + x^d y^e\end{aligned}$$

where $y = 0$ is a (formal) local equation of F, and $x = 0$ is a (formal) local equation of the other component E of D_X containing p. Computing the Jacobian determinant of f at p, we see that $e = 0$. We consider the morphism $f^* : \mathcal{O}_{D,q} \to \mathcal{O}_{F,p}$. $\hat{f}^* : \hat{\mathcal{O}}_{D,q} \to \hat{\mathcal{O}}_{F,p}$ is the $\mathbf{k}$-algebra homomorphism $\hat{f}^* : \mathbf{k}[[\overline{v}, \overline{w}]] \to \mathbf{k}[[x, z]]$ given by $\overline{v} = z$, $\overline{w} = \phi(0, z) + x^d$. Thus $\hat{\mathcal{O}}_{F,p}$ is finite over $\hat{\mathcal{O}}_{D,q}$. It follows that $f^* : \mathcal{O}_{D,q} \to \mathcal{O}_{F,p}$ is quasi-finite, and thus $\mathcal{O}_{F,p} \cong \mathcal{O}_{D,q}$ by Zariski's Main Theorem. In particular, $\{p\} = f^{-1}(q) \cap F$. Let $C = f(E)$. C is an algebraic curve through q which has formal local equations $\overline{u} = \overline{w} - \phi(\overline{u}, \overline{v}) = 0$ at q. Thus C is nonsingular at q.

Suppose that $C' \subset F$ is a curve containing q such that $q \in C'$ and $C' \neq C$. Let C'' be the unique curve on D containing p which dominates C'. A general point of C'' is then not on an exceptional component of f. Thus f is an isomorphism above a general point of C' be Zariski's main theorem.

If E' is a component of the exceptional locus of f such that $q \in f(E')$, we must have that $f(E') = C$. Now let u, v, w be permissible parameters at q such that $u = w = 0$ are local equations of C at q. We see that u, v, w must have a form (4.5) or (4.6) of Definition 4.1 for all $p \in f^{-1}(q)$, since w is divisible by a local equation of a component of D_X.

□

LEMMA 4.9. *Suppose that* X, Y *are projective,* $f : X \to Y$ *is a prepared morphism of 3-folds, and* $q \in D_Y$ *is a 1-point. Then there exist algebraic permissible parameters* u, v, w *at* q *such that a form (4.5) or (4.6) of Definition 4.1 (for this fixed choice of* u, v, w*) holds at* p *for all* $p \in f^{-1}(q)$.

PROOF. Let $q \in Y$ be a 1-point, u, v, w be permissible parameters at q, $\overline{X}$ be the normalization of Y in $\mathbf{k}(X)$. We have a natural commutative diagram

$$\begin{array}{ccc} X & \xrightarrow{f} & Y\\ g & \searrow & \uparrow \lambda\\ & & \overline{X}\end{array}.$$

Let $\lambda^{-1}(q) = \{q_1, \ldots, q_n\}$. Abhyankar's Lemma implies that $\overline{X}$ is nonsingular above a neighborhood of q, and there exist (formal) permissible parameters u_1, v_i, w_i at q_i and $r_i \in \mathbf{N}$ such that

$$u = u_i^{r_i}, v = v_i, w = w_i.$$

Lemma 4.8 (applied to $g : X \to \overline{X}$) implies $u_i, \alpha v_i + \beta w_i$ are toroidal forms at all points of $g^{-1}(q_i)$ for a dense open set of $(\alpha, \beta) \in \mathbf{k}^2$. The condition is that $u_i = \alpha v_i + \beta w_i = 0$ intersects the (nonsingular at q_i) nonfinite locus of g transversally at q_i. Thus there exist $\alpha, \beta \in \mathbf{k}$ such that $u, \alpha v + \beta w$ are toroidal forms at all $p \in f^{-1}(q)$. □

LEMMA 4.10. *Suppose that $f : X \to Y$ is a prepared morphism of 3-folds and E is a component of D_X such that $f(E) = C$ is a curve which is not a 2-curve. Suppose that $q \in C$ is a 1-point. Then all formal branches of C are smooth at q, and there exist permissible parameters u, v, w at q such that $u = w = 0$ are (formal) local equations of C at q.*

PROOF. Suppose that $p \in f^{-1}(q) \cap E$. There exist permissible parameters u, v, w at q such that one of the following forms hold.

p is a 1-point:

$$u = x^a, v = y, w = g(x, y) + x^n z \tag{4.13}$$

where $x = 0$ is a local equation of E, or

p is a 2-point:

$$u = (x^a y^b)^i, v = z, w = g(x^a y^b, z) + x^c y^d \tag{4.14}$$

where $a, b > 0$, $ad - bc \neq 0$ and $x = 0$ is a local equation of E.

In case (4.13), we have an expression $w = \phi(y) + x\Omega$ in $\hat{\mathcal{O}}_{X,p}$ ($n > 0$ since $f(E)$ is a curve). $u = w - \phi(y) = 0$ is a local equation of a branch of C.

In case (4.14), one of the following must hold in $\hat{\mathcal{O}}_{X,p}$:

$$w = \phi(z) + xy\Omega, \tag{4.15}$$

$$w = \phi(z) + xy\Omega + x^c \tag{4.16}$$

where $c > 0$, or

$$w = \phi(z) + xy\Omega + y^d \tag{4.17}$$

where $d > 0$.

In cases (4.15) or (4.16), we have

$$0 = u = w - \phi(z)$$

is a formal local equation of a branch of C at q.

Suppose that case (4.17) holds. $f(E) = C$ a curve implies that there exists an irreducible series $\Psi(\overline{x}, \overline{y})$ in a power series ring $\mathbf{k}[[\overline{x}, \overline{y}]]$ such that x divides $\Psi(v, w)$ in $\hat{\mathcal{O}}_{X,p}$.

$$\Psi(v, w) = \Psi(z, \phi(z) + xy\Omega + y^d)$$

implies $\Psi(0, y^d) = 0$ (set $x = z = 0$), which implies that $\overline{x} \mid \Psi(\overline{x}, \overline{y})$. Since $\Psi(\overline{x}, \overline{y})$ is irreducible, we have that $\Psi = \overline{x}\lambda$ where λ is a unit series. This is a contradiction. □

CHAPTER 5

Toroidal ideals

LEMMA 5.1. *Suppose that X is a nonsingular 3-fold with SNC divisor D_X, defining a toroidal structure on X. Suppose that $\mathcal{I}$ is an ideal sheaf on X which is locally generated by monomials in local equations of components of D_X. Then there exists a sequence of blow ups of 2-curves $\Phi_1 : X_1 \rightarrow X$ such that $\mathcal{I}\mathcal{O}_{X_1}$ is an invertible ideal sheaf. If $\mathcal{I}$ is locally generated by two equations, then Φ_1 is an isomorphism away from the support of $\mathcal{I}$.*

This lemma is an extension of Lemma 18.18 [**C3**], and is generalized to all dimensions in [**Go**].

PROOF. X has a cover by affine open sets $U_1, \ldots, U_n$ such that there exist $g_{i,1}, \ldots, g_{i,l} \in \Gamma(U_i, \mathcal{O}_X)$ such that $g_{i,j} = 0$ are local equations in U_i of irreducible components of D_X, and there exist $f_{i,1}, \ldots, f_{i,m(i)} \in \Gamma(U_i, \mathcal{O}_X)$ such that the $f_{i,j}$ are monomials in the $g_{i,k}$ and $\Gamma(U_i, \mathcal{I}) = (f_{i,1}, \ldots, f_{i,m(i)})$.

Let D_{ij} be an effective divisor supported on the components of D_X such that there is equality of divisors $D_{ij} \cap U_i = (f_{i,j}) \cap U_i$. Let $\mathcal{I}_i \subset \mathcal{O}_X$ be the ideal sheaf which is locally generated by local equations of $D_{i,1}, \ldots, D_{i,m(i)}$. By construction, $\mathcal{I}_i \mid U_i = \mathcal{I} \mid U_i$ for all i.

We will show that for an ideal sheaf of the form $\mathcal{I}_1$, there exists a sequence of blow ups of 2-curves, $\pi : X_1 \rightarrow X$ such that $\mathcal{I}_1\mathcal{O}_{X_1}$ is invertible. Since $\mathcal{I}_i\mathcal{O}_{X_1}$ are locally generated by local equations of $\pi^*(D_{i,1}), \ldots, \pi^*(D_{i,m(i)})$, there exists $\pi_2 : X_2 \rightarrow X$ which is a sequence of blow ups of 2-curves such that $\mathcal{I}_i\mathcal{O}_{X_2}$ is invertible for all i. Since $\mathcal{I}_i\mathcal{O}_{X_2} \mid \pi_2^{-1}(U_i) = \mathcal{I}\mathcal{O}_{X_2} \mid \pi_2^{-1}(U_i)$ for all i, $\mathcal{I}\mathcal{O}_{X_2}$ is invertible.

We may now suppose that there exists $n > 0$ and effective divisors $D_1, \ldots, D_n$ on X whose supports are unions of components of D_X, such that $\mathcal{I}$ is locally generated by local equations of $D_1, \ldots, D_n$.

First suppose that $n = 2$. Suppose that $p \in X$ is a general point of a 2-curve C. Let $x = 0$, $y = 0$ be local equations of the components of D_X containing p. $x = y = 0$ are local equations of C at p. Then there exist $a, b, c, d \in \mathbf{N}$ such that D_1 is defined near p by the divisor of $x^a y^b$, and D_2 is defined near p by the divisor of $x^c y^d$. Define

$$\omega(C) = \begin{cases} \max\{(|a-c|, |b-d|), (|b-d|, |a-c|)\} & \text{if } a-c,\ b-d \text{ are nonzero} \\ & \text{and have opposite signs,} \\ -\infty & \text{otherwise} \end{cases}$$

Here the maximum is computed in the lexicographic order. We see that the stalk $\mathcal{I}_p$ is invertible if and only if $\omega(C) = -\infty$.

Further, if $\omega(C) = -\infty$ for all 2-curves C of X, then $\mathcal{I}$ is invertible, as follows since the divisors D_1 and D_2 are given locally at a 3-point p by the divisors of monomials $x^ay^bz^c$ and $x^dy^ez^f$ where $xyz = 0$ is a local equation of D_X at p. $a-d$ and $b-e$ have the same signs, $a-d$, $c-f$ have the same signs, and $b-e$, $c-f$ have the same signs, so $x^ay^bz^c | x^dy^ez^f$ or $x^dy^ez^f \mid x^ay^bz^c$.

Now define

$$\overline{\omega}(X) = \max\{\omega(C) \mid C \text{ is a 2-curve of } X\}.$$

We have seen that $\mathcal{I}$ is invertible if and only if $\overline{\omega}(X) = -\infty$. Suppose that $\overline{\omega}(X) \neq -\infty$ and C is a 2-curve of X such that $\omega(C) = \overline{\omega}(X)$. Let $\pi : X_1 \to X$ be the blow up of C. Let $D_{X_1} = \pi^{-1}(D_X) = \pi^*(D_X)_{red}$, $D_1' = \pi^*(D_1)$, $D_2' = \pi^*(D_2)$.

We can define the function ω for 2-curves on X_1, relative to D_1' and D_2', and define $\overline{\omega}(X_1)$.

By a local calculation (as shown in the proof of Lemma 18.18 [**C3**]) we see that $\omega(C_1) < \overline{\omega}(X)$ if C_1 is a 2-curve which is contained in the exceptional divisor of π.

Suppose that $C_1, \ldots, C_r$ are the 2-curves C on X such that $\omega(C) = \overline{\omega}(X)$. We obtain a reduction $\overline{\omega}(X_1) < \overline{\omega}(X)$ after blowing up (the strict transforms of) these r curves. By induction on $\overline{\omega}(X)$, we must obtain that $\mathcal{I}\mathcal{O}_{X_2}$ is invertible after an appropriate sequence of blow ups of 2-curves $X_2 \to X$.

Now suppose that $\mathcal{I}$ is locally generated by local equations of $D_1, \ldots, D_n$ (with $n > 2$). Let $\mathcal{I}_1 \subset \mathcal{I}$ be the ideal sheaf which is locally generated by local equations of D_1 and D_2. We have seen that there exists a sequence of blow ups of 2-curves $\pi_1 : X_1 \to X$ such that $\mathcal{I}_1\mathcal{O}_{X_1}$ is invertible. Thus there exists a divisor $\overline{D}$ on X_1 whose support is a union of components of D_{X_1} such that $\mathcal{I}_1\mathcal{O}_{X_1}$ is locally generated by a local equation of $\overline{D}$.

Let $\overline{D}_i = \pi_1^*(D_i)$ for $3 \leq i \leq n$. Then $\mathcal{I}\mathcal{O}_{X_1}$ is locally generated by local equations of the $n-1$ divisors $\overline{D}, \overline{D}_3, \ldots, \overline{D}_n$. By induction, there exists a sequence of blow ups of 2-curves $X_2 \to X$ such that $\mathcal{I}\mathcal{O}_{X_2}$ is invertible.

When $\mathcal{I}$ is locally generated by two equations, we restrict to blowing up 2-curves C for which a general point of C is in U_i to obtain the conclusion that Φ_1 is an isomorphism away from the support of $\mathcal{O}_X/\mathcal{I}$.

□

LEMMA 5.2. *Suppose that $f : X \to Y$ is a prepared morphism of nonsingular 3-folds, and $\overline{C}$ is a 2-curve in Y. Then there exists a sequence of blowups of 2-curves $\Phi : X_1 \to X$ such that $\mathcal{I}_{\overline{C}}\mathcal{O}_{X_1}$ is invertible and Φ is an isomorphism over $f^{-1}(Y - \overline{C})$.*

PROOF. The lemma is a consequence of Lemma 5.1. □

LEMMA 5.3. *Suppose that X is a nonsingular 3-fold with SNC divisor D_X, defining a toroidal structure on X. Suppose that $\mathcal{I}$ is an ideal sheaf on X which is locally generated by monomials in local equations of components of D_X. Then there exists a sequence of blow ups of 2-curves and 3-points $\Phi_1 : X_1 \to X$ such that $\mathcal{I}\mathcal{O}_{X_1}$ is an invertible ideal sheaf and Φ_1 is an isomorphism away from the support of $\mathcal{O}_X/\mathcal{I}$.*

The proof of this lemma follows from the proof of principalization of ideals, as in [**BEV**] or [**BrM**] (cf. the proof of Theorem 6.3 [**C6**]) in the case when the ideal to be principalized is locally generated by monomials in the toroidal structure.

CHAPTER 6

Toroidalization of morphisms from 3-folds to surfaces

The following theorem is Theorem 19.11 [**C3**].

THEOREM 6.1. *Suppose that $\Phi : X \rightarrow S$ is a dominant morphism from a nonsingular 3-fold X to a nonsingular surface S and D_S is a SNC divisor on S such that $D_X = \Phi^{-1}(D_S)$ is a SNC divisor which contains the singular locus of Φ.*

1. *There exists a sequence of blow ups of possible centers $\alpha_1 : X_1 \rightarrow X$ such that*
   (a) *The fundamental locus of α_1^{-1} is contained in the union of irreducible components E of D_X such that E contains a point p such that Φ is not prepared at p (Definition 6.5 [**C3**]).*
   (b) *$\Phi_1 = \Phi \circ \alpha_1 : X_1 \rightarrow S$ is prepared (Definition 6.5 [**C3**])*
2. *Further, there exist sequences of blow ups of possible centers, $\alpha_2 : X_2 \rightarrow X_1$ and $\beta : S_1 \rightarrow S$ such that:*
   (a) *There is a commutative diagram*

$$\begin{array}{ccc} X_2 & \stackrel{\Phi_2}{\rightarrow} & S_1 \\ \alpha_2 \downarrow & & \downarrow \beta \\ X_1 & \stackrel{\Phi_1}{\rightarrow} & S \end{array}$$

   such that Φ_2 is toroidal,
   (b) *β is an isomorphism away from $\beta^{-1}(\Phi_1(Z))$ where Z is the locus where Φ_1 is not toroidal.*
   (c) *α_2 is an isomorphism away from $\alpha_2^{-1}(\Phi_1^{-1}(\Phi_1(Z)))$*

PROOF. The proof is immediate from the proof of Theorem 19.11 [**C3**], which we outline below.

We first prove 1 of the theorem. By Lemma 6.2 [**C3**], there exists a commutative diagram

$$\begin{array}{ccc} X_0 & & \\ \alpha_0 \downarrow & \searrow \Phi_0 & \\ X & \stackrel{\Phi}{\rightarrow} & S \end{array}$$

such that $\Phi_0 : X_0 \rightarrow S$ is a weakly prepared morphism (Definition 6.1 [**C3**]), and the fundamental locus of α_0^{-1} is contained in the locus where Φ is not weakly prepared, (which is contained in the locus where Φ is not prepared). It is not necessary to blow up points on S since D_S, D_X are SNC divisors. Let $D_{X_0} = \alpha_0^{-1}(D_X)$.

By Theorem 17.2 [**C3**] there exists a commutative diagram

$$\begin{array}{ccc} X_1 & & \\ \overline{\alpha}\downarrow & \searrow \Phi_1 & \\ X_0 & \stackrel{\Phi_0}{\to} & S \end{array}$$

such that Φ_1 is prepared. The algorithm consists of a sequence

$$X_1 = Y_n \stackrel{\overline{\alpha}_n}{\to} Y_{n-1} \to \cdots \to Y_1 \stackrel{\overline{\alpha}_1}{\to} Y_0 = X_0$$

of blow ups of points and nonsingular curves which are possible centers. They make SNCs with the preimage D_{Y_i} of D_{X_0} and are contained in a component E of D_{Y_i} such that E contains a point which is not prepared for

$$\overline{\Phi}_i = \Phi_0 \circ \overline{\alpha}_1 \circ \cdots \circ \overline{\alpha}_i : Y_i \to S.$$

If $p \in Y_i$ is prepared for $\overline{\Phi}_i$ then all points of $\overline{\alpha}_{i+1}^{-1}(p)$ are prepared for $\overline{\Phi}_{i+1}$. Thus conditions (a) and (b) of 1 hold.

We now verify 2 of the theorem. We first observe that Φ_1 (from the conclusions of part 1 of this theorem) is strongly prepared (Definition 18.1 [**C3**]). We now examine the monomialization algorithm of Chapter 18 [**C3**] and the toroidalization algorithm of Chapter 19 [**C3**], applied to $\Phi_1 : X_1 \to S$.

The monomialization algorithm of Theorem 18.19 and Theorem 18.21 [**C3**] consists in constructing a commutative diagram

$$\begin{array}{ccc} \tilde{X}_2 & \stackrel{\tilde{\pi}_2}{\to} & X_1 \\ \tilde{\Phi}_2 \downarrow & & \downarrow \Phi_1 \\ \tilde{S}_1 & \stackrel{\tilde{\Psi}_1}{\to} & S \end{array} \tag{6.1}$$

such that $\tilde{\Phi}_2$ is monomial (all points of $\tilde{X}_2$ are good for $\tilde{\Phi}_2$ as defined in Definition 18.5 [**C3**]).

(6.1) has a factorization

$$\begin{array}{ccccccccccc} \tilde{X}_2 & = & Z_l & \stackrel{\epsilon_l}{\to} & \cdots & \to & Z_1 & \stackrel{\epsilon_1}{\to} & Z_0 & = & X_1 \\ \tilde{\Phi}_2 \downarrow & & \Omega_l \downarrow & & & & \Omega_1 \downarrow & & \Omega_0 \downarrow & & \Phi_1 \downarrow \\ \tilde{S}_1 & = & T_l & \stackrel{\delta_l}{\to} & \cdots & \to & T_1 & \stackrel{\delta_1}{\to} & T_0 & = & S \end{array} \tag{6.2}$$

where each Ω_i is strongly prepared, each δ_{i+1} is the blow up of a point q_i such that $\Omega_i^{-1}(q_i)$ contains a point p_i at which Ω_i is not monomial and ϵ_{i+1} is a sequence of blow ups of curves which are exceptional to such q_i. This step is accomplished by performing the algorithms of Lemmas 18.16, 18.17 and 18.18 of [**C3**].

Each map $Z_{i+1} \to Z_i$ has a factorization

$$Z_{i+1} = \overline{Z}_m \stackrel{\lambda_m}{\to} \overline{Z}_{m-1} \to \cdots \stackrel{\lambda_1}{\to} \overline{Z}_0 = Z_i \tag{6.3}$$

where each λ_{j+1} is a blow up of a 2-curve or of a curve C_j which contains a 1-point, makes SNCs with the preimage $D_{\overline{Z}_j}$ of D_X on $\overline{Z}_j$, and is contained in a component of $D_{\overline{Z}_j}$. C_j is contained in the locus where $m_{q_i}\mathcal{O}_{\overline{Z}_j}$ is not invertible. To construct (6.3) we successively apply Lemmas 18.16, 18.17, 18.18 of [**C3**].

The algorithms of Lemma 18.16 and Lemmas 18.18 [**C3**] consist of blow ups of 2-curves.

The algorithm of Lemma 18.17 [**C3**] consists of a sequence of blow ups of curves $\lambda_{j+1} : \overline{Z}_{j+1} \to \overline{Z}_j$ of $C_j \subset D_{\overline{Z}_j}$ which are not 2-curves, and are contained in the locus where $m_{q_i}\mathcal{O}_{\overline{Z}_j}$ is not invertible.

Theorem 19.9 [**C3**] and Theorem 19.10 [**C3**] imply there exists a commutative diagram

$$\begin{array}{ccc} \tilde{X}_3 & \stackrel{\tilde{\pi}_3}{\rightarrow} & \tilde{X}_2 \\ \tilde{\Phi}_3 \downarrow & & \downarrow \tilde{\Phi}_2 \\ \tilde{S}_2 & \stackrel{\tilde{\psi}_2}{\rightarrow} & \tilde{S}_1 \end{array} \tag{6.4}$$

such that 2 of the conclusions of the theorem holds.

□

CHAPTER 7

Preparation above 2 and 3-points

LEMMA 7.1. *Suppose that $f : X \to Y$ is a dominant morphism of nonsingular projective 3-folds with toroidal structures determined by SNC divisors D_Y and $D_X = f^{-1}(D_Y)$ such that D_X contains the singular locus of f. Then there exist a commutative diagram*

$$\begin{array}{ccc} X_1 & \overset{f_1}{\to} & Y_1 \\ \Phi \downarrow & & \downarrow \Psi \\ X & \overset{f}{\to} & Y \end{array}$$

such that Φ and Ψ are products of blow ups of 2-curves and f_1 is toroidal above all 3-points of Y_1.

PROOF. Suppose that ν is a 0-dimensional valuation of $\mathbf{k}(X)$. We will say that ν is resolved for f if the center of ν on Y is not a 3-point or if the center of ν on Y is a 3-point, and f is toroidal at the center of ν on X.

Being resolved is an open condition on the Zariski-Riemann manifold of X, and if ν is resolved for f and

$$\begin{array}{ccc} X_1 & \overset{f_1}{\to} & Y_1 \\ \Phi_1 \downarrow & & \downarrow \Psi_1 \\ X & \overset{f}{\to} & Y \end{array}$$

is a commutative diagram such that Φ_1 and Ψ_1 are products of blow ups of 2-curves, then ν is resolved for f_1.

Suppose that ν is a 0-dimensional valuation of $\mathbf{k}(X)$ such that the center q of ν on Y is a 3-point. Let p be the center of ν on X. Let u, v, w be permissible parameters at q.

Case 1. Suppose that $\nu(u), \nu(v), \nu(w)$ are rationally independent. Since $uvw = 0$ is a local equation of D_X at p, there exist regular parameters x, y, z in $\mathcal{O}_{X,p}$ such that $xyz = 0$ contains the germ of D_X at p, and we have an expression

$$\begin{array}{ll} u & = x^a y^b z^c \gamma_1 \\ v & = x^d y^e z^f \gamma_2 \\ w & = x^g y^h z^i \gamma_3 \end{array}$$

where the γ_i are units in $\mathcal{O}_{X,p}$. Since $\nu(u), \nu(v), \nu(w)$ are rationally independent, $\nu(x), \nu(y), \nu(z)$ are also rationally independent and

$$\mathrm{Det}\begin{pmatrix} a & b & c \\ d & e & f \\ g & h & i \end{pmatrix} \neq 0$$

which implies that p is a 3-point and f is toroidal at p. Thus ν is resolved for f.

Case 2. Suppose that $\nu(u), \nu(v)$ are rationally dependent. After possibly interchanging u, v, w we reduce to this case. Let C be the 2-curve of Y with local equation $u = v = 0$ at q. There exists a sequence of blow ups of 2-curves $\Psi_\nu : Y_\nu \to Y$ which are sections over C such that the center of ν on Y_ν is not a 3-point.

Y_ν is the blow up of a toroidal ideal sheaf $\mathcal{J}_\nu$ of $\mathcal{O}_Y$. Since $f^{-1}(D_Y) = D_X$, $\mathcal{J}_\nu\mathcal{O}_X$ is also a toroidal ideal sheaf. By Lemma 5.1, there exists a sequence of blow ups of 2-curves $\Phi_\nu : X_\nu \to X$ such that there is a commutative diagram of morphisms

$$\begin{array}{ccc} X_\nu & \stackrel{f_\nu}{\to} & Y_\nu \\ \Phi_\nu \downarrow & & \downarrow \Psi_\nu \\ X & \stackrel{f}{\to} & Y. \end{array}$$

Thus ν is resolved for f_ν.

It follows from compactness of the Zariski Riemann manifold of X [**Z**], that there exists a positive integer n and commutative diagrams

$$\begin{array}{ccc} X_i & \stackrel{f_i}{\to} & Y_i \\ \Phi_i \downarrow & & \downarrow \Psi_i \\ X & \stackrel{f}{\to} & Y \end{array}$$

for $1 \le i \le n$ such that Φ_i and Ψ_i are products of blow ups of 2-curves, and every 0-dimensional valuation ν of $\mathbf{k}(X)$ is resolved for some f_i.

Y_i is the blow up of a toroidal ideal sheaf $\mathcal{J}_i$ of $\mathcal{O}_Y$ and X_i is the blow up of a toroidal ideal sheaf $\mathcal{I}_i$ of $\mathcal{O}_X$. Thus there exists a sequence of blow ups of 2-curves $Y^* \to Y$ such that $\prod_i \mathcal{J}_i\mathcal{O}_{Y^*}$ is invertible, by Lemma 5.1. Y^* is thus the blow up of a toroidal ideal sheaf $\mathcal{J} \subset \mathcal{O}_Y$, so that $\mathcal{J}\mathcal{O}_X$ is also a toroidal ideal sheaf. By Lemma 5.1, there exists a sequence of blow ups of 2-curves $X^* \to X$ such that $\mathcal{J}\prod \mathcal{I}_i\mathcal{O}_{X^*}$ is invertible. Thus for $1 \le i \le n$ there exist commutative diagrams of morphisms

$$\begin{array}{ccc} X^* & \stackrel{f^*}{\to} & Y^* \\ \Phi_i^* \downarrow & & \downarrow \Psi_i^* \\ X_i & \stackrel{f_i}{\to} & Y_i \\ \downarrow & & \downarrow \\ X & \to & Y. \end{array}$$

Suppose that ν is a 0-dimensional valuation of $\mathbf{k}(X)$. If the center of ν on Y^* is a 3-point, then the center of ν on Y_i is a 3-point for all i, since Ψ_i^* is toroidal. There exists an i such that ν is resolved for f_i. Thus f_i is toroidal at the center of ν on X_i. Since Φ_i^* and Ψ_i^* are toroidal, f^* is toroidal at the center of ν. Thus ν is resolved for f^*. Since all 0-dimensional valuations of $\mathbf{k}(X)$ are resolved for f^*, it follows that f^* is toroidal above all 3-points of Y^*, and we have achieved the conclusions of the lemma. □

LEMMA 7.2. *Suppose that $f : X \to Y$ is a dominant morphism of nonsingular projective 3-folds, with toroidal structures determined by SNC divisors D_Y and $D_X = f^{-1}(D_Y)$ such that D_X contains the singular locus of f. Further suppose* *that f is toroidal above all 3-points of Y. Then there exists a commutative diagram*

$$\begin{array}{ccc} X_1 & \xrightarrow{f_1} & Y_1 \\ \Phi\downarrow & & \downarrow\Psi \\ X & \xrightarrow{f} & Y \end{array}$$

such that Ψ and Ψ are products of blow ups of 2-curves, f_1 is toroidal above all 3-points of Y_1, and f_1 is prepared (and satisfies 2 (a) of Definition 4.6) above all 2-points of Y_1.

PROOF. Suppose that ν is a 0-dimensional valuation of $\mathbf{k}(X)$. We will say that ν is resolved for f if the center of ν on Y is a 1-point or if the center of ν on Y is a 2-point and f is prepared at the center of ν on X (and satisfies 2 (a) of Definition 4.6), or if the center of ν on Y is a 3-point, and f is toroidal at the center of ν on X.

Being resolved is an open condition on the Zariski-Riemann manifold of X. Suppose that

$$\begin{array}{ccc} X_1 & \xrightarrow{f_1} & Y_1 \\ \Phi\downarrow & & \downarrow\Psi \\ X & \xrightarrow{f} & Y \end{array}$$

is a commutative diagram of morphisms such that Φ and Ψ are products of blow ups of 2-curves. If ν is a 0-dimensional valuation of $\mathbf{k}(X)$ such that ν is resolved for f, then ν is resolved for f_1.

Suppose that $q \in Y$ is a 2-point, and ν is a 0-dimensional valuation of $\mathbf{k}(X)$ such that q is the center of ν on Y. Let p be the center of ν on X. Let u, v, w be permissible parameters at q, so that $u = v = 0$ are local equations of the 2-curve C through q.

Case 1. Suppose that $\nu(u), \nu(v)$ are rationally independent. Since $uv = 0$ is a local equation of D_X at p, there exist regular parameters x, y, z in $\mathcal{O}_{X,p}$ such that $xyz = 0$ contains the germ of D_X in $\mathcal{O}_{X,p}$, and we have an expression

$$\begin{aligned} u &= x^a y^b z^c \gamma_1 \\ v &= x^d y^e z^f \gamma_2 \end{aligned}$$

where the γ_i are units in $\mathcal{O}_{X,p}$. Since $\nu(u), \nu(v)$ are rationally independent,

$$\operatorname{rank}\begin{pmatrix} a & b & c \\ d & e & f \end{pmatrix} = 2$$

which implies that p is a 2 or 3-point and f is prepared at p (and satisfies 2 (a) of Definition 4.6). Thus ν is resolved for f.

Case 2. Suppose that $\nu(u), \nu(v)$ are rationally dependent. There exists a sequence of blow ups of 2-curves which are sections over C, $\Psi_\nu : Y_\nu \to Y$ such that the center of ν on Y_ν is a 1-point.

Y_ν is the blow up of a toroidal ideal sheaf $\mathcal{J}_\nu$ of $\mathcal{O}_Y$. Since $f^{-1}(D_Y) = D_X$, $\mathcal{J}_\nu$ is also a toroidal ideal sheaf. By Lemma 5.1, there exists a sequence of blow ups of

2-curves $\Phi_\nu : X_\nu \to X$ such that there is a commutative diagram of morphisms

$$\begin{array}{lll} X_\nu & \stackrel{f_\nu}{\to} & Y_\nu \\ \Phi_\nu \downarrow & & \downarrow \Psi_\nu \\ X & \stackrel{f}{\to} & Y. \end{array}$$

Thus ν is resolved for f_ν.

It follows from compactness of the Zariski Riemann manifold of X [**Z**] that there exists a positive integer n and commutative diagrams

$$\begin{array}{lll} X_i & \stackrel{f_i}{\to} & Y_i \\ \Phi_i \downarrow & & \downarrow \Psi_i \\ X & \stackrel{f}{\to} & Y \end{array}$$

for $1 \le i \le n$ such that Φ_i and Ψ_i are products of blow ups of 2-curves, and every valuation ν of $\mathbf{k}(X)$ is resolved for some f_i.

Y_i is the blow up of a toroidal ideal sheaf $\mathcal{J}_i$ of $\mathcal{O}_Y$ and X_i is the blow up of a toroidal ideal sheaf $\mathcal{I}_i$ of $\mathcal{O}_X$. Thus there exists a sequence of blow ups of 2-curves $Y^* \to Y$ such that $\prod_i \mathcal{J}_i\mathcal{O}_{Y^*}$ is invertible, by Lemma 5.1. Y^* is thus the blow up of a toroidal ideal sheaf $\mathcal{J} \subset \mathcal{O}_Y$. Thus $\mathcal{J}\mathcal{O}_X$ is also a toroidal ideal sheaf. By Lemma 5.1, there exists a sequence of blow ups of 2-curves $X^* \to X$ such that $\mathcal{J}\prod \mathcal{I}_i\mathcal{O}_{X^*}$ is invertible. Thus for $1 \le i \le n$, there exist commutative diagrams of morphisms

$$\begin{array}{lll} X^* & \stackrel{f^*}{\to} & Y^* \\ \Phi_i^* \downarrow & & \downarrow \Psi_i^* \\ X_i & \stackrel{f_i}{\to} & Y_i \\ \downarrow & & \downarrow \\ X & \to & Y. \end{array}$$

Since Φ_i^* and Ψ_i^* are the blow ups of toroidal ideal sheaves, they are toroidal morphisms.

Suppose that ν is a 0-dimensional valuation of $\mathbf{k}(X)$. If the center of ν on Y is a 3-point then f^* is resolved at the center of ν on X^*. In particular, since $\Psi_i \circ \Psi_i^*$ is toroidal, if the center of ν on Y^* is a 3-point, then the center of ν on Y is a 3-point and ν is resolved for f^*. Suppose that the center of ν on Y^* is 2-point, and the center of ν on Y is not a 3-point. Then the center of ν on Y_i is a 2-point for all i. There exists an i such that ν is resolved for f_i. Thus f_i is prepared (and satisfies 2 (a) of Definition 4.6) at the center of ν on X_i. Since Φ_i^* and Ψ_i^* are toroidal, f^* is prepared (and satisfies 2 (a) of Definition 4.6) at the center of ν. Thus ν is resolved for f^*. Since all 0-dimensional valuations of $\mathbf{k}(X)$ are resolved for f^*, it follows that f^* is toroidal above all 3-points of Y^*, and prepared above all 2-point of Y^*, and we have achieved the conclusions of the lemma. □

LEMMA 7.3. *Suppose that $f : X \to Y$ satisfies the conclusions of Lemma 7.2. Suppose that H is a general hyperplane section of Y. Then f is prepared above all points of H.*

PROOF. Bertini's theorem implies that H is nonsingular and makes SNCs with D_Y. Further, $H' = f^{-1}(H)$ is nonsingular and makes SNCs with D_X. Thus H contains no 3-points of Y and H' contains no 3-points of X.

Suppose that $q \in H \cap D_Y$ is a 1-point, and $p \in f^{-1}(q)$. Let u, v, w be regular parameters in $\mathcal{O}_{Y,q}$ such that $u = 0$ is a local equation of D_Y at q, and $w = 0$ is a local equation of H. Then we have regular paramaters x, y, z in $\mathcal{O}_{X,p}$ such that either p is a 1 point with $x = 0$ a local equation of D_X or p is a 2-point with $xy = 0$ a local equation of D_X at p, and we have an expression $u = x^a\gamma, w = z$ or $u = x^a y^b \gamma, w = z$ where γ is a unit in $\mathcal{O}_{X,p}$. Thus f is prepared at p. □

COROLLARY 7.4. Suppose that $f : X \rightarrow Y$ satisfies the conclusions of Lemma 7.2. Then there exists a finite set of 1-points $\Omega \subset Y$ such that f is prepared above $Y - \Omega$.

PROOF. The locus of points in X where f is prepared is an open set. Since f is proper, the image Ω of the closed set of points where f is not prepared is closed in Y. Since a general hyperplane section of Y is disjoint from Ω by Lemma 7.3, Ω must be a finite set of points. □

LEMMA 7.5. *Suppose that $f : X \rightarrow Y$ is a proper dominant morphism of nonsingular 3-folds and $\pi : Y \rightarrow S$ is a smooth dominant morphism onto a nonsingular surface S. Let $g = \pi \circ f$. Suppose that C is a nonsingular curve of S, $D = \pi^{-1}(C)$ and $D' = f^{-1}(D)$. Suppose that D' is a SNC divisor on X which contains the singular locus of g and the singular locus of f. Suppose that $\overline{q} \in C$ is a point, and that g is toroidal and prepared (with respect to C and D') away from points above finitely many points $\Omega = \{q_1, \ldots, q_m\} \subset \gamma = \pi^{-1}(\overline{q})$. Further suppose that f is finite above a general point of γ. Then there exists a commutative diagram of morphisms*

$$\begin{array}{ccc} X_1 & \stackrel{\Phi_1}{\rightarrow} & X \\ g_1 & \searrow & \downarrow g \\ & & S \end{array}$$

such that Φ_1 is a product of possible blow ups for the preimage of D' supported above Ω and g_1 is prepared (with respect to C and $\Phi_1^{-1}(D')$) in a neighborhood of all components F of $\Phi_1^{-1}(D')$ which dominate D and in a neighborhood of all components F of $\Phi_1^{-1}(D')$ which dominate a curve of Y.

PROOF. Let u, w be regular parameters in $\mathcal{O}_{S,\overline{q}}$ such that $u = 0$ is a local equation of C at $\overline{q}$. Let C' be the curve on S with local equation $w = 0$ at $\overline{q}$. Let $A = \pi^{-1}(C')$.

Since it suffices to prove the lemma above a neighborhood of $\overline{q}$ in S, we may assume that $E = C + C'$ is a SNC divisor on S whose only singular point is $\overline{q}$. Since g is toroidal away from points above Ω, we have that $g^{-1}(E)$ defines a SNC divisor on X away from points above Ω. There exists a morphism $\Phi_1 : X_1 \rightarrow X$ which is a sequence of possible blow ups for the preimage of D' supported above Ω such that with $g_1 = g \circ \Phi_1 : X_1 \rightarrow S$, $g_1^{-1}(E)$ is a SNC divisor, and $(f \circ \Phi_1)^{-1}(q_i)$ are divisors for all $q_i \in \Omega$. We may further assume that the union $\overline{A}$ of codimension 1 subvarieties of X_1 which dominate A are disjoint, since the fact that g is toroidal away from $\overline{q}$ implies they are disjoint away from the preimage of Ω.

Let $\overline{D}$ be the union of codimension 1 subvarieties of X_1 which dominate D, so that $\overline{D}$ is a disjoint union of irreducible components of $D'' = g_1^{-1}(C)$ (by Remark 4.5).

Suppose that $p \in \overline{D}$ and $f \circ \Phi_1(p) = q_i \in \Omega$. Then p must be a 2-point or a 3-point. We have regular parameters x, y, z in $\hat{\mathcal{O}}_{X_1,p}$ such that one of the following cases hold:

1. p is a 2-point and
$$u = x^a y^b, w = y^c$$
where $x = 0$ is a local equation of $\overline{D}$, $u = 0$ is a local equation of D'' and $a, b > 0$.
2. p is a 2-point,
$$u = x^a y^b, w = y^c z$$
where $x = 0$ is a local equation of $\overline{D}$, $u = 0$ is a local equation of D'', $a, b, c > 0$ and $z = 0$ is a local equation of $\overline{A}$.
3. p is a 3-point and
$$u = x^a y^b z^c, w = y^d z^e$$
where $x = 0$ is a local equation of $\overline{D}$, $u = 0$ is a local equation of D'' and $a, b, c > 0$.

Thus g_1 is prepared in a neighborhood of $\overline{D}$.

Now suppose that F is a component of D'' which dominates a curve of Y and $p \in F$ is such that $f \circ \Phi_1(p) = q_i \in \Omega$. Then p must be a 2-point or a 3-point. By our assumption that f is finite above a general point of γ, F dominates the curve C of S. Thus we have regular parameters x, y, z in $\hat{\mathcal{O}}_{X_1,p}$ such that one of the following cases hold:

1. p is a 2-point and
$$u = x^a y^b, w = y^c$$
where $x = 0$ is a local equation of F, $u = 0$ is a local equation of D'' and $a, b > 0$.
2. p is a 2-point,
$$u = x^a y^b, w = y^c z$$
where $x = 0$ is a local equation of F, $u = 0$ is a local equation of D'', $a, b, c > 0$ and $z = 0$ is a local equation of $\overline{A}$.
3. p is a 3-point and
$$u = x^a y^b z^c, w = y^d z^e$$
where $x = 0$ is a local equation of F, $u = 0$ is a local equation of D'' and $a, b, c > 0$.

Thus g_1 is prepared in a neighborhood of F. □

LEMMA 7.6. *Suppose that $f : X \to Y$ is a dominant morphism of nonsingular 3-folds with toroidal structures determined by SNC divisors D_Y and $D_X = f^{-1}(D_Y)$ such that D_X contains the singular locus of f. Further suppose that $f : X \to Y$ is toroidal and $q \in Y$ is a 2-point. Let $\Psi : Y_1 \to Y$ be the blow up of q. Then there exists a commutative diagram of morphisms*

$$\begin{array}{ccc} X_1 & \overset{f_1}{\to} & Y_1 \\ \Phi \downarrow & & \downarrow \Psi \\ X & \overset{f}{\to} & Y \end{array}$$

such that Φ is a sequence of possible blow ups for the preimage of D_X supported above q and f_1 is toroidal with respect to $D_{Y_1} = \Psi^{-1}(D_Y)$ and $D_{X_1} = \Phi^{-1}(D_X)$.

PROOF. There exist permissible parameters u, v, w at q such that if $p \in f^{-1}(q)$ then there exist permissible parameters x, y, z for u, v, w such that if p is a 1-point, then we have a form

$$u = x^a, v = x^b(\alpha + y), w = z \tag{7.1}$$

with $0 \neq \alpha \in \mathbf{k}$, and if p is a 2-point,

$$u = x^a y^b, v = x^c y^d, w = z, \tag{7.2}$$

with $ad - bc \neq 0$. We first show that there exists a sequence of possible blow ups

$$X_m \overset{\Phi_m}{\to} X_{m-1} \to \cdots \to X_1 \overset{\Phi_1}{\to} X \tag{7.3}$$

obtained by blow ups of possible centers supported above q such that the rational map $X_m \to Y_1$ is toroidal wherever it is defined, and if $\mathcal{I}_q\mathcal{O}_{X_m,p}$ is not invertible, then there exist regular parameters x, y, z in $\hat{\mathcal{O}}_{X_m,p}$ such that one of the following forms hold:

p is a 1-point

$$u = x^a, v = x^b(\alpha + y), w = x^c z \tag{7.4}$$

with $0 \neq \alpha \in \mathbf{k}$, and $c = 0$ or 1, or p is a 2-point

$$u = x^a y^b, v = x^c y^d, w = xz \tag{7.5}$$

with $a, c \geq 1$, or p is a 2-point

$$u = x^a y^b, v = x^c y^d, w = xyz \tag{7.6}$$

with $a, c \geq 1$ and $b, d \geq 1$.

After possibly interchanging u and v, the points $p \in f^{-1}(p)$ such that u, v, w do not have a form (7.4), (7.5) or (7.6) at p are 2-points of one of the following forms:

$$u = x^a, v = y^b, w = z, \tag{7.7}$$

in which case $V(x, y, z)$ is the locus in $\text{spec}(\hat{\mathcal{O}}_{X,p})$ where $\mathcal{I}_q\hat{\mathcal{O}}_{X,p}$ is not invertible,

$$u = x^a, v = x^b y^c, w = z \tag{7.8}$$

with $b, c > 0$, in which case $V(x, z)$ is the locus in $\text{spec}(\hat{\mathcal{O}}_{X,p})$ where $\mathcal{I}_q\hat{\mathcal{O}}_{X,p}$ is not invertible,

$$u = x^a y^b, v = x^c y^d, w = z \tag{7.9}$$

with $a, b, c, d > 0$ in which case $V(x, z) \cup V(y, z)$ is the locus in $\text{spec}(\hat{\mathcal{O}}_{X,p})$ where $\mathcal{I}_q\hat{\mathcal{O}}_{X,p}$ is not invertible,

Let Z be the closed locus of points r in X such that $\mathcal{I}_q\mathcal{O}_{X,r}$ is not invertible. The isolated points p in Z have a form (7.7). If p is a non isolated point in Z which is a 2-point, then p has a form (7.8) or (7.9).

Suppose that E is a curve in Z such that E contains a 2-point p satisfying (7.8) or (7.9). Then a generic point of E satisfies (7.1) and all 2-points of E must have a form (7.8) or (7.9).

Let $\Phi_1 : X_1 \to X$ be the blow up of the finitely many points $p \in X$ of the form (7.7). Suppose that $p \in X$ is such a point, and $p_1 \in \Phi_1^{-1}(p)$. After possibly interchanging u and v, we may assume that $a \leq b$ in (7.7). There are regular parameters x_1, y_1, z_1 in $\hat{\mathcal{O}}_{X_1,p_1}$ of one of the following forms:

$$x = x_1, y = x_1(y_1 + \alpha), z = x_1(z_1 + \beta) \tag{7.10}$$

with $\alpha, \beta \in \mathbf{k}$,

$$x = x_1 y_1, y = y_1, z = y_1(z_1 + \alpha) \tag{7.11}$$

with $\alpha \in \mathbf{k}$ or

$$x = x_1 z_1, y = y_1 z_1, z = z_1. \tag{7.12}$$

Suppose that (7.10) holds. Then u, v, w have a form

$$u = x_1^a, v = x_1^b (y_1 + \alpha)^b, w = x_1(z_1 + \beta)$$

at p_1. If $a = 1$, then $f \circ \Phi_1$ factors through Y_1 at p_1 and we have one of the following toroidal forms:

1-point maps to 2-point:

$$u_1 = u = x_1, v_1 = \frac{v}{u} = x_1^{b-1}(y_1 + \alpha)^b, w_1 = \frac{w}{u} - \beta = z_1$$

if $b > a = 1$ and $\alpha \neq 0$,

1-point maps to 1-point:

$$u_1 = u = x_1, v_1 = \frac{v}{u} - \alpha = y_1, w_1 = \frac{w}{u} - \beta = z_1$$

if $b = a = 1$, $\alpha \neq 0$,

2-point maps to 2-point:

$$u_1 = u = x_1, v_1 = \frac{v}{u} = x_1^{b-1} y_1^b, w_1 = \frac{w}{u} - \beta = z_1$$

if $a = 1$ and $\alpha = 0$.

Suppose that (7.10) holds and $a > 1$. If $\beta \neq 0$, we have that $\Phi_1 \circ f$ factors through Y_1 at p_1 and we have a toroidal form, obtained from a change of variable in

$$u_1 = \frac{u}{w} = x_1^{a-1}(z_1 + \beta)^{-1}, v_1 = \frac{v}{w} = x_1^{b-1}(z_1 + \beta)^{-1}(y_1 + \alpha)^b, w_1 = w = x_1(z_1 + \beta)$$

where p_1 is 1-point mapping to a 3-point if $\alpha \neq 0$ and p_1 is 2-point mapping to a 3-point if $\alpha = 0$.

If $\beta = 0$ (and $a > 1$) then we have

$$u = x_1^a, v = x_1^b(y_1 + \alpha), w = x_1 z_1$$

of the form (7.4) if $\alpha \neq 0$ and of the form (7.5) if $\alpha = 0$.

Suppose that (7.11) holds. Then at p_1, u, v, w have a form:

$$u = x_1^a y_1^a, v = y_1^b, w = y_1(z_1 + \alpha).$$

Assume $b = 1$ (which implies $a = 1$). then $f \circ \Phi_1$ factors through Y_1 at p_1, and there is a toroidal form:

$$u_1 = \frac{u}{v} = x_1, v_1 = v = y_1, w_1 = \frac{w}{v} - \alpha = z_1$$

where p_1 is 2-point mapping to a 2-point.

Assume that $b > 1$ and $\alpha \neq 0$. Then $f \circ \Phi_1$ factors through Y_1 at p_1, and there is a toroidal form, obtained from a change of variable in

$$u_1 = \frac{u}{w} = x_1^a y_1^{a-1}(z_1 + \alpha)^{-1}, v_1 = \frac{v}{w} = y_1^{b-1}(z_1 + \alpha)^{-1}, w_1 = w = y_1(z_1 + \alpha)$$

where p_1 is a 2-point mapping to a 3-point.

If $b > 1$ and $\alpha = 0$, then we have a form:

$$u = x_1^a y_1^a, v = y_1^b, w = y_1 z_1$$

of the form (7.5).

Suppose that (7.12) holds. Then p_1 is a 3-point and u, v, w have a form

$$u = x_1^a z_1^a, v = y_1^b z_1^b, w = z_1.$$

Thus $\Phi_1 \circ f$ factors through Y_1 at p_1 by

$$u_1 = \frac{u}{w} = x_1^a z_1^{a-1}, v_1 = \frac{v}{w} = y_1^b z_1^{b-1}, w_1 = w = z_1,$$

where p_1 is a 3-point mapping to a 3-point.

We have thus completed the analysis of Φ_1.

We now construct (7.3) by induction. Each X_i will be such that the rational map $X_i \to Y_1$ is toroidal wherever it is defined, and if $p \in X_i$ is a 2-point such that $\mathcal{I}_q\mathcal{O}_{X_1,p}$ is not invertible, then there exist regular parameters x, y, z at p such that u, v, w have one of the forms (7.4), (7.5), (7.6), (7.8) or (7.9) at p. If a form (7.8) or (7.9) holds at p, then $\Phi_1 \circ \cdots \circ \Phi_i$ is an isomorphism near p.

Each $\Phi_{i+1} : X_{i+1} \to X_i$ for $i \geq 1$ will be the blow up of a curve E_i which is a possible center and is the strict transform of a component of $Z \subset X$.

Suppose that we have constructed (7.3) out to X_i, and $p \in X_i$ is a 2-point such that $\mathcal{I}_q\mathcal{O}_{X_i,p}$ is not invertible, and u, v, w do not have a form (7.4), (7.5) or (7.6) at p. Then u, v, w have a form (7.8) or (7.9) at p. Let $E = E_i$ be a curve in the locus where $\mathcal{I}_q\mathcal{O}_{X_i}$ is not invertible which contains p. Let F be the component of D_{X_i} containing E_i We necessarily have $\mathrm{ord}_F w = 0$ and $\mathrm{ord}_F u > 0$, $\mathrm{ord}_F v > 0$. Further, $\Phi_1 \circ \cdots \circ \Phi_i$ is an isomorphism near p. Thus E is the strict transform of a component of Z.

Suppose that $p' \in E_i$ is another 2-point. Then at p', since $\mathrm{ord}_F w = 0$, u, v, w must have a form (7.8), (7.9) or (7.5), where in this last case, $y = z = 0$ is a local equation of E at p' and $b, d \geq 1$ (since $\mathrm{ord}_F w = 0$, $\mathrm{ord}_F u > 0$ and $\mathrm{ord}_F v > 0$). If $p' \in E_i$ is a 1-point, then u, v, w have a form (7.1) at p', since $\mathrm{ord}_F w = 0$.

Let $\Phi_{i+1} : X_{i+1} \to X_i$ be the blow up of E and $\overline{\Phi}_{i+1} = \Phi_1 \circ \cdots \circ \Phi_{i+1}$.

Suppose that $p \in E$ is a 1-point and $p_1 \in \Phi_{i+1}^{-1}(p)$. We have that there is a form (7.4) at p with $c = 0$. Then $f \circ \overline{\Phi}_{i+1}$ is toroidal whenever it is defined, and points above p where $f \circ \overline{\Phi}_{i+1}$ does not factor through Y_i have a form (7.4) (with $c = 1$). A detailed analysis of a case including this one is given later in the proof, after (7.19).

Suppose that $p \in E$ is a 2-point of the form (7.9) and $p_1 \in \Phi_{i+1}^{-1}(p)$.

There are regular parameters x_1, y_1, z_1 in $\hat{\mathcal{O}}_{X_i,p_1}$ of one of the following forms:

(7.13) $$x = x_1, z = x_1(z_1 + \alpha)$$

with $\alpha \in \mathbf{k}$ or

(7.14) $$x = x_1 z_1, z = z_1.$$

Suppose that (7.13) holds. We have that p_1 is a 2-point, and

$$u = x_1^a y^b, v = x_1^c y^d, w = x_1(z_1 + \alpha).$$

If $\alpha \neq 0$, we have that $f \circ \overline{\Phi}_{i+1}$ factors through Y_1 at p_1. We have a form:

$$u_1 = \frac{u}{w} = x_1^{a-1} y^b (z_1 + \alpha)^{-1}, v_1 = \frac{v}{w} = x_1^{c-1} y^d (z_1 + \alpha)^{-1}, w_1 = x_1(z_1 + \alpha)$$

at the 2-point p_1, which maps to a 3-point, and thus is toroidal, after a change of variables.

If $\alpha = 0$ in (7.13), we have

$$u = x_1^a y^b, v = x_1^c y^d, w = x_1 z_1$$

of the form (7.5).

If (7.14) holds, then p_1 is a 3-point and

$$u = x_1^a y^b z_1^a, v = x_1^c y^d z_1^c, w = z_1.$$

Thus $f \circ \overline{\Phi}_{i+1}$ factors through Y_1 at p_1, and we have a toroidal form:

$$u_1 = \frac{u}{w} = x_1^a y^b z_1^{a-1}, v_1 = \frac{v}{w} = x_1^c y^d z_1^{c-1}, w_1 = w = z_1$$

at the 3-point p_1, which maps to a 3-point.

The analysis of Φ_{i+1} above points (7.8) and above points satisfying (7.5) where $y = z = 0$ are local equations of E (and $b, d \geq 1$) is similar. This last case will lead to a form (7.6). Since Z has only finitely many components, we inductively construct (7.3).

There now exists a sequence of blow ups of 2-curves $X_r \rightarrow X_m$ which are supported above q such that the rational map $X_r \rightarrow Y_1$ is toroidal where ever it is defined, and if $\mathcal{I}_q \mathcal{O}_{X_r,p}$ is not invertible, then there there exist permissible parameters x, y, z at p for u, v, w such that one of the following forms hold:

p a 1-point

$$u = x^a, v = x^b(\alpha + y), w = x^d z \tag{7.15}$$

with $0 \neq \alpha \in \mathbf{k}$ and $d < \min\{a, b\}$ or

p a 2-point

$$u = x^a y^b, v = x^c y^d, w = x^e y^f z \tag{7.16}$$

with $(e, f) < (a, b) < (c, d)$ or $(e, f) < (c, d) < (a, b)$.

We accomplish this as follows. We first consider u and v. Suppose that $p \in X_m$ is a 2-point such that $\mathcal{I}_q \mathcal{O}_{X_m,p}$ is not invertible. We have forms

$$u = x^a y^b, v = x^c y^d, w = x^e y^f z \tag{7.17}$$

with $e + f > 0$ at 2-points p_i above p in the construction of the sequence $X_r \rightarrow X_m$. At p_i we have an invariant $(a - c)(b - d)$. This is a nonnegative integer if and only if $(a, b) \leq (c, d)$ or $(c, d) \leq (a, b)$. Further, if $(a - c)(b - d) < 0$, then after blowing up the 2-curve E which has local equations $x = y = 0$ at p_i, we obtain that all 2-points above p_i have a form (7.17), but $(a - c)(b - d)$ has increased. Further E contracts to q on Y since $e + f > 0$.

After a finite number of blow ups of 2-curves above X_m (which must contract to q) we achieve that all 2-points p_i above a 2-point $p \in X_m$ such that $\mathcal{I}_q \mathcal{O}_{X_m,p}$ is not invertible have a form (7.17) with $(a, b) \leq (c, d)$ or $(c, d) \leq (a, b)$.

We now apply this algorithm to the pairs $u, x^e y^f$ and $v, x^e y^f$ in (7.17) to construct $X_r \rightarrow X_m$.

We will now inductively construct $X_n \rightarrow X_r$ so that $\mathcal{I}_q \mathcal{O}_{X_n}$ is invertible everywhere and the morphism $X_n \rightarrow Y_1$ is toroidal. We will construct a sequence of blow ups

$$X_n \rightarrow X_{n-1} \rightarrow \cdots \rightarrow X_{r+1} \rightarrow X_r \tag{7.18}$$

so that each $\Phi_{i+1} : X_{i+1} \rightarrow X_i$ is the blow up of a nonsingular curve λ_i which is a possible center and is contained in the locus where $\mathcal{I}_q \mathcal{O}_{X_i}$ is not invertible. We will have that the rational map $f_i : X_i \to Y_1$ is toroidal where ever it is defined, and all points $p \in X_i$ where $\mathcal{I}_q\mathcal{O}_{X_i,p}$ is not invertible have a form (7.15) or (7.16).

Suppose that we have inductively constructed (7.18) up to X_i and $\mathcal{I}_q\mathcal{O}_{X_i}$ is not invertible. We will construct $\Phi_{i+1} : X_{i+1} \to X_i$.

Inspection of the forms (7.15) and (7.16) shows that the locus in X_i where $\mathcal{I}_q\mathcal{O}_{X_i}$ is not invertible is a union of nonsingular curves which are possible centers. For such a curve λ, let η be a general point of λ, so that a form (7.15) holds at η. Let $A(\lambda) = \min\{a, b\} - d > 0$.

Choose a curve λ_i which maximizes $A(\lambda)$ on X_i. Let $\Phi_{i+1} : X_{i+1} \to X_i$ be the blow up of λ_i. Suppose that $p_i \in \lambda_i$ and $p_{i+1} \in \Phi_{i+1}^{-1}(\lambda_i)$.

First suppose that p_i has the form (7.15). After possibly interchanging u and v, we may assume that $a \le b$. There are regular parameters x_1, y, z_1 in $\hat{\mathcal{O}}_{X_{i+1},p_{i+1}}$ satisfying

$$x = x_1, z = x_1(z_1 + \beta) \tag{7.19}$$

for some $\beta \in \mathbf{k}$, or

$$x = x_1 z_1, z = z_1. \tag{7.20}$$

Suppose that (7.19) holds. p_{i+1} is then a 1-point, and

$$u = x_1^a, v = x_1^b(\alpha + y), w = x_1^{d+1}(z_1 + \beta). \tag{7.21}$$

If $d + 1 = a = b$ in (7.21), then $X_{i+1} \to Y_1$ is a morphism near p_{i+1}, which maps p_{i+1} to a 1-point, and has a toroidal form

$$u_1 = u = x_1^a, v_1 = \frac{v}{u} - \alpha = y_1, w_1 = \frac{w}{u} - \beta = z_1.$$

If $d + 1 = a < b$ in (7.21), then $X_{i+1} \to Y_1$ is a morphism near p_{i+1}, which maps p_{i+1} to a 2-point, and has a toroidal form

$$u_1 = u = x_1^a, v_1 = \frac{v}{u} = x_1^{b-a}(\alpha + y_1), w_1 = \frac{w}{u} - \beta = z_1.$$

If $d + 1 < a \le b$ and $\beta \ne 0$ in (7.21) then $X_{i+1} \to Y_1$ is a morphism near p_{i+1}, which maps p_{i+1} to a 3-point, and has a toroidal form obtained from a change of variable in

$$u_1 = \frac{u}{w} = x_1^{a-d-1}(z_1 + \beta)^{-1}, v_1 = \frac{v}{w} = x_1^{b-d-1}(\alpha + y_1)(z_1 + \beta)^{-1},$$

$$w_1 = w = x_1^{d+1}(z_1 + \beta).$$

If $d + 1 < a \le b$ and $\beta = 0$ then (7.21) has a form (7.15) with $d < d + 1 < \min\{a, b\}$.

Suppose that (7.20) holds. p_{i+1} is then a 2-point, and

$$u = x_1^a z_1^a, v = x_1^b z_1^b(\alpha + y), w = x_1^d z_1^{d+1}.$$

$X_{i+1} \to Y_1$ is thus a morphism near p_{i+1}, which maps p_{i+1} to a 3-point, and has a toroidal form

$$u_1 = \frac{u}{w} = x_1^{a-d} z_1^{a-d-1}, v_1 = \frac{v}{w} = x_1^{b-d} z_1^{b-d-1}(\alpha + y_1), w_1 = w = x_1^d z_1^{d+1}.$$

Now suppose that p_i has the form (7.16). After possibly interchanging u and v, we may assume that $(a, b) < (c, d)$. After possibly interchanging x and y, we may assume that there are regular parameters x_1, y, z_1 in $\hat{\mathcal{O}}_{X_{i+1},p_{i+1}}$ satisfying (7.19) or (7.20) (so that $e < a$).

Suppose that (7.19) holds. Then p_{i+1} is a 2-point. We have

$$u = x_1^a y^b, v = x_1^c y^d, w = x_1^{e+1} y^f (z_1 + \beta). \tag{7.22}$$

If $(e+1, f) = (a, b)$ in (7.22), then $X_{i+1} \to Y_1$ is a morphism near p_{i+1}, which maps p_{i+1} to a 2-point, and has a toroidal form

$$u_1 = u = x_1^a y^b, v_1 = \frac{v}{u} = x_1^{c-a} y_1^{d-b}, w_1 = \frac{w}{u} - \beta = z_1.$$

If $(e+1, f) < (a, b)$ and $\beta \neq 0$ in (7.22), then $X_{i+1} \to Y_1$ is a morphism near p_{i+1}, which maps p_{i+1} to a 3-point, and has a toroidal form obtained from a change of variable in

$$u_1 = \frac{u}{w} = x_1^{a-e-1} y^{b-f} (z_1 + \beta)^{-1}, v_1 = \frac{v}{w} = x_1^{c-e-1} y_1^{d-f} (z_1 + \beta)^{-1},$$

$$w_1 = w = x_1^{e+1} y^f (z_1 + \beta).$$

If $(e+1, f) < (a, b)$ and $\beta = 0$ then (7.22) has a form (7.16) with $(e, f) < (e+1, f) < (a, b)$.

Suppose that (7.20) holds. p_{i+1} is then a 3-point, and

$$u = x_1^a y^b z_1^a, v = x_1^c y^d z_1^c, w = x_1^e y^f z_1^{e+1}.$$

$X_{i+1} \to Y_1$ is thus a morphism near p_{i+1}, which maps p_{i+1} to a 3-point, and has a toroidal form

$$u_1 = \frac{u}{w} = x_1^{a-e} y^{b-f} z_1^{a-e-1}, v_1 = \frac{v}{w} = x_1^{c-e} y^{d-f} z_1^{c-e-1}, w_1 = w = x_1^e y^f z_1^{e+1}.$$

In summary, we have that all points where $\mathcal{I}_q \mathcal{O}_{X_{i+1}}$ is not invertible have a form (7.15) or (7.16) and if $\lambda_{i+1} \subset \Phi_i^{-1}(\lambda_i)$ is a curve such that $\mathcal{I}_q \mathcal{O}_{X_{i+1}}$ is not invertible along λ_i, we have $0 < A(\lambda_{i+1}) < A(\lambda_i)$. Thus after a finite number of blow ups, we construct the desired sequence (7.18), completing the proof of the lemma. □

LEMMA 7.7. *Suppose that $f : X \to Y$ is a dominant morphism of nonsingular 3-folds with toroidal structures determined by SNC divisors D_Y and $D_X = f^{-1}(D_Y)$ such that D_X contains the singular locus of f. Further suppose that $f : X \to Y$ is toroidal and $C \subset Y$ is a possible center for D_Y which is a curve and contains a 1-point. Let $\Psi : Y_1 \to Y$ be the blow up of C. Then there exists a commutative diagram of morphisms*

$$\begin{array}{ccc} \overline{X}_1 & \xrightarrow{f_1} & Y_1 \\ \Phi \downarrow & & \downarrow \Psi \\ X & \xrightarrow{f} & Y \end{array}$$

such that Φ is a sequence of possible blow ups for the preimage of D_X supported above C and f_1 is toroidal with respect to $D_{Y_1} = \Psi^{-1}(D_Y)$ and $D_{\overline{X}_1} = \Phi^{-1}(D_X)$.

Further, Φ has a factorization

$$\overline{X}_1 = X_n \to X_{n-1} \to \cdots \to X_1 \to X$$

where each $\Phi_{i+1} : X_{i+1} \to X_i$ is either the blow up of a section E_i over C such that $\mathcal{I}_C \mathcal{O}_{X_i}$ is not invertible, or $\Phi_{i+1} : X_{i+1} \to X_i$ is the blow up of a curve E_i which maps to a 2-point of Y and such that E_i is contained in the locus where $\mathcal{I}_C \mathcal{O}_{X_i}$ is invertible.

PROOF. We follow the algorithm of Lemma 18.17 [**C3**] to construct Φ.

Suppose that $q \in C$ and $p \in f^{-1}(q)$. Then there are permissible parameters u, v, w for D_Y in $\mathcal{O}_{Y,q}$ and regular parameters x, y, z in $\hat{\mathcal{O}}_{X,p}$ such that one of the following cases holds:

q is a 2-point and p is a 2-point,

$$u = x^a y^b, v = x^d y^e, w = z \tag{7.23}$$

where $uv = 0$ is a local equation of D_Y and $u = w = 0$ is a local equation of C.

q is a 2-point and p is a 1-point,

$$u = x^a, v = x^b(y + \alpha), w = z \tag{7.24}$$

where $0 \neq \alpha \in \mathbf{k}$, $uv = 0$ is a local equation of D_Y and $u = w = 0$ is a local equation of C.

q is a 1-point and p is a 1-point,

$$u = x^a, v = y, w = z \tag{7.25}$$

where $u = 0$ is a local equation of D_Y and $u = w = 0$ is a local equation of C.

We will construct a sequence of morphisms

$$\cdots \rightarrow X_n \overset{\Phi_n}{\rightarrow} X_{n-1} \overset{\Phi_{n-1}}{\rightarrow} \cdots \rightarrow X_1 \overset{\Phi_1}{\rightarrow} X \tag{7.26}$$

where each Φ_{i+1} is the blow up of a nonsingular curve E_i contained in the locus where $\mathcal{I}_C\mathcal{O}_{X_i}$ is not invertible, and for each $q \in C$ and $p \in (f \circ \Phi_1 \circ \cdots \circ \Phi_i)^{-1}(q)$ such that $\mathcal{I}_C\mathcal{O}_{X_i,p}$ is not invertible, there are permissible parameters u, v, w for D_Y in $\mathcal{O}_{Y,q}$ and permissible parameters x, y, z in $\hat{\mathcal{O}}_{X,p}$ such that one of the forms (7.27) - (7.29) below hold.

q a 2-point, p a 2-point

$$u = x^a y^b, v = x^d y^e, w = x^g y^h z \tag{7.27}$$

with $ae - ba \neq 0$, $(g, h) < (a, b)$.

q a 2-point, p a 1-point

$$u = x^a, v = x^b(y + \alpha), w = x^d z \tag{7.28}$$

with $0 \neq \alpha \in \mathbf{k}$, $d < a$,

q a 1-point, p a 1-point

$$u = x^a, v = y, w = x^d z \tag{7.29}$$

with $d < a$. Further in the locus where the rational map $X_i \rightarrow Y_1$ is a morphism, $X_i \rightarrow Y_1$ is toroidal.

Observe that the forms (7.23), (7.24) and (7.25) are special cases of (7.27), (7.28) and (7.29) respectively.

The locus of points where $\mathcal{I}_C\mathcal{O}_{X_i}$ is not invertible is a union of nonsingular curves which intersect transversally. If E is a curve in this locus, and $p' \in E$ is a general point, then u, v, w have a form (7.28) or (7.29) at p'. In either case, we define an invariant

$$\Omega(E) = a - d > 0.$$

Let $\Phi_{i+1} : X_{i+1} \rightarrow X_i$ be the blow up of a curve E_i such that $\Omega(E_i)$ is maximal. Suppose that $p_1 \in E_i$, $p_2 \in \Phi_{i+1}^{-1}(p_1)$ and $q = (f \circ \Phi_1 \circ \cdots \circ \Phi_i)(p_1)$.

Suppose that p_1 has a form (7.28). Then $\hat{\mathcal{O}}_{X_{i+1},p_2}$ has regular parameters x_1, y, z_1 such that

$$x = x_1, z = x_1(z_1 + \beta) \tag{7.30}$$

with $\beta \in \mathbf{k}$ or

$$x = x_1 z_1, z = z_1. \tag{7.31}$$

Suppose that (7.30) holds. Then p_2 is a 1-point,

$$u = x_1^a, v = x_1^b(y + \alpha), w = x_1^{d+1}(z_1 + \beta). \tag{7.32}$$

If $d + 1 = a$ in (7.32), then $X_{i+1} \to Y_1$ is a morphism near p_2, mapping p_2 to a 2-point, and at p_2, we have a toroidal form

$$u_1 = u = x_1^a, v = x_1^b(y + \alpha), w_1 = \frac{w}{u} - \beta = z_1.$$

If $d + 1 < a$ and $\beta \neq 0$ in (7.32) then $X_{i+1} \to Y_1$ is a morphism near p_2, mapping p_2 to a 3-point, and at p_2, we have a toroidal form obtained from a change of variable in

$$u_1 = \frac{u}{w} = x_1^{a-d-1}(z_1 + \beta)^{-1}, v = x_1^b(y_1 + \alpha), w_1 = w = x_1^{d+1}(z_1 + \beta).$$

If $d + 1 < a$ and $\beta = 0$ in (7.32), then we have a form (7.28) with d increased to $d + 1$. The curve E' containing p_2 in the locus where $\mathcal{I}_C \mathcal{O}_{X_{i+1}}$ is not invertible satisfies

$$0 < \Omega(E') = a - (d + 1) < \Omega(E).$$

Suppose that (7.31) holds. Then p_2 is a 2-point.

$$u = x_1^a z_1^a, v = x_1^b z_1^b(y + \alpha), w = x_1^d z_1^{d+1}. \tag{7.33}$$

Further, $X_{i+1} \to Y_1$ is a morphism near p_2, mapping p_2 to a 3-point, and at p_2, we have a toroidal form

$$u_1 = \frac{u}{w} = x_1^{a-d} z_1^{a-d-1}, v = x_1^b z_1^b(y + \alpha), w_1 = w = x_1^d z_1^{d+1}.$$

There is a similar argument if p_1 satisfies (7.29).

Suppose that p_1 has a form (7.27) and $x = z = 0$ are local equations of E_i (so that $g < a$). $\hat{\mathcal{O}}_{X_{i+1},p_2}$ has regular parameters x_1, y_1, z_1 satisfying (7.30) or (7.31).

Suppose that (7.30) holds. Then p_2 is a 2-point,

$$u = x_1^a y^b, v = x_1^d y^e, w = x_1^{g+1} y^h(z_1 + \beta). \tag{7.34}$$

If $(g + 1, h) = (a, b)$ in (7.34), then $X_{i+1} \to Y_1$ is a morphism near p_2, mapping p_2 to a 2-point, and at p_2, we have a toroidal form

$$u_1 = u = x_1^a y^b, v_1 = v = x_1^d y^e, w_1 = \frac{w}{u} - \beta = z_1.$$

If $(g + 1, h) < (a, b)$ and $\beta \neq 0$ in (7.34), then $X_{i+1} \to Y_1$ is a morphism near p_2, mapping p_2 to a 3-point, and we have a toroidal form obtained from a change of variable in

$$u_1 = \frac{u}{w} = x_1^{a-g-1} y^{b-h}(z_1 + \beta)^{-1}, v = x_1^d y^e, w_1 = w = x_1^{g+1} y^h(z_1 + \beta).$$

If $(g + 1, h) < (a, b)$ and $\beta = 0$ in (7.34), then (7.34) has the form (7.27) with g increased to $g + 1$.

Suppose that (7.31) holds. Then p_2 is a 3-point,

$$u = x_1^a y^b z_1^a, v = x_1^d y^e z_1^d, w = x_1^g y_1^h z_1^{g+1}.$$

$X_{i+1} \to Y_1$ is a morphism near p_2, mapping p_2 to a 3-point, and at p_2, we have a toroidal form

$$u_1 = \frac{u}{w} = x_1^{a-g} y_1^{b-h} z_1^{a-g-1}, v = x_1^d y^e z_1^d, w_1 = w = x_1^g y_1^h z_1^{g+1}.$$

By descending induction on $\max(\Omega(E))$, we see that the sequence (7.26) must terminate after a finite number of blow ups, and we complete the proof of the lemma. □

CHAPTER 8

Preparation

In this chapter we prove Theorem 1.3 stated in the introduction, giving the construction of a prepared morphism of 3-folds.

To prove this theorem, we may assume by Lemmas 7.1 and 7.2 that f is prepared (of type 2 (a) of Definition 4.6) above 2 points and toroidal above 3-points of Y. By Corollary 7.4, f only fails to be prepared above a finite set of 1-points $\Sigma \subset Y$.

Suppose that $q \in \Sigma$. Let D be the component of D_Y containing q. There exists a very ample effective divisor L on Y such that $q \notin L$ and $D + L \sim H$ where H is a very ample effective divisor such that $q \notin H$. Let $\alpha : Z \rightarrow Y$ be the blow up of q, with exceptional divisor E. We may replace L with a high multiple of L so that $\alpha^* H - E$ is very ample on Z. Let N be a general member of $\alpha^* H - E$. By Bertini's theorem, N is nonsingular, makes SNCs with $D_Z = \alpha^{-1}(D_Y)$, intersects 2-curves of D_Z transversally at general points, does not contain a component of the strict transform on Z of the nonfinite locus of f, and is disjoint from $\alpha^{-1}(\Sigma - \{q\})$. Let $M = \alpha(N)$. Then $M \sim H$, M is nonsingular and intersects D transversally in a nonsingular curve $\overline{\gamma}$ which contains q, M contains no other points of Σ, contains no 3-points of D_Y, intersects 2-curves of D_Y transversally at general points, does not contain a component of the nonfinite locus of f and by Bertini's theorem, at points which are not above q, $f^*(M)$ is nonsingular and $f^*(M) + D_X$ is a SNC dvisor. After possibly replacing L and H with effective divisors linearly equivalent to L and H respectively, we may assume that $\overline{\gamma} \cap (L + H)$ consists of 1-points of D_Y and is disjoint from the nonfinite locus of f.

$U = Y - (L + H)$ is an affine neighborhood of q. Let $\gamma = \overline{\gamma} \cap U$. There exist $\overline{f}, \overline{g} \in \Gamma(Y, \mathcal{O}_Y(H))$ such that $(\overline{f}) = D + L - H$ and $(\overline{g}) = M - H$. We can thus define a morphism $\pi : U \rightarrow S = \mathbf{A}^2$ by $\pi(a) = (\overline{f}(a), \overline{g}(a))$ for $a \in U$. Let $\overline{q} = \pi(q)$. $\pi^{-1}(\overline{q}) = \gamma$ (scheme theoretically) so π is smooth in a neighborhood of γ. We may thus replace U with an open neighborhood of γ so that π is smooth.

Let $\overline{X} = f^{-1}(U)$, and $\overline{f} = f \mid \overline{X}$. Let $D_U = D_Y \cap U$, $D_U^* = D \cap U$, $D_S^* = \pi(D_U^*)$, $g = \pi \circ \overline{f} : \overline{X} \rightarrow S$, $D_{\overline{X}}^* = g^{-1}(D_S^*)$, $D_{\overline{X}} = D_X \cap \overline{X}$. The map π is toroidal with respect to D_S^* and D_U^*.

Let $D_1, \ldots, D_m$ be the components of D_Y other than D which intersect γ. Since γ intersects these components transversally, we may assume then that $\pi \mid D_i \cap U$ is étale onto its image for $1 \leq i \leq m$. We further may assume that $\Sigma \cap U = \{q\}$, and (by Bertini's theorem) for $\overline{q}' \in D_S^* - \{\overline{q}\}$, there exist regular parameters u, w at $\overline{q}'$ such that $u = 0$ is a local equation of D_S^*, and if E is the curve $w = 0$ on S, then E is nonsingular, $D_S^* + E$ is a SNC divisor, $D_U^* + \pi^{-1}(E)$ is a SNC divisor on U, $g^{-1}(E)$ is nonsingular, and $g^{-1}(E) + D_{\overline{X}}$ is a SNC divisor on $\overline{X}$. Thus if $q' \in \pi^{-1}(\overline{q}')$, there exist permissible parameters u, v, w in $\mathcal{O}_{U,q'}$ (for D_U) such that if $p\in \overline{f}^{-1}(q')$ then there exist regular parameters x,y,z in $\hat{\mathcal{O}}_{\overline{X},p}$ such that

$$u=x^ay^b, w=z \tag{8.1}$$

where $u=x^ay^b=0$ is a local equation of $D_{\overline{X}}$ at p (with $a>0$, $b\ge 0$) if $q'\in D-\cup D_i$ and

$$u=x^ay^b, v=x^cy^d\gamma_1, w=z \tag{8.2}$$

where $\gamma_1\in\hat{\mathcal{O}}_{\overline{X},p}$ is a unit and $uv=x^{a+c}y^{b+d}=0$ is a local equation of $D_{\overline{X}}$ at p if $q'\in D\cap D_i$ for some i.

Since γ intersects the 2-curves $D_i\cap D\cap U$ of U at general points of the 2-curves, after possibly replacing U with a smaller open neighborhood of γ, we have that the intersection of the nonfinite locus of f with U is contained in $D\cap U$.

We will now establish that g is toroidal and prepared with respect to D_S^* and $D_{\overline{X}}^*$ away from the preimages of finitely many 1-points $\Omega\subset\gamma$ of D_U.

Suppose that $q'\in(D_i-D)\cap U$, $p\in\overline{f}^{-1}(q')$, and $\overline{q}'=\pi(q')$, which implies that there exist regular parameters u,w at $\overline{q}'$, u,v,w at q' such that $v=0$ is a local equation of D_i. q' is not in the nonfinite locus of $\overline{f}$, and q' is a 1-point of D_U, so by Abhyankar's lemma ([**Ab4**] and XIII 5.3 [**G**]) there exist regular parameters x,y,z in $\hat{\mathcal{O}}_{\overline{X},p}$ such that

$$u=x, v=y^b, w=z. \tag{8.3}$$

g is defined by $u=x, w=z$ near p, which implies that g is smooth, and thus prepared and toroidal for D_S^* and $D_{\overline{X}}^*$ at p.

Suppose that $q'\in(D-\gamma)\cap U$, $p\in\overline{f}^{-1}(q')$, $\overline{q}'=\pi(q')$. Then we have a form (8.1) or (8.2) at p, so that g is prepared and toroidal for D_S^* and $D_{\overline{X}}^*$ at p.

Let $\delta=D\cap D_i\cap U$ for some $1\le i\le m$. Suppose that $q'\in\delta\cap\gamma$ and $p\in\overline{f}^{-1}(q')$. Then $\pi(q')=\overline{q}$. There exist regular parameters u,w in $\mathcal{O}_{S,\overline{q}}$ such that $u=0$ is a local equation of D on U, $w=0$ is a local equation of M on U, there exists $v\in\mathcal{O}_{U,q'}$ such that $v=0$ is a local equation of D_i and u,v,w are regular parameters in $\mathcal{O}_{U,q'}$. By our choice of M, $\overline{f}^*(M)$ is nonsingular and makes SNCs with $D_{\overline{X}}$ at p. Since $uv=0$ is a local equation of $D_{\overline{X}}$ at p, and $w=0$ is a local equation of $\overline{f}^*(M)$ at p, there exist regular parameters x,y,z in $\mathcal{O}_{\overline{X},p}$ such that

$$u=x^ay^b\gamma_1, v=x^cy^d\gamma_2, w=z$$

with γ_1,γ_2 units in $\mathcal{O}_{\overline{X},p}$. Thus g is prepared and toroidal for D_S^* and $D_{\overline{X}}^*$ at p.

Suppose that $q'\in\gamma$ is a general point. Then q' is a 1-point of D_U and $\overline{f}$ is finite above q'. There exist regular parameters u,v,w in $\mathcal{O}_{U,q'}$ such that u,w are permissible parameters for D_S^* at $\overline{q}=\pi(q')$, and if $p\in\overline{f}^{-1}(q')$, then there exist permissible parameters x,y,z in $\hat{\mathcal{O}}_{\overline{X},p}$ such that

$$u=x^a, v=y, w=z \tag{8.4}$$

by Abhyankar's lemma, which implies that g is prepared and toroidal at p for D_S^* and $D_{\overline{X}}^*$.

We conclude that g is toroidal and prepared with respect to D_S^* and $D_{\overline{X}}^*$ away from points above finitely many 1-points $\Omega\subset\gamma$ of D_U.

By Lemma 7.5 (applied to a neighborhood of Ω), there exists a morphism $\Phi_1:\overline{X}_1\to\overline{X}$ such that Φ_1 is a sequence of possible blow ups for the preimage of $D_{\overline{X}}^*$ of points and nonsingular curves supported above Ω such that if $g_1 = g\circ\Phi_1 : \overline{X}_1 \to S$ and $f_1 = \overline{f}\circ\Phi_1 : \overline{X}_1 \to U$, then in a neighborhood of all components of $D^*_{\overline{X}_1}$ which do not map to a point of Ω, g_1 is prepared for D^*_S and $D^*_{\overline{X}_1} = \Phi_1^{-1}(D^*_{\overline{X}})$

By 1 of Theorem 6.1, there exists a morphism $\Phi_2 : \overline{X}_2 \to \overline{X}_1$ which is a sequence of possible blow ups for the preimage of $D^*_{\overline{X}_1}$ of points and nonsingular curves supported above Ω, such that $g_2 = \pi\circ f_1\circ\Phi_2 : \overline{X}_2 \to S$ is prepared for D^*_S and $D^*_{\overline{X}_2} = \Phi_2^{-1}(D^*_{\overline{X}_1})$. Let $f_2 = f_1\circ\Phi_2 : \overline{X}_2 \to U$.

Now by 2 of Theorem 6.1, there exists a commutative diagram

$$\begin{array}{ccc} \overline{X}_3 & \stackrel{g_3}{\to} & S_1 \\ \overline{\Phi}_3 \downarrow & & \downarrow \lambda_1 \\ \overline{X}_2 & \stackrel{g_2}{\to} & S \end{array}$$

such that λ_1 is a sequence of possible blow ups for the preimage of D^*_S of points supported above $\overline{q}$, $\overline{\Phi}_3$ is a sequence of possible blow ups for the preimage of $D^*_{\overline{X}_2}$ of points and nonsingular curves supported above γ, and g_3 is toroidal with respect to $D^*_{S_1} = \lambda_1^{-1}(D^*_S)$ and $D^*_{\overline{X}_3} = \overline{\Phi}_3^{-1}(D^*_{\overline{X}_2})$. Let $f_3 = f_2\circ\overline{\Phi}_3 : \overline{X}_3 \to U$.

Consider the commutative diagram

$$\begin{array}{ccccc} \overline{X}_3 & \stackrel{\overline{f}_3}{\to} & \overline{Y}_1 & \stackrel{\pi_1}{\to} & S_1 \\ \overline{\Phi} \downarrow & & \overline{\Psi} \downarrow & & \downarrow \lambda_1 \\ \overline{X} & \stackrel{\overline{f}}{\to} & U & \stackrel{\pi}{\to} & S \end{array}$$

where $\overline{\Phi} = \Phi_1\circ\Phi_2\circ\overline{\Phi}_3$, $\overline{Y}_1 = U\times_S S_1$ and $\overline{\Psi} : \overline{Y}_1 \to U$, $\pi_1 : \overline{Y}_1 \to S_1$ are the natural projections, and $\overline{f}_3 = f_3\times g_3$. $D^*_{\overline{Y}_1} = \overline{\Psi}^{-1}(D^*_U)$ and $D_{\overline{Y}_1} = \overline{\Psi}^{-1}(D_U)$ are SNC divisors. $\overline{Y}_1$ is nonsingular, and is obtained from U by possible blow ups for the preimage of D_U of sections over γ. Since g_3 is toroidal with respect to $D^*_{S_1}$ and $D^*_{\overline{X}_3}$, $\overline{f}_3$ is prepared with respect to $D^*_{\overline{Y}_1}$ and $D^*_{\overline{X}_3}$.

Suppose that $q' \in \gamma$ is a general point, and $p \in (\overline{f}\circ\overline{\Phi})^{-1}(q')$. Then $\overline{f}$ has a toroidal and prepared form (8.4) at $\overline{\Phi}(p)$. By Theorem 6.1, and the form of the factorization (6.3), we see that $\overline{f}_3$ is toroidal and prepared at p with respect to $D^*_{\overline{Y}_3}$ and $D^*_{\overline{X}_3}$. Also, over a general point of γ, $\overline{\Phi}$ is a sequence of possible blow ups for the preimages of $D^*_{\overline{X}}$ of sections over γ.

Recall that $D_Y = D + D_1 + \cdots + D_m + G$, where G consists of the components of D_Y disjoint from U, and that the $D_i\cap U$ are étale over their images in S. Let $\overline{D}_i$ be the strict transform of D_i on $\overline{Y}_1$ for $1\le i\le m$. Let

$$D_{\overline{Y}_1} = \overline{\Psi}^{-1}(D_U) = D^*_{\overline{Y}_1} + \overline{D}_1 + \cdots + \overline{D}_m,$$

$D_{\overline{X}_3} = \overline{\Phi}^{-1}(D_{\overline{X}})$.

We will now verify that $D_{\overline{X}_3}$ is a SNC divisor on $\overline{X}_3$ and that $\overline{f}_3$ is prepared for $D_{\overline{Y}_1}$ and $D_{\overline{X}_3}$. Since $\overline{f}_3$ is prepared for $D^*_{\overline{Y}_1}$ and $D^*_{\overline{X}_3}$, we need only verify that $\overline{f}_3$ is prepared for $D_{\overline{Y}_1}$ and $D_{\overline{X}_3}$ at points $p' \in \overline{X}_3$ such that $q' = \overline{f}\circ\overline{\Phi}(p') \in D_i$ for some i.

First suppose that $q' \in D_i - \gamma$ for some i. Then $\overline{\Phi}$ and $\overline{\Psi}$ are isomorphisms near p' and q' respectively. Suppose that $q' \notin D$. Then we have permissible parameters v, u, w for D_U at q' which have an expression (8.3) at p'. Thus $\overline{f}_3$ has an expression

3 of Definition 4.6 at p'. Suppose that $q' \in D \cap D_i - \gamma$. Then q' is a 2-point of D_U, so that $\overline{f}$ is prepared above q' for D_U and $D_{\overline{X}}$. Thus $\overline{f}_3$ is prepared above q' (for $D_{\overline{Y}_1}$ and $D_{\overline{X}_3}$).

Suppose that $q' = \overline{f} \circ \overline{\Phi}(p') \in \gamma \cap D_i$ for some i. Without loss of generality, we may assume that $D_i = D_1$. Recall that $q' \in \gamma \cap D_1$ is a general point of the 2-curve $D \cap D_1$, $\overline{f}$ is prepared above q' and $\overline{f}^*(M)$ is nonsingular and makes SNCs with $D_{\overline{X}}$ above q'. Since q' is a general point of $D \cap D_1$, there are no 3-points in $\overline{f}^{-1}(q')$. Let D_1' be the reduced divisor on $\overline{X}$ whose components dominate D_1. The irreducible components of D_1' are disjoint by Remark 4.5.

There exist permissible parameters u, v, w in $\mathcal{O}_{U,q'}$ for the two point q' of D_U such that $u = 0$ is a local equation of D, $v = 0$ is a local equation of D_1, $w = 0$ is a local equation of M on U, and u, w are regular parameters in $\mathcal{O}_{S,\overline{q}}$ such that if $p = \overline{\Phi}(p') \in \overline{f}^{-1}(q')$, then there exist regular parameters x, y, z in $\hat{\mathcal{O}}_{\overline{X},p}$ such that one of the following prepared forms for $\overline{f}$ hold at p. u, v are toroidal forms for D_U and $D_{\overline{X}}$ in all cases.

1. p is a 1-point of $D_{\overline{X}}$

$$\begin{aligned} u &= x^a \\ v &= x^b\gamma \\ w &= z \end{aligned} \tag{8.5}$$

where $\gamma \in \hat{\mathcal{O}}_{\overline{X},p}$ is a unit and $x = 0$ is a local equation of $D_{\overline{X}}$.

2. p is a 2-point of $D_{\overline{X}}$ which is not on D_1'

$$\begin{aligned} u &= x^a y^b \\ v &= x^c y^d \gamma \\ w &= z \end{aligned} \tag{8.6}$$

with $a, b > 0$, $\gamma \in \hat{\mathcal{O}}_{\overline{X},p}$ is a unit and $xy = 0$ is a local equation of $D_{\overline{X}}$.

3. p is a 2-point which is on D_1'

$$\begin{aligned} u &= x^a \\ v &= x^b y^c \\ w &= z \end{aligned} \tag{8.7}$$

where $xy = 0$ is a local equation of $D_{\overline{X}}$ and $y = 0$ is a local equation of D_1'.

$\overline{\Psi}$ is the sequence of monodial transforms induced by a sequence of quadratic transforms,

$$S_1 = \overline{S}_n \to \cdots \to \overline{S}_0 = S.$$

Each map $\overline{S}_{j+1} \to \overline{S}_j$ is the blow up of the ideal sheaf m_j of a point $\overline{q}_j$ above $\overline{q}$. Let

$$\overline{Y}_1 = \tilde{Y}_n \to \cdots \to \tilde{Y}_0 = U \tag{8.8}$$

be the induced factorization of $\overline{\Psi}$, where $\tilde{\Psi}_{j+1} : \tilde{Y}_{j+1} = U \times_S \overline{S}_{j+1} \to \tilde{Y}_j = U \times_S \overline{S}_j$, is the blow up of a curve C_j. Let $\overline{\pi}_j : \tilde{Y}_j \to \overline{S}_j$ be the natural projection.

$\overline{\Phi}$ is a sequence of morphisms

$$\overline{X}_3 = \tilde{X}_n \to \cdots \to \tilde{X}_0 = \overline{X}_2 \to \overline{X}.$$

where $\tilde{\Phi}_{j+1} : \tilde{X}_{j+1} \rightarrow \tilde{X}_j$ is a principalization of $m_j\mathcal{O}_{\tilde{X}_j}$, with natural morphism $\tilde{f}_j : \tilde{X}_j \rightarrow \tilde{Y}_j$. Let $D_{\tilde{X}_j}$, $D_{\tilde{Y}_j}$ be the respective preimages of D_U, and let $D^*_{\tilde{X}_j}$, $D^*_{\tilde{Y}_j}$ be the respective preimages of D^*_U. Let $D^*_{\overline{S}_j}$ be the preimage of D^*_S in $\overline{S}_j$. The principalizations $\tilde{\Phi}_j$ are explicitly described in the proof of Theorem 6.1. We have a factorization

$$\tilde{X}_{j+1} = \hat{X}_{n_j,j} \rightarrow \cdots \rightarrow \hat{X}_{0,j} = \tilde{X}_j. \tag{8.9}$$

where each $\hat{\Phi}_{i+1,j} : \hat{X}_{i+1,j} \rightarrow \hat{X}_{i,j}$ is the blow up of a single curve E_{ij} which is a possible center for the preimage $D^*_{\hat{X}_{ij}}$ of D^*_U on $\hat{X}_{i,j}$. E_{ij} is in the locus where $m_j\mathcal{O}_{\hat{X}_{i,j}}$ is not locally principal. Let $D_{\hat{X}_{ij}}$ be the preimage of D_U on $\hat{X}_{ij}$.

Recall that $\overline{X}_2 \rightarrow \overline{X}$ is an isomorphism above q'.

We will prove that $\overline{f}_3 : \overline{X}_3 \rightarrow \overline{Y}_1$ is prepared for $D_{\overline{Y}_1}$ and $D_{\overline{X}_3}$ above q' by induction on j in the morphisms $\tilde{f}_j : \tilde{X}_j \rightarrow \tilde{Y}_j$.

Recall that we have a fixed choice of regular parameters $u = u_0, v, w = w_0$ in $\mathcal{O}_{U,q'}$, which are permissible parameters for D_Y at the 2-point q', and one of the forms (8.5) - (8.7) holds at all points of $\overline{X}$ above q'.

Suppose by induction that $D_{\tilde{X}_j}$ is a SNC divisor, $\tilde{f}_j : \tilde{X}_j \rightarrow \tilde{Y}_j$ is prepared for $D_{\tilde{Y}_j}$ and $D_{\tilde{X}_j}$, and if $q_j \in \tilde{Y}_j$ and $\tilde{\Psi}_1 \circ \cdots \circ \tilde{\Psi}_j(q_j) = q'$, then

1. q_j is a 2-point or a 3-point of $D_{\tilde{Y}_j}$ and there exist regular parameters u_j, w_j in $\mathcal{O}_{\overline{S}_j,\overline{q}'_j}$, where $\overline{\pi}_j(q_j) = \overline{q}'_j$, such that u_j, v, w_j are permissible parameters for $D_{\tilde{Y}_j}$ in $\mathcal{O}_{\tilde{Y}_j,q_j}$.
2. If $q_j \in C_j$, then $u_j = w_j = 0$ are local equations of C_j at q_j.
3. If $p_j \in \tilde{f}_j^{-1}(q_j)$, then there exist regular parameters x_j, y_j, z_j in $\hat{\mathcal{O}}_{\tilde{X}_j,p_j}$ such that one of the following forms hold:

Case 1. q_j is a 2-point of $D_{\tilde{Y}_j}$, and $u_jv = 0$ is a local equation of $D_{\tilde{Y}_j}$ (so that $\overline{\pi}_j(q_j) = \overline{q}'_j$ is a 1-point), and p_j is a 1-point of $D_{\tilde{X}_j}$ with

$$u_j = x_j^a, v = x_j^b\gamma_j, w_j = z_j \tag{8.10}$$

where $x_j = 0$ is a local equation of $D_{\tilde{X}_j}$, γ_j is a unit in $\hat{\mathcal{O}}_{\tilde{X}_j,p_j}$ or p_j is a 2-point of $D_{\tilde{X}_j}$ with

$$u_j = x_j^a y_j^b, v = x_j^c y_j^d \gamma_j, w_j = z_j \tag{8.11}$$

where $x_j y_j = 0$ is a local equation of $D_{\tilde{X}_j}$, γ_j is a unit in $\hat{\mathcal{O}}_{\tilde{X}_j,p_j}$.

Case 2. q_j is a 3-point of $D_{\tilde{Y}_j}$, and $u_jvw_j = 0$ is a local equation of $D_{\tilde{Y}_j}$ (so that $\overline{\pi}_j(q_j) = \overline{q}'_j$ is a 2-point), and p_j is a 1-point of $D_{\tilde{X}_j}$ with

$$u_j = x_j^a, v = x_j^b\gamma_j, w_j = x_j^c(z_j + \beta) \tag{8.12}$$

where $x_j = 0$ is a local equation of $D_{\tilde{X}_j}$, γ_j is a unit in $\hat{\mathcal{O}}_{\tilde{X}_j,p_j}$ and $0 \neq \beta \in \mathbf{k}$ or p_j is a 2-point of $D_{\tilde{X}_j}$ with

$$u_j = x_j^a z_j^b, v = x_j^c z_j^d \gamma_j, w_j = x_j^e z_j^f \tag{8.13}$$

where $af - be \neq 0$, $x_jz_j = 0$ is a local equation of $D_{\tilde{X}_j}$, γ_j is a unit in $\hat{\mathcal{O}}_{\tilde{X}_j,p_j}$ or p_j is a 2-point of $D_{\tilde{X}_j}$ with

$$u_j = (x_j^a y_j^b)^k, v = x_j^d y_j^e \gamma_j, w_j = (x_j^a y_j^b)^t (z_j + \beta) \tag{8.14}$$

where $x_jy_j = 0$ is a local equation of $D_{\tilde{X}_j}$, γ_j is a unit in $\hat{\mathcal{O}}_{\tilde{X}_j,p_j}$, $\gcd(a,b) = 1$ and $0 \neq \beta \in \mathbf{k}$ or p_j is a 3-point of $D_{\tilde{X}_j}$ with

$$u_j = x_j^a y_j^b z_j^c, v = x_j^d y_j^e z_j^f \gamma_j, w_j = x_j^g y_j^h z_j^i \tag{8.15}$$

where $x_jy_jz_j = 0$ is a local equation of $D_{\tilde{X}_j}$, γ_j is a unit in $\hat{\mathcal{O}}_{\tilde{X}_j,p_j}$ and

$$\operatorname{rank}\begin{pmatrix} a & b & c \\ g & h & i \end{pmatrix} = 2.$$

All of the forms of Case 1 and Case 2 are prepared for $D_{\tilde{Y}_j}$ and $D_{\tilde{X}_j}$. In case 1, this follows from Remark 4.7. In case 2, u_j and w_j are toroidal forms.

We will prove that the above statements hold for $\tilde{f}_{j+1} : \tilde{X}_{j+1} \to \tilde{Y}_{j+1}$. The argument is essentially the same as in the proof of Lemma 7.7.

Suppose that $q_j \in C_j$ is a 2-point (and $\tilde{\Psi}_1 \circ \cdots \circ \tilde{\Psi}_j(q_j) = q'$), so that Case 1 holds, and $p_j \in \tilde{f}_j^{-1}(q_j)$.

$\mathcal{I}_{C_j}\mathcal{O}_{\tilde{X}_j,p_j}$ is not invertible, and u_j, w_j satisfy (185) [**C3**] at p_j if (8.10) holds, u_j, w_j satisfy (190) [**C3**] at p_j if (8.11) holds and $a, b > 0$ (so that p_j is a 2-point of $D^*_{\tilde{X}_j}$), u_j, w_j satisfy (185) [**C3**] at p_j if (8.11) holds and $b = 0$ (so that p_j is a 1-point of $D^*_{\tilde{X}_j}$).

The algorithm of Lemma 18.17 [**C3**] (as used in the proof of Theorem 6.1) is applied to construct $\tilde{\Phi}_{j+1} : \tilde{X}_{j+1} \to \tilde{X}_j$ and $\tilde{f}_{j+1} : \tilde{X}_{j+1} \to \tilde{Y}_{j+1}$ above q_j. Suppose that $q_{j+1} \in \tilde{\Psi}_{j+1}^{-1}(q_j)$, and $\overline{\pi}_{j+1}(q_{j+1}) = \overline{q}'_{j+1} \in \overline{S}_{j+1}$. Then there exist regular parameters u_{j+1}, w_{j+1} in $\mathcal{O}_{\overline{S}_{j+1},\overline{q}'_{j+1}}$ such that u_{j+1}, v, w_{j+1} are regular parameters in $\mathcal{O}_{\tilde{Y}_{j+1},q_{j+1}}$ and one of the following forms hold:

$\overline{q}'_{j+1}$ is a 1-point of $D^*_{\overline{S}_{j+1}}$

$$u_j = u_{j+1}, w_j = u_{j+1}(w_{j+1} + \alpha) \tag{8.16}$$

with $\alpha \in \mathbf{k}$, or $\overline{q}'_{j+1}$ is a 2-point for $D^*_{\overline{S}_{j+1}}$

$$u_j = u_{j+1}w_{j+1}, w_j = w_{j+1}. \tag{8.17}$$

If (8.16) holds at $\overline{q}'_{j+1}$ and $p_{j+1} \in \tilde{f}_{j+1}^{-1}(q_{j+1})$, then an analysis of the algorithm of Lemma 18.17 [**C3**] and Theorem 6.1 shows that u_{j+1}, v, w_{j+1} satisfy one of the forms (8.10) or (8.11) at p_{j+1}.

If (8.17) holds at $\overline{q}'_{j+1}$, and $p_{j+1} \in \tilde{f}_{j+1}^{-1}(q_{j+1})$, then u_{j+1}, v, w_{j+1} satisfy one of the forms (8.12) - (8.15) at p_{j+1}.

If $q_{j+1} \in C_{j+1}$, then $u_{j+1} = w_{j+1} = 0$ are local equations of C_{j+1}.

Now suppose that $q_j \in C_j$ is a 3-point (and $\tilde{\Psi}_1 \circ \cdots \circ \tilde{\Psi}_j(q_j) = q'$), so that Case 2 holds, and $p_j \in \tilde{f}_j^{-1}(q_j)$.

If $\mathcal{I}_{C_j}\mathcal{O}_{\tilde{X}_j,p_j}$ is not invertible, then after possibly interchanging u_j and w_j, then we have one of the following forms. u_j, w_j satisfy (187) [**C3**] at p_j if (8.13) holds and $a, b > 0$, u_j, w_j satisfy (191) [**C3**] at p_j if (8.13) holds and $b = e = 0$ (in both cases, p_j is a 2-point of $D^*_{\tilde{X}_j}$). u_j, w_j satisfy (187) or (191) [**C3**] if (8.15) holds and p_j is a 2 point of $D^*_{\tilde{X}_j}$. u_j, w_j satisfy (193), (194) or (195) [**C3**] at p_j if (8.15) holds and p_j is a 3-point of $D^*_{\tilde{X}_j}$.

The algorithm of Lemma 18.18 [**C3**] is then applied to construct $\tilde{\Phi}_{j+1} : \tilde{X}_{j+1} \rightarrow \tilde{X}_j$ and $\tilde{f}_{j+1} : \tilde{X}_{j+1} \rightarrow \tilde{Y}_{j+1}$ above q_j. Suppose that $q_{j+1} \in \tilde{\Psi}^{-1}_{j+1}(q_j)$, and $\overline{\pi}(q_{j+1}) = \overline{q}'_{j+1} \in \overline{S}_{j+1}$. Then there exist regular parameters u_{j+1}, w_{j+1} in $\mathcal{O}_{\overline{S}_{j+1}, \overline{q}'_{j+1}}$ such that u_{j+1}, v, w_{j+1} are regular parameters in $\mathcal{O}_{\tilde{Y}_{j+1}, q_{j+1}}$ and one of the following forms hold:

$\overline{q}'_{j+1}$ is a 1-point of $D^*_{\overline{S}_{j+1}}$

$$u_j = u_{j+1}, w_j = u_{j+1}(w_{j+1} + \alpha) \tag{8.18}$$

with $0 \neq \alpha \in \mathbf{k}$, or $\overline{q}'_{j+1}$ is a 2-point for $D^*_{\overline{S}_{j+1}}$

$$u_j = u_{j+1}, w_j = u_{j+1}w_{j+1}, \tag{8.19}$$

or $\overline{q}'_{j+1}$ is a 2-point for $D^*_{\overline{S}_{j+1}}$

$$u_j = u_{j+1}w_{j+1}, w_j = w_{j+1}. \tag{8.20}$$

If (8.18) holds at $\overline{q}'_{j+1}$ and $p_{j+1} \in \tilde{f}^{-1}_{j+1}(q_{j+1})$, then an analysis of the algorithm of Lemma 18.18 [**C3**] shows that u_{j+1}, v, w_{j+1} satisfy one of the forms (8.10) or (8.11) at p_{j+1}.

If (8.19) or (8.20) holds at $\overline{q}'_{j+1}$, and $p_{j+1} \in \tilde{f}^{-1}_{j+1}(q_{j+1})$, then u_{j+1}, v, w_{j+1} satisfy one of the forms (8.12) - (8.15) at p_{j+1}.

We have shown that $D_{\tilde{X}_{j+1}} = \tilde{f}^{-1}_{j+1}(D_{\tilde{Y}_{j+1}})$ is a SNC divisor above q_j and that $\tilde{f}_{j+1}$ is prepared for $D_{\tilde{Y}_{j+1}}$ and $D_{\tilde{X}_{j+1}}$.

We have thus established that $\overline{f}_3$ is prepared for $D_{\overline{Y}_1}$ and $D_{\overline{X}_3}$.

Recall that $\tilde{\Phi}_{i+1} : \tilde{X}_{i+1} \rightarrow \tilde{X}_i$ is a principalization of $m_i\mathcal{O}_{\tilde{X}_i}$ which in a neighborhood of a general point of γ is a sequence of blow ups of sections over γ where $m_i\mathcal{O}_{\tilde{X}_i}$ is not invertible. Each $\tilde{\Psi}_{i+1} : \tilde{Y}_{i+1} \rightarrow \tilde{Y}_i$ is the blow up of a curve C_i which is a section over γ and is a possible center for $D_{\tilde{Y}_i}$.

We will construct $\Psi_1 : Y_1 \rightarrow Y$ such that $\Psi_1^{-1}(U) \cong \overline{Y}_1$, $\Psi_1 \mid \Psi_1^{-1}(U) = \overline{\Psi}$ and $\Psi_1^{-1}(D_Y)$ is a SNC divisor by constructing a sequence of morphisms

$$Y_1 = \hat{Y}_n \stackrel{\hat{\Psi}_n}{\rightarrow} \hat{Y}_{n-1} \rightarrow \cdots \stackrel{\hat{\Psi}_1}{\rightarrow} Y \tag{8.21}$$

where each $\hat{\Psi}_{i+1}$ is a product of blow ups of possible centers for the preimage of D_Y, and $\hat{\Psi}^{-1}_{i+1}(\tilde{Y}_i) \cong \tilde{Y}_{i+1}$, $\hat{\Psi}_{i+1} \mid \tilde{Y}_{i+1} = \tilde{\Psi}_{i+1}$ for all i.

We will inductively construct (8.21). Suppose that we have constructed $\hat{\Psi}_i : \hat{Y}_i \rightarrow \hat{Y}_{i-1}$.

Let γ_i be the Zariski closure of C_i in $\hat{Y}_i$. Then γ_i is a section over $\overline{\gamma}$, and is thus a nonsingular curve. We construct $\hat{\Psi}_{i+1}$ by first blowing up points on (the strict transform of) γ_i above $\overline{\gamma} - \gamma$ where (the strict transform of) γ_i does not make SNCs with (the preimage of) $D_{\hat{Y}_i}$, and then blowing up the strict transform of γ_i.

$\overline{X}_2 \rightarrow \overline{X}$ is an isomorphism away from the preimage of Ω. Thus the sequence of blow ups $\overline{X}_2 \rightarrow X$ extends trivially to a morphism $X_2 \rightarrow X$, so that $X_2 \rightarrow X$ is an isomorphism away from the preimage of Ω.

Now we construct $\Phi_3 : X_3 \to X_2$ such that $\Phi_3^{-1}(\overline{X}_2) = \overline{X}_3$, $\Phi_3 \mid \overline{X}_3 = \overline{\Phi}_3$ and $\Phi_3^{-1}(D_{X_3})$ is a SNC divisor by constructing a sequence of morphisms

$$X_3 = \hat{X}_n \stackrel{\hat{\Phi}_n}{\to} \hat{X}_{n-1} \to \cdots \stackrel{\hat{\Phi}_1}{\to} \hat{X}_0 = X_2$$

where $\hat{\Phi}_i^{-1}(\tilde{X}_{i-1}) \cong \tilde{X}_i$ and $\hat{\Phi}_i \mid \tilde{X}_i = \tilde{\Phi}_i$ for all i, and so that there are morphisms $\hat{X}_i \to \hat{Y}_i$ which are toroidal (with respect to the preimages of D_Y and D_X) over points of $\overline{\gamma} - \gamma$. This follows from application of Lemmas 7.6 and 7.7, and the fact that the case when γ_i is a 2-curve (or a 3-point is blown up) extends directly to a toroidal morphism.

The resulting morphism Φ_3 is an isomorphism away from the preimage of $\overline{\gamma}$.

We have constructed a diagram

$$\begin{array}{ccc} X_3 & \stackrel{f_3}{\to} & Y_1 \\ \Phi \downarrow & & \downarrow \Psi_1 \\ X & \stackrel{f}{\to} & Y \end{array}$$

such that Φ and Ψ_1 are isomorphisms away from the preimage of $\overline{\gamma}$ and f_3 is prepared with respect to $D_{Y_1} = \Psi_1^{-1}(D_Y)$ and $D_{X_3} = \Phi^{-1}(D_X)$ away from the points in $\Sigma - \{q\}$.

By induction on $|\Sigma|$, we repeat this construction to prove Theorem 1.3.

CHAPTER 9

The τ invariant

Throughout this chapter, we assume that $f : X \to Y$ is a dominant morphism of nonsingular 3-folds, with toroidal structures D_Y, $D_X = f^{-1}(D_Y)$ such that D_X contains the singular locus of f.

DEFINITION 9.1. Suppose that $f : X \to Y$ is prepared.

Suppose that $p \in D_X$. Define $\tau_f(p) = \tau(p) = -\infty$ if there exist permissible parameters u, v, w at $q = f(p)$ such that u, v, w are toroidal forms at p. (The existence of such formal parameters u, v, w implies the existence of algebraic permissible parameters which are toroidal forms at p).

Suppose that $p \in X$ is a 3-point, and $\tau(p) \neq -\infty$. Suppose that u, v, w are permissible parameters at $q = f(p)$. Then there is an expression (after possibly permuting u, v, w if q is a 3-point)

$$
\begin{aligned}
u &= x^a y^b z^c \\
v &= x^d y^e z^f \\
w &= \textstyle\sum_{i\geq 0} \alpha_i M_i + N
\end{aligned}
\tag{9.1}
$$

where x, y, z are permissible parameters at p for u, v, w, $\text{rank}(u, v) = 2$, the sum in w is over (possibly infinitely many) monomials $M_i = x^{a_i} y^{b_i} z^{c_i}$ in x, y, z such that $\text{rank}(u, v, M_i) = 2$ for all i, $\deg(M_i) \leq \deg(M_j)$ if $i < j$, N is a monomial in x, y, z such that $\text{rank}(u, v, N) = 3$, $0 \neq \alpha_i \in \mathbf{k}$ for all i and $N \not| M_i$ for any M_i in the series $\sum \alpha_i M_i$.

If q is a 3-point and u, v, w is not a toroidal form at p, we necessarily have (since f is prepared) that

$$
\sum \alpha_i M_i = M_0 \gamma
\tag{9.2}
$$

where γ is a unit series in the monomials $\frac{M_i}{M_0}$ in x, y, z such that

$$
\text{rank}(u, v, M_0) = \text{rank}(u, v, \frac{M_i}{M_0}) = 2
$$

for all i, and $M_0 \mid N$.

Define a group $H_p = H_{f,p}$ as follows. The Laurent monomials in x, y, z form a group under multiplication. We define $H_p = H_{f,p}$ to be the subgroup of $\mathbf{Z}^3$ defined by

$$
H_p = H_{f,p} = (a, b, c)\mathbf{Z} + (d, e, f)\mathbf{Z} + \sum_{i\geq 0} (a_i, b_i, c_i)\mathbf{Z}.
$$

Define a subgroup A_p of H_p by:

$$
A_p = A_{f,p} = \begin{cases} (a, b, c)\mathbf{Z} + (d, e, f)\mathbf{Z} + (a_0, b_0, c_0)\mathbf{Z} & \text{if } q \text{ is 3-point} \\ (a, b, c)\mathbf{Z} + (d, e, f)\mathbf{Z} & \text{if } q \text{ is a 2-point.} \end{cases}
$$

Define

$$L_p = L_{f,p} = H_p/A_p,$$
$$\tau(p) = \tau_f(p) = |L_p|.$$

Suppose that $p \in X$ is a 2-point, and $\tau(p) \neq -\infty$. Suppose that u, v, w are permissible parameters at $q = f(p)$ (which satisfy the conclusions of Lemma 4.9 if $f(p) = q$ is a 1-point). Then there is an expression (after possibly permuting u, v, w if $q = f(p)$ is a 3-point) of one the following forms:

q is 2-point or a 3-point:

$$\begin{aligned} u &= x^a y^b \\ v &= x^c y^d \\ w &= \textstyle\sum_{i \geq 0} \alpha_i x^{a_i} y^{b_i} + x^e y^f (z + \beta) \end{aligned} \tag{9.3}$$

where x, y, z are permissible parameters at p for u, v, w, $0 \neq \alpha_i \in \mathbf{k}$ for all i and $x^e y^f \not| x^{a_i} y^{b_i}$ for all i, or

q is a 2-point or a 3-point:

$$\begin{aligned} u &= (x^a y^b)^k \\ v &= (x^a y^b)^t (z + \beta) \\ w &= \textstyle\sum_{i \geq 0} \alpha_i(z) (x^a y^b)^i + x^c y^d \end{aligned} \tag{9.4}$$

where $\gcd(a, b) = 1$, $0 \neq \beta \in \mathbf{k}$, x, y, z are permissible parameters at p for u, v, w, $\alpha_i(z) \in \mathbf{k}[[z]]$ for all i and $x^c y^d \not| (x^a y^b)^i$ for all i, or

q is a 2-point:

$$\begin{aligned} u &= (x^a y^b)^k \\ v &= \textstyle\sum_{i \geq 0} \alpha_i(z) (z^a y^b)^i + x^c y^d \\ w &= z \end{aligned} \tag{9.5}$$

where $\gcd(a, b) = 1$, x, y, z are permissible parameters at p for u, v, w, $\alpha_i(z) \in \mathbf{k}[[z]]$ for all i and $x^c y^d \not| (x^a y^b)^i$ for all i, or

q is a 1-point:

$$\begin{aligned} u &= (x^a y^b)^k \\ v &= z \\ w &= \textstyle\sum_{i \geq 0} \alpha_i(z) (x^a y^b)^i + x^c y^d \end{aligned} \tag{9.6}$$

where $\gcd(a, b) = 1$, x, y, z are permissible parameters at p for u, v, w, $\alpha_i(z) \in \mathbf{k}[[z]]$ for all i and $x^c y^d \not| (x^a y^b)^i$ for all i.

If q is a 3-point, we necessarily have (since f is prepared) that

$$w = M_0 \gamma \tag{9.7}$$

where γ is a unit series and M_0 is a monomial in x and y.

In equation (9.5), we have

$$v = (x^a y^b)^{i_0} \gamma \tag{9.8}$$

where γ is a unit series.

Suppose that (9.3) holds. Then define

$$\tau(p) = |((a, b)\mathbf{Z} + (c, d)\mathbf{Z} + \sum_{i \geq 0} (a_i, b_i)\mathbf{Z})/((a, b)\mathbf{Z} + (c, d)\mathbf{Z})|$$

if q is a 2-point,

$$\tau(p) = |((a,b)\mathbf{Z} + (c,d)\mathbf{Z} + \sum_{i\ge 0}(a_i,b_i)\mathbf{Z})/((a,b)\mathbf{Z} + (c,d)\mathbf{Z} + (a_0,b_0)\mathbf{Z})|$$

if q is a 3-point, and $M_0 = x^{a_0}y^{b_0}$ in (9.7).

Suppose that (9.4) holds. Then define

$$\tau(p) = |(k\mathbf{Z} + t\mathbf{Z} + \sum_{\alpha_i\ne 0} i\mathbf{Z})/(k\mathbf{Z}+t\mathbf{Z})|$$

if q is a 2-point,

$$\tau(p) = |(k\mathbf{Z} + t\mathbf{Z} + \sum_{\alpha_i\ne 0} i\mathbf{Z})/(k\mathbf{Z}+t\mathbf{Z}+i_0\mathbf{Z})|$$

if q is a 3-point, and $M_0 = x^{i_0}\gamma$ in (9.7).

Suppose that (9.5) (and (9.8)) hold. Then define

$$\tau(p) = |(k\mathbf{Z} + \sum_{\alpha_i\ne 0} i\mathbf{Z})/(k\mathbf{Z}+i_0\mathbf{Z})|.$$

Suppose that (9.6) holds. Then define

$$\tau(p) = |(k\mathbf{Z} + \sum_{\alpha_i\ne 0} i\mathbf{Z})/k\mathbf{Z}|.$$

Suppose that $p\in X$ is a 1-point, and $\tau(p)\ne -\infty$. Suppose that u,v,w are permissible parameters at $q=f(p)$ which satisfy the conclusions of Lemma 4.9 if $f(p)=q$ is a 1-point. Then there is an expression (after possibly permuting u,v,w if $q=f(p)$ is a 3-point) of one the following forms:

q is 2-point or a 3-point

$$\text{(9.9)}\qquad \begin{aligned} u &= x^a\\ v &= x^b(\alpha+y)\\ w &= \textstyle\sum_{i<c}\alpha_i(y)x^i + x^c(z+\beta)\end{aligned}$$

where x,y,z are permissible parameters at p for u,v,w, $0\ne\alpha\in\mathbf{k}$, $\beta\in\mathbf{k}$ and $\alpha_i(y)\in\mathbf{k}[[y]]$ or

q is 2-point

$$\text{(9.10)}\qquad \begin{aligned} u &= x^a\\ v &= \textstyle\sum_{i<c}\alpha_i(y)x^i + x^c(z+\beta)\\ w &= y\end{aligned}$$

where x,y,z are permissible parameters at p for u,v,w, $\beta\in\mathbf{k}$ and $\alpha_i(y)\in\mathbf{k}[[y]]$, or

q is a 1-point

$$\text{(9.11)}\qquad \begin{aligned} u &= x^a\\ v &= y\\ w &= \textstyle\sum_{i<c}\alpha_i(y)x^i + x^c(z+\beta)\end{aligned}$$

where x,y,z are permissible parameters at p for u,v,w, $\beta\in\mathbf{k}$, $\alpha_i(y)\in\mathbf{k}[[y]]$.

In case (9.9) we have

$$w = x^{i_0}\gamma \tag{9.12}$$

where γ is a unit series if q is a 3-point. In case (9.10) we have

$$v = x^{i_0}\gamma \tag{9.13}$$

where γ is a unit series.

Suppose that (9.9) holds. Then define

$$\tau(p) = |(a\mathbf{Z} + b\mathbf{Z} + \sum_{\alpha_i \neq 0} i\mathbf{Z})/(a\mathbf{Z} + b\mathbf{Z})|$$

if q is a 2-point,

$$\tau(p) = |(a\mathbf{Z} + b\mathbf{Z} + \sum_{\alpha_i \neq 0} i\mathbf{Z})/(a\mathbf{Z} + b\mathbf{Z} + i_0\mathbf{Z})|$$

if q is a 3-point.

Suppose that (9.10) holds. Then define

$$\tau(p) = |(a\mathbf{Z} + \sum_{\alpha_i \neq 0} i\mathbf{Z})/(a\mathbf{Z} + i_0\mathbf{Z})|.$$

Suppose that (9.11) holds. Then define

$$\tau(p) = |(a\mathbf{Z} + \sum_{\alpha_i \neq 0} i\mathbf{Z})/a\mathbf{Z}|.$$

Observe that $\tau(p) < \infty$ in Definition 9.1, since $\tau(p)$ is the order of a finitely generated group.

We define

$$\tau(X) = \tau_f(X) = \sup\{\tau_f(p) \mid p \in X\}.$$

It will follow from Lemma 9.3 that $\tau(X) < \infty$.

LEMMA 9.2. *Suppose that $f : X \to Y$ is prepared and $p \in D_X$. Then $\tau_f(p)$ is independent of choice of permissible parameters u, v, w at $q = f(p)$ and permissible parameters x, y, z at p for u, v, w.*

PROOF. First suppose that $q \in Y$ is a 3-point, and $p \in X$ is a 3-point.

The condition $\tau(p) = -\infty$ is independent of permuting u, v, w, multiplying u, v, w by units in $\hat{\mathcal{O}}_{Y,q}$ and multiplying x, y, z by units in $\hat{\mathcal{O}}_{X,p}$ so that the conditions of (9.1) hold. Thus $\tau(p) = -\infty$ is independent of choice of permissible parameters at (the 3-points) q and p.

Suppose that u, v, w are permissible parameters at q and x, y, z are permissible parameters at p for u, v, w satisfying (9.1). Let τ be the computation of $\tau(p)$ for these variables.

Suppose that $\tilde{u}, \tilde{v}, \tilde{w}$ is another set of permissible parameters at q, and $\tilde{x}, \tilde{y}, \tilde{z}$ are permissible parameters for $\tilde{u}, \tilde{v}, \tilde{w}$ at p, satisfying (9.1). Let τ_1 be the computation for $\tau(p)$ with respect to these variables. We must show that $\tau = \tau_1$. We may assume that $\tau \geq 1$ and $\tau_1 \geq 1$.

Since p and q are 3-points, $\tilde{u}, \tilde{v}, \tilde{w}$ can be obtained from u, v, w by permuting the variables u, v, w and then multiplying u, v, w by unit series (in u, v, w). $\tilde{x}, \tilde{y}, \tilde{z}$ can be obtained from x, y, z by permuting x, y, z and then multiplying x, y, z by unit series (in x, y, z).

We then reduce to proving the following:

1. Suppose that $\tilde{u}, \tilde{v}, \tilde{w}$ is a permutation of u, v, w such that $\tilde{u}, \tilde{v}$ are toroidal forms at p. Then there exist permissible parameters $\tilde{x}, \tilde{y}, \tilde{z}$ at p for $\tilde{u}, \tilde{v}, \tilde{w}$ such that a form (9.1) holds, and $\tau_1 = \tau$.
2. Suppose that $\tilde{u}, \tilde{v}, \tilde{w}$ are obtained from u, v, w by multiplying u, v, w by unit series. Then there exist permissible parameters $\tilde{x}, \tilde{y}, \tilde{z}$ at p for $\tilde{u}, \tilde{v}, \tilde{w}$ such that a form (9.1) holds, and $\tau_1 = \tau$.
3. Suppose that $u = \tilde{u}$, $v = \tilde{v}$ and $w = \tilde{w}$ and x, y, z, $\tilde{x}, \tilde{y}, \tilde{z}$ are two sets of permissible parameters for u, v, w. Then $\tau_1 = \tau$.

We now verify 1. The case when

$$\tilde{u} = v, \tilde{v} = u, \tilde{w} = w$$

is immediate. We will verify the case when

$$\tilde{u} = w, \tilde{v} = v, \tilde{w} = u.$$

Since the symmetric group S_3 is generated by the permutations (1, 2) and (1,3), the remaining cases of 1 will follow.

Since $\tau \geq 1$, and $\tilde{u}, \tilde{v}, \tilde{w}$ have a form (9.1) at p, we have $\operatorname{rank}(v, M_0) = 2$. Since w, v are (by assumption) toroidal forms at p, there exist permissible parameters $\tilde{x}, \tilde{y}, \tilde{z}$ at p such that w, v, u (in this order) have an expression of the form of (9.1) in terms of $\tilde{x}, \tilde{y}, \tilde{z}$. We will show that there is an isomorphism of the corresponding group $\tilde{H}$ (computed for these variables) and the group H computed for u, v, w and x, y, z which takes the corresponding group $\tilde{A}$ to A.

With the notation of (9.2), we have

$$w = M_0(\gamma + \overline{N}_0)$$

where $\overline{N}_0 = \frac{N}{M_0}$ is a monomial in x, y, z.

Set $\overline{M}_0 = M_0$, $\overline{M}_i = \frac{M_i}{M_0}$ for all M_i appearing in the series $\sum_{i \geq 1} \alpha_i M_i$. Thus $\gamma = \alpha_0 + \sum_{i \geq 1} \alpha_i \overline{M}_i$. There exist $a_i, b_i, c_i \in \mathbf{N}$ such that

$$\overline{M}_i = x^{a_i} y^{b_i} z^{c_i}$$

for $0 \leq i$. Let $\overline{m}_i = (a_i, b_i, c_i)$ for $i \geq 0$,

$$H = (a, b, c)\mathbf{Z} + (d, e, f)\mathbf{Z} + \sum_{i \geq 0} \overline{m}_i \mathbf{Z}.$$

Define a finite sequence

$$1 = \mu(1) < \mu(2) < \cdots < \mu(\overline{r})$$

(for appropriate $\overline{r}$) so that

$$\sum_{i=1}^{\mu(\overline{r})} \overline{m}_i \mathbf{Z} = \sum_{i=1}^{\infty} \overline{m}_i \mathbf{Z},$$

$$\sum_{i=1}^{\mu(j)} \overline{m}_i \mathbf{Z} = \sum_{i=1}^{n} \overline{m}_i \mathbf{Z}$$

if $\mu(j) < n < \mu(j+1)$ and

$$\sum_{i=1}^{\mu(j)} \overline{m}_i \mathbf{Z} \neq \sum_{i=1}^{\mu(j+1)} \overline{m}_i \mathbf{Z}.$$

Set

$$G_j = \sum_{i=1}^{\mu(j)} \overline{m}_i \mathbf{Z}.$$

There exist $k_i, l_i \in \mathbf{Z}$ and $e_i \in \mathbf{N}$ with $\gcd(e_i, k_i, l_i) = 1$ such that $\overline{M}_i^{e_i} = u^{k_i} v^{l_i}$ for $i \in \{0, \mu(1), \ldots, \mu(\overline{r})\}$, and there exist $g, h, i \in \mathbf{N}$ such that $\overline{N}_0 = x^g y^h z^i$. Thus H is the subgroup of $\mathbf{Z}^3$ generated by

$$\begin{array}{c}\delta_1 = (a, b, c), \delta_2 = (d, e, f), \epsilon_0 = (a_0, b_0, c_0),\\ \epsilon_{\mu(1)} = (a_{\mu(1)}, b_{\mu(1)}, c_{\mu(1)}), \ldots, \epsilon_{\mu(\overline{r})} = (a_{\mu(\overline{r})}, b_{\mu(\overline{r})}, c_{\mu(\overline{r})}).\end{array}$$

A is the subgroup with generators δ_1, δ_2 and ϵ_0.

Since $\operatorname{rank}(v, M_0) = 2$, we can make a change of variables

$$x = \overline{x}\lambda_1, y = \overline{y}\lambda_2, z = \overline{z}\lambda_3$$

where $\lambda_i = (\gamma + \overline{N}_0)^{\beta_i}$ for some $\beta_i \in \mathbf{Q}$, so that

$$\begin{array}{ll} \lambda_1^{a_0}\lambda_2^{b_0}\lambda_3^{c_0} & = (\gamma + \overline{N}_0)^{-1} \\ \lambda_1^{d}\lambda_2^{e}\lambda_3^{f} & = 1 \\ \lambda_1^{a}\lambda_2^{b}\lambda_3^{c} & = (\gamma + \overline{N}_0)^{t} \end{array}$$

for some $t \in \mathbf{Q}$. Since u, v, w are algebraically independent in $\hat{\mathcal{O}}_{X,p}$ (by Zariski's subspace theorem, Theorem 10.14 [**Ab3**]), and

$$\operatorname{rank}(u, v, M_0) = 2,$$

we have that $t \neq 0$,

$$\begin{array}{ll} w & = \overline{x}^{a_0}\overline{y}^{b_0}\overline{z}^{c_0} \\ v & = \overline{x}^{d}\overline{y}^{e}\overline{z}^{f} \\ u & = \overline{x}^{a}\overline{y}^{b}\overline{z}^{c}(\gamma^t + \overline{x}^g\overline{y}^h\overline{z}^i\gamma_2) \end{array}$$

where γ_2 is a unit series in $\overline{x}, \overline{y}, \overline{z}$.

Set $\tilde{M}_i = \overline{x}^{a_i}\overline{y}^{b_i}\overline{z}^{c_i}$ for $i \geq 1$. In the Taylor's series expansion of $h(\zeta) = (\alpha_0 + \zeta)^t$ we substitute $\zeta = \sum_{i=1}^{\infty} \alpha_i \overline{M}_i$ to see that

$$\gamma(x, y, z)^t \equiv \sum_{j=0}^{\infty} q_j \mod (\overline{x}, \overline{y}, \overline{z})\overline{N}_0$$

where each q_j is a series in monomials of degree j in $\{\tilde{M}_i \mid i \geq 1\}$ with coefficients in $\mathbf{k}$, and

$$q_0 = \alpha_0^t, q_1 = \sum_{i=1}^{\infty} \sigma_i \tilde{M}_i$$

with

$$\sigma_i = t\alpha_0^{t-1+a_i\beta_1+b_i\beta_2+c_i\beta_3}\alpha_i$$

for $i \geq 1$. Let

$$\omega = \sum_{j=0}^{\infty} q_j.$$

There exists a unit series γ_3 such that

$$u = \overline{x}^a\overline{y}^b\overline{z}^c(\omega + \overline{x}^g\overline{y}^h\overline{z}^i\gamma_3).$$

We now see that the coefficient of $\tilde{M}_{\mu(j)}$ for $1 \leq j \leq \overline{r}$ in the expansion of ω as a series in $\overline{x}, \overline{y}, \overline{z}$ is $\sigma_{\mu(j)}$. If not, there would exist a relation $\tilde{M}_{\mu(j)} = \tilde{M}_{i_1} \cdots \tilde{M}_{i_n}$ for some $n > 1$. Thus $i_1, \ldots, i_n < \mu(j)$ and $G_j = G_{j-1}$, a contradiction.

Since

$$\mathrm{Det}\begin{pmatrix} a & b & c\\ d & e & f\\ g & h & i\end{pmatrix}\neq 0,$$

there exists a change of variables

$$\overline{x}=\tilde{x}\phi_1,\overline{y}=\tilde{y}\phi_2,\overline{z}=\tilde{z}\phi_3$$

where ϕ_1,ϕ_2,ϕ_3 are unit series in $\overline{x},\overline{y},\overline{z}$ (fractional powers of γ_3) such that

$$\begin{array}{ll}\phi_1^{\frac{a}{e_0}}\phi_2^{\frac{b}{e_0}}\phi_3^{\frac{c}{e_0}} & =1\\ \phi_1^{\frac{d}{e_0}}\phi_2^{\frac{e}{e_0}}\phi_3^{\frac{f}{e_0}} & =1\\ \phi_1^g\phi_2^h\phi_3^i & =\gamma_3^{-1}.\end{array}$$

Since $e_i(a_i,b_i,c_i)=k_i(a,b,c)+l_i(d,e,f)$ for $i\geq 0$, we have an expression

$$\begin{array}{ll} w & =\tilde{x}^{a_0}\tilde{y}^{b_0}\tilde{z}^{c_0}\\ v & =\tilde{x}^d\tilde{y}^e\tilde{z}^f\\ u & =\tilde{x}^a\tilde{y}^b\tilde{z}^c(\omega+\tilde{x}^g\tilde{y}^h\tilde{z}^i)\end{array}$$

and $\tilde{M}_i=\tilde{x}^{a_i}\tilde{y}^{b_i}\tilde{z}^{c_i}\eta_i$ for some e_i-th root of unity η_i in $\mathbf{k}$ for all $1\leq i$.

$\tilde{H}$ is thus the subgroup of $\mathbf{Z}^3$ with generators

$$\overline{\delta}_1,\overline{\delta}_2,\overline{\epsilon}_0,\overline{\epsilon}_{\mu(1)},\ldots,\overline{\epsilon}_{\mu(\overline{r})},$$

defined by

$$\begin{array}{l}\overline{\delta}_1=(a_0,b_0,c_0),\overline{\delta}_2=(d,e,f),\overline{\epsilon}_0=(a,b,c),\\ \overline{\epsilon}_{\mu(1)}=(a_{\mu(1)},b_{\mu(1)},c_{\mu(1)}),\ldots,\overline{\epsilon}_{\mu(\overline{r})}=(a_{\mu(\overline{r})},b_{\mu(\overline{r})},c_{\mu(\overline{r})}).\end{array}$$

$\tilde{A}$ is the subgroup of $\tilde{H}$ generated by $\overline{\delta}_1,\overline{\delta}_2$ and $\overline{\epsilon}_0$. Thus $H=\tilde{H}$ and $A=\tilde{A}$.

We have thus completed the verification of 1. The verification of 2 and 3 follow from simpler calculations. We thus obtain the conclusions of the lemma when p is a 3-point and q is a 3-point.

Now assume that $q\in Y$ is a 2-point, and p is a 3-point. $\tau(p)$ is independent of interchanging u and v, multiplying u and v by unit series in $\hat{\mathcal{O}}_{Y,q}$, and permuting x,y,z and multiplying x,y,z by unit series in $\hat{\mathcal{O}}_{X,p}$, so that the conditions of (9.1) hold. If we replace w by $w'\in\hat{\mathcal{O}}_{Y,q}$ so that u,v,w' are permissible parameters at q, then by the Weierstrass preparation theorem, there exists a unit series $\alpha(u,v,w)\in\hat{\mathcal{O}}_{Y,q}$ and a series $\beta(u,v)\in\mathbf{k}[[u,v]]$ such that $w=\alpha^{-1}(w'-\beta(u,v))$.

There exists a series ϕ in x,y,z such that

$$\alpha(u,v,w)=\alpha(u,v,\sum\alpha_iM_i+N)=\alpha(u,v,\sum\alpha_iM_i)+N\phi.$$

Then

$$w'=\beta(u,v)+\alpha(u,v,\sum\alpha_iM_i)(\sum\alpha_iM_i)+N[\alpha(u,v,\sum\alpha_iM_i)+(\sum\alpha_iM_i)\phi+N\phi].$$

Now as in the calculation we make in the verification that $\tau_1=\tau$ in the case when p is a 3-point and q is a 3-point, we see that there exist permissible parameters $\tilde{x},\tilde{y},\tilde{z}$ for u,v,w' at p such that $\tau_1=\tau$ (where τ is computed for u,v,w and x,y,z and τ' is computed for u,v,w' and $\tilde{x},\tilde{y},\tilde{z}$). Thus $\tau(p)$ is independent of choice of permissible parameters at q and p when p is a 3-point and q is a 2-point.

Now assume that p is a 1-point and $q=f(p)$ is a 2-point or a 3-point.

Invariance of $\tau_f(p)$ is not difficult to prove when there exists a computation of $\tau_f(p)$ which gives $-\infty$, so we will assume that $\tau_f(p) \geq 1$ for all computations of $\tau_f(p)$.

Assume that p is a 1-point, and that $q = f(p)$ is a 2-point or a 3-point. Suppose that u, v, w are permissible parameters at q, such that u, v, w have an expression of the form (4.1) and (9.9), or of the form 2 (b) of Definition 4.6 and (9.10).

We will first show that for fixed permissible parameters u, v, w at q, $\tau_f(p)$ is independent of choice of permissible parameters x, y, z for u, v, w at p which have an expression of the form (4.1) and (9.9).

We have that u, v, w have an expression

$$
\begin{aligned}
u &= x^a \\
v &= x^b(\alpha + y) \\
w &= \textstyle\sum_{j=0}^{n} \alpha_{i_j}(y)x^{i_j} + x^c(z + \beta)
\end{aligned}
\tag{9.14}
$$

where $i_j < c$ for all j, and $\alpha_{i_j}(y) \neq 0$ for all j.

If $\overline{x}, \overline{y}, \overline{z}$ are other permissible parameters for u, v, w at q, which also give an expansion of the form (4.1) and (9.9), then we have $x = \omega\overline{x}$ where $\omega^a = 1$, and thus $y = \omega^{-b}\overline{y}$, $z = \omega^{-c}\overline{z}$. Thus the computation of $\tau_f(p)$ for $\overline{x}, \overline{y}, \overline{z}$ is the same as for x, y, z.

By a similar calculation, for fixed permissible parameters u, v, w at q, τ is independent of choice of permissible parameters x, y, z for u, v, w at p which have an expression of type 2 (b) of Definition 4.6 and (9.10) at p.

Now suppose that q is a 2-point, and u, v, w admit expressions of both the form 2 (b) of Definition 4.6 and of the form (4.1) and (9.9) at p. Let τ_1 be the computation of $\tau_f(p)$ from a representation of u, v, w in the form 2 (b) of Definition 4.6, and let τ_2 be the computation of $\tau_f(p)$ from a representation of u, v, w in the form of (4.1) and (9.9).

Since u, v, w have an expression of type 2 (b) at p, there are permissible parameters x, y, z at p such that there is an expression

$$
\begin{aligned}
u &= x^a \\
v &= \textstyle\sum_{j=0}^{n} \alpha_{i_j}(y)x^{i_j} + x^c(z + \beta) \\
w &= y
\end{aligned}
\tag{9.15}
$$

of the form (9.10), where α_{ij} are all nonzero and $i_j < c$ for all j. τ_1 is the computation of $\tau_f(p)$ for this expression. Since u, v, w also have an expression of type (4.1) at p, u, v are toroidal forms at p. Since $\tau_1 \geq 1$, we have

$$v = (\overline{\alpha} + y\lambda(y))x^{i_0} + \alpha_{i_1}(y)x^{i_1} + \cdots + \alpha_{i_n}(y)x^{i_n} + x^c(z + \beta)$$

where $0 \neq \overline{\alpha} = \alpha_{i_0}(0) \in \mathbf{k}$ and $\lambda(y)$ is a unit series. Set

$$\overline{y} = y\lambda(y) + \alpha_{i_1}(y)x^{i_1 - i_0} + \cdots + \alpha_{i_n}(y)x^{i_n - i_0} + x^{c - i_0}(z + \beta).$$

Then

$$y = \lambda(y)^{-1}\overline{y} - \lambda(y)^{-1}\alpha_{i_1}(y)x^{i_1 - i_0} - \cdots - \lambda(y)^{-1}\alpha_{i_n}(y)x^{i_n - i_0} - x^{c - i_0}\lambda(y)^{-1}(z + \beta). \tag{9.16}$$

Iterate, substituting (9.16) into successive iterations of (9.16), to get a series $\overline{P}$ such that

$$y = \overline{P}(\overline{y}, x^{i_1 - i_0}, \ldots, x^{i_n - i_0}) + x^{c - i_0}\Omega$$

for some $\Omega \in \hat{\mathcal{O}}_{X,p}$.

We compute the Jacobian determinant

$$\text{Det}\left(\frac{\partial(u,v,w)}{\partial(x,y,z)}\right) = -ax^{a+c-1}. \tag{9.17}$$

We have

$$\begin{aligned} u &= x^a \\ v &= x^{i_0}(\overline{y}+\overline{\alpha}) \\ w &= \overline{P}(\overline{y}, x^{i_1-i_0}, \ldots, x^{i_n-i_0}) + x^{c-i_0}\Omega, \end{aligned}$$

and

$$\text{Det}\left(\frac{\partial(u,v,w)}{\partial(x,\overline{y},z)}\right) = ax^{a+c-1}\frac{\partial\Omega}{\partial z}. \tag{9.18}$$

Since (9.17) and (9.18) differ by multiplication of a unit series, $\frac{\partial\Omega}{\partial z}(0,0,0) \neq 0$. Set $\overline{\beta} = \Omega(0,0,0)$, $\overline{z} = \Omega - \overline{\beta}$. $x, \overline{y}, \overline{z}$ are regular parameters in $\hat{\mathcal{O}}_{X,p}$. We have

$$\begin{aligned} u &= x^a \\ v &= x^{i_0}(\overline{y}+\alpha) \\ w &= \overline{P}(\overline{y}, x^{i_1-i_0}, \ldots, x^{i_n-i_0}) + x^{c-i_0}(\overline{\beta}+\overline{z}). \end{aligned}$$

τ_2 is the value of $\tau_f(p)$ for this expression, computed from (9.9). We have

$$\begin{aligned} \tau_2 &\leq |(a\mathbf{Z} + i_0\mathbf{Z} + \textstyle\sum_{j=1}^n (i_j - i_0)\mathbf{Z})/(a\mathbf{Z} + i_0\mathbf{Z})| \\ &= |(a\mathbf{Z} + \textstyle\sum_{j=0}^n i_j\mathbf{Z})/(a\mathbf{Z} + i_0\mathbf{Z})| \\ &= \tau_1. \end{aligned} \tag{9.19}$$

We now show that $\tau_1 \leq \tau_2$. Since u, v, w have an expression of type (4.1) at p, we have an expression

$$\begin{aligned} u &= x^a \\ v &= x^b(\alpha + y) \\ w &= P(x,y) + x^c(z+\beta) \end{aligned} \tag{9.20}$$

of the form (9.9), where

$$P = \alpha_{i_0}(y) + \alpha_{i_1}(y)x^{i_1} + \cdots + \alpha_{i_n}(y)x^{i_n},$$

$\alpha_{i_j}(y) \neq 0$ for all j, and $i_n < c$.

τ_2 is the computation of $\tau_f(p)$ for this expression. Since $\tau_2 \geq 1$, and u, v, w also have an expression of type 2 (b) of Definition 4.6, we have $\frac{\partial P}{\partial y}(0,0) \neq 0$. We then have

$$\alpha_{i_0}(y) = y\lambda(y)$$

where λ is a unit series. Set $\overline{z} = P + x^c(z+\beta)$. $\overline{z}, x, z$ are regular parameters in $\hat{\mathcal{O}}_{X,p}$. We have

(9.21)

$$\begin{aligned} y &= \lambda(y)^{-1}[\overline{z} - \alpha_{i_1}(y)x^{i_1} - \cdots - \alpha_{i_n}(y)x^{i_n} - x^c(z+\beta)] \\ &= \lambda(y)^{-1}\overline{z} - \lambda(y)^{-1}\alpha_{i_1}(y)x^{i_1} - \cdots - \lambda(y)^{-1}\alpha_{i_n}(y)x^{i_n} - \lambda(y)^{-1}x^c(z+\beta). \end{aligned}$$

Iterate, substituting (9.21) into successive iterations of (9.21), to get series $\overline{P}$ and Ω such that

$$y = \overline{P}(\overline{z}, x^{i_1}, \ldots, x^{i_n}) + x^c\Omega(\overline{z}, x, z).$$

Thus we have an expression

$$\begin{aligned} u &= x^a \\ v &= x^b(\alpha + \overline{P}(\overline{z}, x^{i_1}, \ldots, x^{i_n})) + x^{b+c}\Omega \\ w &= \overline{z}. \end{aligned}$$

We compute the Jacobian determinant from (9.20),

$$\mathrm{Det}\left(\frac{\partial(u,v,w)}{\partial(x,y,z)}\right) = ax^{a+b+c-1}.$$

We compare with

$$\mathrm{Det}\left(\frac{\partial(u,v,w)}{\partial(x,z,\overline{z})}\right) = ax^{a+b+c-1}\frac{\partial\Omega}{\partial z},$$

to see that $\frac{\partial\Omega}{\partial z}(0,0,0)\neq 0$. Set $\overline{\beta}=\Omega(0,0,0)$, $\overline{y}=\Omega-\overline{\beta}$. $x,\overline{y},\overline{z}$ are regular parameters in $\hat{\mathcal{O}}_{X,p}$. We have

$$\begin{aligned} u &= x^a\\ v &= x^b(\alpha+\overline{P}(\overline{z},x^{i_1},\ldots,x^{i_n}))+x^{b+c}(\overline{\beta}+\overline{y})\\ w &= \overline{z}\end{aligned}$$

in the form of 2 (b) of Definition 4.6. τ_1 is the value of $\tau_f(p)$ for this expression. We compute from (9.10),

$$\begin{aligned} \tau_1 &\leq |(a\mathbf{Z}+b\mathbf{Z}+\textstyle\sum_{j=0}^n i_j\mathbf{Z})/(a\mathbf{Z}+b\mathbf{Z})|\\ &= \tau_2.\end{aligned} \tag{9.22}$$

By (9.19) and (9.22) we see that $\tau_1=\tau_2$.

From our calculations so far, we conclude that for fixed u,v,w, assuming that p is a 1-point and q is a 2-point or a 3-point, $\tau_f(p)$ is independent of choice of permissible parameters x,y,z for u,v,w at p.

Now we will show that if u_1,v_1,w_1 are a permutation of u,v,w such that u_1,v_1,w_1 have one of the forms (4.1) and (9.9) or 2 (b) of Definition 4.6 and (9.10) at p, then we obtain the same value of $\tau_f(p)$.

First suppose that $u_1=v$, $v_1=u$, $w_1=w$, and u,v are toroidal forms at p (so that (4.1) holds). Then u,v,w have an expression (9.14) at p. We have

$$\begin{aligned} u_1=v &= \overline{x}^b\\ v_1=u &= \overline{x}^a(\overline{\alpha}+\overline{y})\end{aligned}$$

where $x=\overline{x}(\alpha+y)^{-\frac{1}{b}}$, $\overline{\alpha}=\alpha^{-\frac{a}{b}}$, and $y=\lambda(\overline{y})\overline{y}$ for an appropriate unit series $\lambda(\overline{y})$.

Set

$$\overline{\beta}=\beta\alpha^{-\frac{c}{b}},\overline{z}=(z+\beta)(\alpha+\lambda(\overline{y})\overline{y})^{-\frac{c}{b}}-\overline{\beta}.$$

We have

$$\begin{aligned} w &= \textstyle\sum_{j=0}^n \alpha_{i_j}(\lambda(\overline{y})\overline{y})(\alpha+\lambda(\overline{y})\overline{y})^{-\frac{i_j}{b}}\overline{x}^{i_j}+\overline{x}^c(z+\beta)(\alpha+\lambda(\overline{y})\overline{y})^{-\frac{c}{b}}\\ &= \textstyle\sum_{j=0}^n \overline{\alpha}_{ij}(\overline{y})\overline{x}^{i_j}+\overline{x}^c(\overline{\beta}+\overline{z})\end{aligned}$$

where $\overline{\alpha}_{ij}(\overline{y})\neq 0$ for all j.

Let τ_1 be the calculation of $\tau_f(p)$ for the variables u,v,w, and let τ_2 be the calculation of τ for the variables u_1,v_1,w_1. We see that $\tau_1=\tau_2$.

Now suppose that $u_1=v,v_1=u,w_1=w$, and u,v,w have an expression of the form 2 (b) of Definition 4.6 and (9.10). u,v,w have an expression (9.15) at p. We have

$$\begin{aligned} u &= x^a\\ v &= x^{i_0}(\textstyle\sum_{j=0}^n \alpha_{i_j}(y)x^{i_j-i_0}+x^{c-i_0}(z+\beta))\\ w &= y\end{aligned}$$

where $\alpha_{i_0}(y)$ is a unit series. Set

$$\gamma = \sum_{j=0}^{n} \alpha_{i_j}(y)x^{i_j-i_0} + x^{c-i_0}(z+\beta).$$

Define $\overline{x}$ by

$$x = \overline{x}\gamma^{-\frac{1}{i_0}}. \tag{9.23}$$

We have a series P and $\Omega \in \hat{\mathcal{O}}_{X,p}$ such that

$$\begin{aligned} u_1 &= v = \overline{x}^{i_0} \\ v_1 &= u = \overline{x}^a\gamma^{-\frac{a}{i_0}} = \overline{x}^a(P(y, x^{i_1-i_0}, \ldots, x^{i_n-i_0}) + \overline{x}^{c-i_0}\Omega) \\ w &= y \end{aligned}$$

By iterating substitution of (9.23) in $P(y, x^{i_1-i_0}, \ldots, x^{i_n-i_0})$, we see that there is a series $\overline{P}(y, \overline{x}^{i_1-i_0}, \ldots, \overline{x}^{i_n-i_0})$ and $\overline{\Omega} \in \hat{\mathcal{O}}_{X,p}$ such that

$$v_1 = \overline{x}^a(\overline{P}(y, \overline{x}^{i_1-i_0}, \ldots, \overline{x}^{i_n-i_0}) + \overline{x}^{c-i_0}\overline{\Omega}).$$

Since the Jacobians

$$\mathrm{Det}\left(\frac{\partial(u,v,w)}{\partial(x,y,z)}\right) = -ax^{a+c-1}$$

and

$$\mathrm{Det}\left(\frac{\partial(u_1,v_1,w_1)}{\partial(\overline{x},y,z)}\right) = -i_0\overline{x}^{a+c-1}\frac{\partial\overline{\Omega}}{\partial z}$$

differ by multiplication by a unit, $\frac{\partial\Omega}{\partial z}(0,0,0) \neq 0$.

Set

$$\overline{\beta} = \overline{\Omega}(0,0,0),\ \overline{z} = \Omega - \overline{\beta}.$$

Then $\overline{x}, y, \overline{z}$ are regular parameters in $\hat{\mathcal{O}}_{X,p}$, and

$$\begin{aligned} u_1 &= \overline{x}^{i_0} \\ v_1 &= \overline{x}^a\overline{P}(y, \overline{x}^{i_1-i_0}, \ldots, \overline{x}^{i_n-i_0}) + \overline{x}^{a+c-i_0}(\overline{\beta} + \overline{z}) \\ w &= y. \end{aligned}$$

Let τ_1 be the calculation of $\tau_f(p)$ for the variables u, v, w, and let τ_2 be the calculation of $\tau_f(p)$ for the variables u_1, v_1, w_1 (from (9.10). We see that $\tau_2 \leq \tau_1$. From this argument for the change of variables u_1, v_1, w_1 to u, v, w, we see that $\tau_1 \leq \tau_2$, so that $\tau_1 = \tau_2$.

Now suppose that q is a 3-point, u, v, w have an expression (4.1) and (9.9), and $u_1 = w, v_1 = v, w_1 = u$ also have an expression (4.1) and (9.9).

We have

$$\begin{aligned} u &= x^a \\ v &= x^b(\alpha + y) \\ w &= x^{i_0}(\textstyle\sum_{j=0}^{n} \alpha_{i_j}(y)x^{i_j-i_0} + x^{c-i_0}(z+\beta)) \end{aligned} \tag{9.24}$$

where $i_n < c$, $\alpha_{i_j}(y) \neq 0$ for all j, and $\alpha_{i_0}(y)$ is a unit series. Set

$$\gamma = \sum_{j=0}^{n} \alpha_{i_j}(y)x^{i_j-i_0} + x^{c-i_0}(z+\beta).$$

Set $x = \overline{x}\gamma^{-\frac{1}{i_0}}$. We have

$$\begin{aligned} u_1 = w &= \overline{x}^{i_0} \\ v_1 = v &= \overline{x}^b \gamma^{-\frac{b}{i_0}}(\alpha + y) \\ w_1 = u &= \overline{x}^a \gamma^{-\frac{a}{i_0}}. \end{aligned}$$

As we are assuming that all calculations of $\tau_f(p)$ are ≥ 1, we have that $c > 0$. Since w, v are assumed to be toroidal forms at p, we then have

$$\frac{\partial}{\partial y}(\gamma^{-\frac{b}{i_0}}(\alpha + y))(0,0,0) \neq 0.$$

Set

$$\overline{\alpha} = \gamma^{-\frac{b}{i_0}}(0,0,0)\alpha, \ \overline{y} = \gamma^{-\frac{b}{i_0}}(\alpha + y) - \overline{\alpha}.$$

We have

$$x = \overline{x}[(\alpha + y)^{-1}(\overline{y} + \overline{\alpha})]^{\frac{1}{b}}. \tag{9.25}$$

There exists a unit series $\lambda(y)$, series $\overline{\alpha}_{r_1, \ldots r_n}(y)$, and $\overline{\Omega} \in \hat{\mathcal{O}}_{X,p}$, such that

$$\overline{y} = y\lambda(y) + \sum_{r_1, \ldots, r_n > 0} \overline{\alpha}_{r_1, \ldots, r_n}(y) x^{(i_1 - i_0)r_1} \cdots x^{(i_n - i_0)r_n} + x^{c - i_0}\overline{\Omega}.$$

Thus

(9.26)

$$y = \overline{y}\lambda(y)^{-1} - \sum_{r_1, \ldots, r_n > 0} \lambda(y)^{-1}\overline{\alpha}_{r_1, \ldots, r_n}(y) x^{(i_1 - i_0)r_1} \cdots x^{(i_n - i_0)r_n} - x^{c - i_0}\lambda(y)^{-1}\overline{\Omega}.$$

There exists a series

$$P(y, x^{i_1 - i_0}, \ldots, x^{i_n - i_0}) \tag{9.27}$$

such that $u = \overline{x}^a(P + \overline{x}^{c - i_0}\Omega)$ for some $\Omega \in \hat{\mathcal{O}}_{X,p}$.

Substituting (9.25) and (9.26) into successive iterations of (9.27), we see that there is a series $\overline{P}(\overline{y}, \overline{x}^{i_1 - i_0}, \ldots, \overline{x}^{i_n - i_0})$ and $\Omega' \in \hat{\mathcal{O}}_{X,p}$ such that

$$w_1 = u = \overline{x}^a(\overline{P}(\overline{y}, \overline{x}^{i_1 - i_0}, \ldots, \overline{x}^{i_n - i_0}) + \overline{x}^{c - i_0}\Omega').$$

Thus there is an expression

$$\begin{aligned} u_1 = w &= \overline{x}^{i_0} \\ v_1 = v &= \overline{x}^b(\overline{\alpha} + \overline{y}) \\ w_1 = u &= \overline{x}^a(\overline{P}(\overline{y}, \overline{x}^{i_1 - i_0}, \ldots, \overline{x}^{i_n - i_0}) + \overline{x}^{c - i_0}\Omega'). \end{aligned} \tag{9.28}$$

We compute the Jacobian determinants

$$\text{Det}\left(\frac{\partial(u, v, w)}{\partial(x, y, z)}\right) = ax^{a+b+c-1}$$

and

$$\text{Det}\left(\frac{\partial(u_1, v_1, w_1)}{\partial(\overline{x}, \overline{y}, z)}\right) = i_0\overline{x}^{a+b+c-1}\frac{\partial\Omega'}{\partial z}.$$

Thus $\frac{\partial\Omega'}{\partial z}(0,0,0) \neq 0$. Set

$$\overline{\beta} = \Omega'(0,0,0), \ \overline{z} = \Omega' - \overline{\beta}.$$

$\overline{x}, \overline{y}, \overline{z}$ are regular parameters in $\hat{\mathcal{O}}_{X,p}$. We have an expression

$$
(9.29) \qquad \begin{aligned}
u_1 = w &= \overline{x}^{i_0}\\
v_1 = v &= \overline{x}^{b}(\overline{\alpha}+\overline{y})\\
w_1 = u &= \overline{x}^{a}\overline{P}(\overline{y}, \overline{x}^{i_1-i_0}, \ldots, \overline{x}^{i_n-i_0}) + \overline{x}^{a+c-i_0}(\overline{\beta}+\overline{z})
\end{aligned}
$$

of the form (9.9).

Let τ_1 be the computation of $\tau_f(p)$ from (9.24), τ_2 be the computation of $\tau_f(p)$ from (9.29). We see that $\tau_2 \le \tau_1$.

By this argument applied to the change of variables $u_1 = w, v_1 = v, w_1 = u$ to u, v, w, we see that $\tau_1 \le \tau_2$, so that $\tau_1 = \tau_2$.

There is a similar calculation if $u_1 = u, v_1 = w, w_1 = v$.

Now all other permutations u_1, v_1, w_1 of u, v, w such that u_1, v_1, w_1 have a form (4.1) and (9.9) at p (so that u_1, v_1 are toroidal forms at p) are a composition of change of variables of the three types analyzed above. We have shown that $\tau_f(p)$ is invariant under such change of variables. Hence we get the same evaluation of $\tau_f(p)$ for all possible permuations of the variables u, v, w and u_1, v_1, w_1.

Now assume that $u_1 = u, v_1 = v$ and w_1 are permissible parameters at q.

Suppose that u, v, w have an expression of the form (4.1) at p. We then have an expression (9.14) of u, v, w. Let τ_1 be the computation of $\tau_f(p)$ for this expression. Set

$$P = \sum_{j=0}^{n} \alpha_{ij}(y)x^{i_j}.$$

We have an expansion

$$(9.30) \qquad w_1 = \sum_{i=0}^{\infty} g_i(u,v)w^i$$

and $w_1 = w\gamma$ where $\gamma(u, v, w)$ is a unit series if $q = f(p)$ is a 3-point.

Substitute $w = P + x^c(z+\beta)$ into (9.30) to get

$$w_1 = \sum_{i=0}^{\infty}\sum_{j=0}^{i} g_i(u,v)\binom{i}{j}P^{i-j}x^{jc}(z+\beta)^j.$$

We see that there exists a series $\overline{P}(y, x^a, x^b, x^{i_0}, \ldots, x^{i_n})$ such that

$$w_1 = \overline{P}(y, x^a, x^b, x^{i_0}, \ldots, x^{i_n}) + x^c\Omega$$

for some series Ω. Comparing the Jacobian determinants

$$\mathrm{Det}\left(\frac{\partial(u,v,w)}{\partial(x,y,z)}\right) = ax^{a+b+c-1}$$

and

$$\mathrm{Det}\left(\frac{\partial(u,v,w_1)}{\partial(x,y,z)}\right) = ax^{a+b+c-1}\frac{\partial\Omega}{\partial z},$$

we see that $\frac{\partial\Omega}{\partial z}(0,0,0) \ne 0$. Set

$$\overline{\beta} = \Omega(0,0,0),\ \overline{z} = \Omega - \overline{\beta}.$$

We have

$$
\begin{aligned}
u &= x^a\\
v &= x^b(\alpha+y)\\
w_1 &= \overline{P}(y, x^a, x^b, x^{i_0}, \ldots, x^{i_n}) + x^c(\overline{z}+\overline{\beta})
\end{aligned}
$$

of the form of (9.9). Further, we have $w_1 = x^{i_0}\overline{\gamma}$ where $\overline{\gamma}(\overline{x},\overline{y},\overline{z})$ is a unit series if $q = f(p)$ is a 3-point. Let τ_2 be the computation of $\tau_f(p)$ for this expression. We see that $\tau_2 \le \tau_1$. Now applying this argument to the change of variables from u, v, w_1 to u, v, w, we see that $\tau_1 = \tau_2$.

Now suppose that u, v, w have an expression of the form 2 (b) of Definition 4.6 and (9.10) (so that $q = f(p)$ is a 2-point). We have

$$\begin{aligned} u &= x^a \\ v &= \textstyle\sum_{j=0}^n \alpha_{i_j}(y)x^{i_j} + x^c(z+\beta) \\ w &= y \end{aligned}$$

where $\alpha_{i_j}(y)$ are all nonzero and $i_j < c$ for all j.

Let τ_1 be the computation of $\tau_f(p)$ for u, v, w.

By the Weierstrass preparation theorem, there exists a unit series $\gamma(u,v,w)$ and a series $\phi(u,v)$ with $\phi(0,0) = 0$ such that $w_1 = \gamma(w-\phi)$. We have

$$w_1 = \gamma(0,0,y)y + x\Omega'$$

for some series $\Omega'(x,y,z)$. We may thus set $\overline{y} = w_1$, and have that $x, \overline{y}, z$ are regular parameters in $\hat{\mathcal{O}}_{X,p}$.

There exist a unit series $\overline{\alpha}_0(y)$ and series $\overline{\alpha}_{r_0,\ldots,r_n}(y)$ such that there is an expression

$$\overline{y} = \overline{\alpha}_0(y)y + \sum_{s+r_0+\cdots+r_n>0} \overline{\alpha}_{r_0,\ldots,r_n}(y)x^{as+i_0r_0+\cdots+i_nr_n} + x^c\overline{\Omega}$$

with $\overline{\alpha}_0(0) \ne 0$.

We thus have an expression

(9.31)
$$y = \overline{\alpha}_0(y)^{-1}\overline{y} - \sum_{s+r_0+\cdots+r_n>0} \overline{\alpha}_0(y)^{-1}\overline{\alpha}_{r_0,\ldots,r_n}(y)x^{as+i_0r_0+\cdots+i_nr_n} + x^c\overline{\alpha}_0(y)^{-1}\overline{\Omega}.$$

We substitute (9.31) into successive iterations of

$$v = \sum_{j=0}^n \alpha_{i_j}(y)x^{i_j} + x^c(z+\beta)$$

to obtain an expression

$$v = P(\overline{y}, x^a, x^{i_0}, \ldots, x^{i_n}) + x^c\Omega.$$

Comparing the Jacobian determinants

$$\operatorname{Det}\left(\frac{\partial(u,v,w)}{\partial(x,y,z)}\right) = -ax^{a+c-1}$$

and

$$\operatorname{Det}\left(\frac{\partial(u,v,w_1)}{\partial(x,\overline{y},z)}\right) = -ax^{a+c-1}\frac{\partial\Omega}{\partial z},$$

we see that $\frac{\partial\Omega}{\partial z}(0,0,0) \ne 0$. Set

$$\overline{\beta} = \Omega(0,0,0),\ \overline{z} = \Omega - \overline{\beta}.$$

$x, \overline{y}, \overline{z}$ are regular parameters in $\hat{\mathcal{O}}_{X,p}$, and we have an expression

$$\begin{aligned} u &= x^a \\ v &= P(y, x^a, x^{i_0}, \ldots, x^{i_n}) + x^c(\overline{z}+\overline{\beta}) \\ w_1 &= \overline{y} \end{aligned}$$

of the form of (9.10). Let τ_2 be the computation of $\tau_f(p)$ for u, v, w_1. We have $\tau_2 \leq \tau_1$. Now applying this argument to the change of variables from u, v, w_1 to u, v, w, we see that $\tau_1 = \tau_2$.

Using the above techniques, we can show that we have the same computation of $\tau_f(p)$ for a change of variables

$$u_1 = \lambda_1 u, v_1 = \lambda_2 v, w_1 = w$$

where λ_1, λ_2 are unit series.

Finally, suppose that u_1, v_1, w_1 are arbitrary permissible parameters at q. Then u_1, v_1, w_1 may be obtained from u, v, w be a series of changes of variables of the types considered above. It follows that the computation of $\tau_f(p)$ for the variables u, v, w and u_1, v_1, w_1 are the same.

We leave to the reader the verification of the remaining cases (p a 1-point and q a 1-point, p a 2-point).

□

LEMMA 9.3. *Suppose that $f : X \rightarrow Y$ is prepared. Then τ_f is upper semi continuous on D_X.*

PROOF. This follows from a local calculation, computing a deformation in etale local coordinates of the possible prepared forms.

Suppose that $p^* \in X$. We must find a Zariski open neighborhood U of p^* in X such that $p \in U \cap D_X$ implies $\tau_f(p) \leq \tau_f(p^*)$.

We work out the case when p^* is a 3-point. The other cases are similar. Suppose that $p* \in X$ is a 3-point, $\tau_f(p^*) \geq 1$ and there exist (algebraic) permissible parameters u, v, w at $q* = f(p*)$ such that u, v are toroidal forms at $p*$. Then there exist regular parameters $\tilde{x}, \tilde{y}, \tilde{z} \in \mathcal{O}_{X,p*}$, and units $\gamma_1, \gamma_2 \in \mathcal{O}_{X,p*}$ such that

$$\begin{aligned} u &= \tilde{x}^a \tilde{y}^b \tilde{z}^c \gamma_1 \\ v &= \tilde{x}^d \tilde{y}^e \tilde{z}^f \gamma_2 \end{aligned}$$

with

$$\text{rank} \begin{pmatrix} a & b & c \\ d & e & f \end{pmatrix} = 2.$$

There exist rational numbers $\overline{a}_{ij}$ such that if we set

$$x = \tilde{x}\gamma_1^{\overline{a}_{11}}\gamma_2^{\overline{a}_{12}}, y = \tilde{x}\gamma_1^{\overline{a}_{21}}\gamma_2^{\overline{a}_{22}}, z = \tilde{x}\gamma_1^{\overline{a}_{31}}\gamma_2^{\overline{a}_{32}},$$

then

$$u = x^a y^b z^c, \ v = x^d y^e z^f.$$

From the Jacobian determinant

$$\text{Det}\left(\frac{\partial(u, v, w)}{\partial(x, y, z)} \right)$$

we see that we have an expansion

$$\begin{aligned} u &= x^a y^b z^c \\ v &= x^d y^e z^f \\ w &= \textstyle\sum_{i \geq 0} \alpha_i x^{a_i} y^{b_i} z^{c_i} + x^g y^h z^i \gamma \end{aligned} \tag{9.32}$$

where γ is a unit series,

$$\text{Det} \begin{pmatrix} a & b & c \\ d & e & f \\ g & h & i \end{pmatrix} \neq 0,$$

and for all i, $0 \neq \alpha_i \in \mathbf{k}$,

$$\mathrm{Det}\begin{pmatrix} a & b & c \\ d & e & f \\ a_i & b_i & c_i \end{pmatrix} = 0,$$

and $(a_i, b_i, c_i) \not\geq (g, h, i)$.

We can compute $\tau_f(p^*)$ by changing variables, multiplying x, y, z by appropriate rational powers of γ. We obtain that

$$\begin{aligned} u &= \hat{x}^a \hat{y}^b \hat{z}^c \\ v &= \hat{x}^d \hat{y}^e \hat{z}^f \\ w &= \textstyle\sum_{i \geq 0} \alpha_i \hat{x}^{a_i} \hat{y}^{b_i} \hat{z}^{c_i} + \hat{x}^g \hat{y}^h \hat{z}^i. \end{aligned}$$

Thus

$$\tau_f(p^*) = |((a,b,c)\mathbf{Z} + (d,e,f)\mathbf{Z} + \sum_{i \geq 0} (a_i, b_i, c_i)\mathbf{Z})/((a,b,c)\mathbf{Z} + (d,e,f)\mathbf{Z})|$$

if q^* is a 2-point,

$$\tau_f(p^*) = |((a,b,c)\mathbf{Z} + (d,e,f)\mathbf{Z} + \sum_{i \geq 0} (a_i, b_i, c_i)\mathbf{Z})/((a,b,c)\mathbf{Z} + (d,e,f)\mathbf{Z} + (a_0, b_0, c_0)\mathbf{Z})|$$

if q^* is a 3-point.

We compute the Jacobian

$$\mathrm{Det}\left(\frac{\partial(u,v,w)}{\partial(\hat{x},\hat{y},\hat{z})}\right) = \mathrm{Det}\begin{pmatrix} a & b & c \\ d & e & f \\ g & h & i \end{pmatrix} \hat{x}^{a+d+g-1} \hat{y}^{b+e+h-1} \hat{z}^{c+f+i-1}. \tag{9.33}$$

Let $\bar{e}$ be a common denominator of the $\bar{a}_{ij}$. There exists an affine neighborhood $U = \mathrm{spec}(A)$ of p^* such that $\tilde{x}, \tilde{y}, \tilde{z}$ are uniformizing parameters on U, and γ_1, γ_2 are units on U. Let

$$B = A[\gamma_1^{\frac{1}{\bar{e}}}, \gamma_2^{\frac{1}{\bar{e}}}],$$

$V = \mathrm{spec}(B)$. After possibly shrinking U to a smaller neighorhood of p^*, we may assume that V is an etale cover of U and x, y, z are uniformizing parameters on V. Let $\Lambda : V \to U$ be the natural morphism.

Suppose that $p \in D_X \cap U$ is a 2-point. We will show that $\tau_f(p) \leq \tau_f(p^*)$. After possibly interchanging x, y, z, we may assume that $\hat{\mathcal{O}}_{U,p}$ has regular parameters $x, y, \tilde{z} = z - \alpha$ for some $0 \neq \alpha \in \mathbf{k}$. After interchanging u, v if necessary, we have three possible cases:

1. $ae - bd \neq 0$,
2. $ae - bd = 0$, $(a,b) \neq (0,0)$ and $(d,e) \neq (0,0)$,
3. $(d,e) = 0$.

We will analyze these three cases in turn.

Suppose that $ae - bd \neq 0$, Define (formal) regular parameters $\overline{x}, \overline{y}, \overline{z}$ at p by choosing $\lambda_1, \lambda_2 \in \mathbf{Q}$ so that

$$x = \overline{x} z^{\lambda_1}, y = \overline{y} z^{\lambda_2}, z = \overline{z} + \alpha$$

satisfy

$$u = \overline{x}^a \overline{y}^b, v = \overline{x}^d \overline{y}^e.$$

We thus have an expression

$$u=\overline{x}^a\overline{y}^b, v=\overline{x}^d\overline{y}^e, w=P(\overline{x},\overline{y})+\overline{x}^m\overline{y}^n\Omega \tag{9.34}$$

for some series $P(\overline{x},\overline{y})$ and $\Omega\in\hat{\mathcal{O}}_{X,p}$, where $\frac{\partial\Omega}{\partial\overline{z}}(p)\neq 0$.

We compute the Jacobian determinant

$$\mathrm{Det}\left(\frac{\partial(u,v,w)}{\partial(\overline{x},\overline{y},\overline{z})}\right)=(ae-bd)\frac{\partial\Omega}{\partial\overline{z}}\overline{x}^{a+d+m-1}\overline{y}^{b+e+n-1}$$

to see that $\frac{\partial\Omega}{\partial\overline{z}}(p)\neq 0$. Comparing with (9.33), we see that $m=g$ and $n=h$.

We compute that

$$\frac{\partial}{\partial\overline{x}}=z^{\lambda_1}\frac{\partial}{\partial x},\ \frac{\partial}{\partial\overline{y}}=z^{\lambda_2}\frac{\partial}{\partial y},\ \frac{\partial}{\partial\overline{z}}=\lambda_1\frac{x}{z}\frac{\partial}{\partial x}+\lambda_2\frac{y}{z}\frac{\partial}{\partial y}+\frac{\partial}{\partial z}.$$

Thus for $m,n\in\mathbf{N}$,

$$\frac{\partial^{m+n}w}{\partial\overline{x}^m\partial\overline{y}^n}=z^{m\lambda_1+n\lambda_2}\frac{\partial^{m+n}w}{\partial x^m\partial y^n}.$$

From (9.32) we see that $\frac{\partial^{m+n}w}{\partial x^m\partial y^n}\in(x,y)$ if $(m,n)\neq(a_i,b_i)$ for any i, and $(m,n)\not\geq(g,h)$. We have an expansion

$$P(\overline{x},\overline{y})=\sum_{(m,n)\not\geq(g,h)}\frac{1}{m!n!}\frac{\partial^{m+n}w}{\partial\overline{x}^m\partial\overline{y}^n}(p)\overline{x}^m\overline{y}^n.$$

Thus there is an expansion

$$P=\sum_{i\geq 0}\beta_i\overline{x}^{a_i}\overline{y}^{b_i}$$

with

$$\beta_i=\alpha^{m\lambda_1+n\lambda_2}\frac{\partial^{m+n}w}{\partial x^m\partial y^n}(p)$$

if $(m,n)=(a_i,b_i)$.

We conclude that we may make a formal change of variables in (9.34), setting $\beta=\Omega(p)$ and $z^*=\Omega-\beta$, to get

$$\begin{aligned} u &=\overline{x}^a\overline{y}^b\\ v &=\overline{x}^d\overline{y}^e\\ w &=\textstyle\sum\beta_i\overline{x}^{a_i}\overline{y}^{b_i}+\overline{x}^g\overline{y}^h(z^*+\beta).\end{aligned}\tag{9.35}$$

We now compare $\tau_f(p)$ to $\tau_f(p^*)$. Let $q=f(p)$.

Suppose that q^* is a 3-point. If q is a 2-point, then either $(a_0,b_0)=(0,0)$, or $g=h=0$ in (9.35). $(a_0,b_0)=(0,0)$ implies $ae-bd=0$ which is not possible. Thus $g=h=0$, and we see that $\tau_f(p)=-\infty\leq\tau_f(p^*)$. Suppose that q is a 3-point. Then we compute $\tau_f(p)$ from (9.35) and (9.3), to get

$$\begin{aligned}\tau_f(p) &=|((a,b)\mathbf{Z}+(d,e)\mathbf{Z}+\textstyle\sum_{\beta_i\neq 0,(a_i,b_i)\not\geq(g,h)}(a_i,b_i)\mathbf{Z})\\ &\quad/((a,b)\mathbf{Z}+(d,e)\mathbf{Z}+(a_0,b_0)\mathbf{Z})|\\ &\leq\tau_f(p^*).\end{aligned}$$

Suppose that q^* is a 2-point. Then q is a 2-point, and we have from (9.35) and (9.3) that

$$\begin{aligned}\tau_f(p) &=|((a,b)\mathbf{Z}+(d,e)\mathbf{Z}+\textstyle\sum_{\beta_i\neq 0,(a_i,b_i)\not\geq(g,h)}(a_i,b_i)\mathbf{Z})/((a,b)\mathbf{Z}+(d,e)\mathbf{Z})|\\ &\leq\tau_f(p^*).\end{aligned}$$

Suppose that $ae - bd = 0$**,** $(a,b) \neq (0,0)$ **and** $(d,e) \neq (0,0)$**.**

There exist $\overline{a}, \overline{b}, t, k$ such that $(a,b) = k(\overline{a}, \overline{b})$, $(d,e) = t(\overline{a}, \overline{b})$ with $\gcd(\overline{a}, \overline{b}) = 1$, $k, t \neq 0$.

Define (formal) regular parameters $\overline{x}, \overline{y}, \overline{z}$ at p by choosing $\lambda_1, \lambda_2, \lambda_3 \in \mathbf{Q}$ so that

$$x = \overline{x} z^{\lambda_1}, y = \overline{y} z^{\lambda_2}, z^{\lambda_3} = \overline{z} + \overline{\alpha}$$

satisfy

$$u = (\overline{x}^{\overline{a}} \overline{y}^{\overline{b}})^k, v = (\overline{x}^{\overline{a}} \overline{y}^{\overline{b}})^t (\overline{z} + \overline{\alpha}).$$

We thus have an expression

$$u = (\overline{x}^{\overline{a}} \overline{y}^{\overline{b}})^k, v = (\overline{x}^{\overline{a}} \overline{y}^{\overline{b}})^t (\overline{z} + \overline{\alpha}), w = P(\overline{x}^{\overline{a}} \overline{y}^{\overline{b}}, \overline{z}) + \overline{x}^m \overline{y}^n \Omega \tag{9.36}$$

where Ω is a unit series and $\overline{a}n - \overline{b}m \neq 0$.

We compute the Jacobian determinant

$$\operatorname{Det}\left(\frac{\partial(u,v,w)}{\partial(\overline{x}, \overline{y}, \overline{z})} \right) = (\overline{x}^{\overline{a}} \overline{y}^{\overline{b}})^{t+k-1} \overline{x}^{m+\overline{a}-1} \overline{y}^{n+\overline{b}-1} \gamma,$$

where

$$\gamma = k(\overline{b}m - \overline{a}n)\Omega + k\overline{b}\overline{x}\frac{\partial \Omega}{\partial \overline{x}} - k\overline{a}\overline{y}\frac{\partial \Omega}{\partial \overline{y}}$$

is a unit series. Comparing with (9.33), we see that $m = g$ and $n = h$.

We compute that

$$\frac{\partial}{\partial \overline{x}} = z^{\lambda_1} \frac{\partial}{\partial x}, \frac{\partial}{\partial \overline{y}} = z^{\lambda_2} \frac{\partial}{\partial y}, \frac{\partial}{\partial \overline{z}} = \frac{\lambda_1}{\lambda_3} x z^{-\lambda_3} \frac{\partial}{\partial x} + \frac{\lambda_2}{\lambda_3} y z^{-\lambda_3} \frac{\partial}{\partial y} + \frac{1}{\lambda_3} z^{1-\lambda_3} \frac{\partial}{\partial z}.$$

Thus for $m, n \in \mathbf{N}$,

$$\frac{\partial^{m+n} w}{\partial \overline{x}^m \partial \overline{y}^n} = z^{m\lambda_1 + n\lambda_2} \frac{\partial^{m+n} w}{\partial x^m \partial y^n}.$$

From (9.32) we see that $\frac{\partial^{m+n} w}{\partial x^m \partial y^n} \in (x,y)$ if $(m,n) \neq (a_i, b_i)$ for any i, and $(m,n) \not\geq (g,h)$. There is thus an expansion

$$P(\overline{x}^{\overline{a}} \overline{y}^{\overline{b}}, z) = \sum_{j,k} \beta_{j,k} (\overline{x}^{\overline{a}} \overline{y}^{\overline{b}})^j \overline{z}^k$$

with

$$\beta_{j,k} = \frac{1}{k!(j\overline{a})!(j\overline{b})!} \frac{\partial^k}{\partial \overline{z}^k} \left(\frac{\partial^{j(\overline{a}+\overline{b})} w}{\partial \overline{x}^{j\overline{a}} \partial \overline{y}^{j\overline{b}}} \right)(p).$$

Thus $\beta_{j,k} = 0$ if $(j\overline{a}, j\overline{b}) \neq (a_i, b_i)$ for any i, and $(j\overline{a}, j\overline{b}) \not\geq (g,h)$.

We conclude that we may make a formal change of variables in (9.36), setting

$$\overline{x} = x^* \gamma_1 \text{ and } \overline{y} = y^* \gamma_2,$$

with appropriate unit series γ_1 and γ_2, to get

$$\begin{aligned} u &= ((x^*)^{\overline{a}} (y^*)^{\overline{b}})^k \\ v &= (x^*)^{\overline{a}} (y^*)^{\overline{b}})^t (\overline{z} + \overline{\alpha}) \\ w &= P((x^*)^{\overline{a}} (y^*)^{\overline{b}}, \overline{z}) + (x^*)^g (y^*)^h. \end{aligned} \tag{9.37}$$

We now compare $\tau_f(p)$ to $\tau_f(p^*)$. Let $q = f(p)$.

Suppose that q^* is a 3-point and q is a 3-point. We compute $\tau_f(p)$ from (9.37) and (9.4), to get

$$\tau_f(p) = |(k\mathbf{Z} + t\mathbf{Z} + \sum_{\beta_{j,k}\neq 0,(j\overline{a},j\overline{b})\not\geq(g,h)} j\mathbf{Z})/(k\mathbf{Z}+t\mathbf{Z}+j_0\mathbf{Z})|,$$

where $j_0(\overline{a},\overline{b}) = (a_0, b_0)$. We have a surjection

$$((a,b,c)\mathbf{Z} + (d,e,f)\mathbf{Z} + \textstyle\sum(a_i,b_i,c_i)\mathbf{Z})/((a,b,c)\mathbf{Z}+(d,e,f)\mathbf{Z}+(a_0,b_0,c_0)\mathbf{Z})$$
$$\to (k\mathbf{Z} + t\mathbf{Z} + \textstyle\sum_{j(\overline{a},\overline{b})=(a_i,b_i)\text{ for some } i} j\mathbf{Z})/(k\mathbf{Z}+t\mathbf{Z}+j_0\mathbf{Z})$$

defined by $j(\overline{a},\overline{b},0) + l(0,0,1) \mapsto j$. Thus $\tau_f(p) \le \tau_f(p^*)$.

If q^* is a 3-point and q is a 2-point, then we must have that $(a_0, b_0) = (0,0)$ or $g = h = 0$. But we cannot have $g = h = 0$ since

$$\mathrm{Det}\begin{pmatrix} a & b & c\\ d & e & f\\ g & h & i\end{pmatrix} \neq 0.$$

If $(a_0,b_0) = (0,0)$, then we have $(a_0,b_0,c_0) = (0,0,c_0)$, from which we conclude by comparison with (9.37) and (9.4) that $\tau_f(p) \le \tau_f(p^*)$.

Suppose that q^* is a 2-point. Then q is a 2-point.

We compute $\tau_f(p)$ from (9.37) and (9.4), to get

$$\tau_f(p) = |(k\mathbf{Z} + t\mathbf{Z} + \sum_{\beta_{j,k}\neq 0,(j\overline{a},j\overline{b})\not\geq(g,h)} j\mathbf{Z})/(k\mathbf{Z}+t\mathbf{Z})|.$$

As in the case when q^* is a 3-point, we conclude that $\tau_f(p) \le \tau_f(p^*)$.

Suppose that $d = e = 0$. We have $(a,b) = k(\overline{a},\overline{b})$ with $\overline{a},\overline{b} > 0$ and $\gcd(\overline{a},\overline{b}) = 1$.

Define regular parameters $\overline{x},\overline{y},\overline{z}$ in $\hat{\mathcal{O}}_{X,p}$ by choosing $\lambda_1,\lambda_2 \in \mathbf{Q}$ and $\overline{\alpha} = \alpha^f$, so that

$$x = \overline{x}z^{\lambda_1}, y = \overline{y}z^{\lambda_2}, z^f = \overline{z} + \overline{\alpha}$$

satisfy

$$u = (\overline{x}^{\overline{a}}\overline{y}^{\overline{b}})^k, v = \overline{z} + \overline{\alpha}.$$

We thus have an expression

$$u = (\overline{x}^{\overline{a}}\overline{y}^{\overline{b}})^k, \overline{v} = v - \overline{\alpha} = \overline{z}, w = P(\overline{x}^{\overline{a}}\overline{y}^{\overline{b}}, \overline{z}) + \overline{x}^m\overline{y}^n\Omega \tag{9.38}$$

where $u, \overline{v}$ are toroidal forms at $q = f(p)$, P is a series, Ω is a unit series and $\overline{a}n - \overline{b}m \neq 0$.

We compute the Jacobian

$$\mathrm{Det}\left(\frac{\partial(u,v,w)}{\partial(\overline{x},\overline{y},\overline{z})}\right) = \overline{x}^{a+m-1}\overline{y}^{b+n-1}\gamma.$$

where

$$\gamma = (mb - an)\Omega + b\overline{x}\frac{\partial\Omega}{\partial\overline{x}} - a\overline{y}\frac{\partial\Omega}{\partial\overline{y}}$$

is a unit series. Comparing with (9.33), we see that $m = g$ and $n = h$.

We compute that

$$\frac{\partial}{\partial\overline{x}} = z^{\lambda_1}\frac{\partial}{\partial x}, \frac{\partial}{\partial\overline{y}} = z^{\lambda_2}\frac{\partial}{\partial y}, \frac{\partial}{\partial\overline{z}} = \frac{\lambda_1}{f}\frac{x}{z^f}\frac{\partial}{\partial x} + \frac{\lambda_2}{f}\frac{y}{z^f}\frac{\partial}{\partial y} + \frac{1}{f}z^{1-f}\frac{\partial}{\partial z}.$$

Thus for $m, n \in \mathbf{N}$,

$$\frac{\partial^{m+n} w}{\partial \overline{x}^m \partial \overline{y}^n} = z^{m\lambda_1 + n\lambda_2} \frac{\partial^{m+n} w}{\partial x^m \partial y^n}.$$

From (9.32) we see that $\frac{\partial^{m+n} w}{\partial x^m \partial y^n} \in (x, y)$ if $(m, n) \neq (a_i, b_i)$ and $(m, n) \not\geq (g, h)$. There is an expansion

$$P = \sum_{j,k} \beta_{j,k} (\overline{x}^{\overline{a}} \overline{y}^{\overline{b}})^j \overline{z}^k$$

with

$$\beta_{j,k} = \frac{1}{k!(j\overline{a})!(j\overline{b})!} \frac{\partial^k}{\partial \overline{z}^k} \left(\frac{\partial^{j(\overline{a}+\overline{b})} w}{\partial \overline{x}^{j\overline{a}} \partial \overline{y}^{j\overline{b}}} \right) (p).$$

Thus $\beta_{j,k} = 0$ if $(j\overline{a}, j\overline{b}) \neq (a_i, b_i)$ for any i, and $(j\overline{a}, j\overline{b}) \not\geq (g, h)$.

We conclude that we may make a formal change of variables in (9.38), setting

$$\overline{x} = x^* \gamma_1 \text{ and } \overline{y} = y^* \gamma_2,$$

with appropriate unit series γ_1 and γ_2, to get

$$(9.39) \qquad \begin{aligned} u &= ((x^*)^{\overline{a}} (y^*)^{\overline{b}})^k \\ \overline{v} &= v - \overline{\alpha} = \overline{z} \\ w &= P((x^*)^{\overline{a}} (y^*)^{\overline{b}}, \overline{z}) + (x^*)^g (y^*)^h. \end{aligned}$$

We now compare $\tau_f(p)$ to $\tau_f(p^*)$. Let $q = f(p)$.

Suppose that q^* is a 3-point. Then q is a 1-point or a 2-point. If q is a 2-point, then $u, w, \overline{v}$ are permissible parameters at q such that there is an expression

$$\begin{aligned} u &= ((x^*)^{\overline{a}} (y^*)^{\overline{b}})^k \\ w &= P((x^*)^{\overline{a}} (y^*)^{\overline{b}}, \overline{z}) + (x^*)^g (y^*)^h \\ \overline{v} &= \overline{z} \end{aligned}$$

of the form of 2 (c) of Definition 4.6, and (9.5).

We have

$$\tau_f(p) = |(k\mathbf{Z} + \sum_{\beta_{j,k} \neq 0, (j\overline{a}, j\overline{b}) \not\geq (g,h)} j\mathbf{Z}) / (k\mathbf{Z} + j_0\mathbf{Z})|,$$

where $j_0(\overline{a}, \overline{b}) = (a_{i_0}, b_{i_0})$.

We thus have a surjection

$$\begin{aligned} &((a, b, c)\mathbf{Z} + (d, e, f)\mathbf{Z} + \textstyle\sum (a_i, b_i, c_i)\mathbf{Z}) / ((a, b, c)\mathbf{Z} + (d, e, f)\mathbf{Z} + (a_0, b_0, c_0)\mathbf{Z}) \\ &\to (k\mathbf{Z} + \textstyle\sum_{j(\overline{a}, \overline{b}) = (a_i, b_i) \text{ for some } i} j\mathbf{Z}) / (k\mathbf{Z} + j_0\mathbf{Z}) \end{aligned}$$

defined by $j(\overline{a}, \overline{b}, 0) + l(0, 0, 1) \mapsto j$. Thus $\tau_f(p) \leq \tau_f(p^*)$.

If q^* is a 3-point and q is a 1-point, then we have permissible parameters $u, \overline{v}, \overline{w}$ at q, defined by

$$\begin{aligned} u &= ((x^*)^{\overline{a}} (y^*)^{\overline{b}})^k \\ \overline{v} &= \overline{z} \\ \overline{w} &= w - \beta = P((x^*)^{\overline{a}} (y^*)^{\overline{b}}, \overline{z}) - \beta + (x^*)^g (y^*)^h \end{aligned}$$

of the form (4.6) and (9.6), where $0 \neq \beta = P(0, 0)$ so that $(a_{j_0}, b_{j_0}) = (0, 0)$.

We have

$$\tau_f(p) = |(k\mathbf{Z} + \sum_{\beta_{j,k} \neq 0, (j\overline{a}, j\overline{b}) \not\geq (g,h)} j\mathbf{Z}) / k\mathbf{Z}|.$$

We have a surjection

$$((a,b,c)\mathbf{Z}+(d,e,f)\mathbf{Z}+\sum(a_i,b_i,c_i)\mathbf{Z})/((a,b,c)\mathbf{Z}+(d,e,f)\mathbf{Z}+(a_0,b_0,c_0)\mathbf{Z})$$
$$\to (k\mathbf{Z}+\textstyle\sum_{j(\overline{a},\overline{b})=(a_i,b_i)\text{ for some }i} j\mathbf{Z})/(k\mathbf{Z}+j_0\mathbf{Z})$$

defined by $j(\overline{a},\overline{b},0)+l(0,0,1)\mapsto j$. Thus $\tau_f(p)\le\tau_f(p^*)$.

There is a similar analysis if q^* is a 2-point.

We have completed the analysis for a 2-point $p\in D_X\cap U$.

Now suppose that $p\in D_X\cap U$ is a 1-point. Recall that we are assuming that p^* is a 3-point. We will show that $\tau_f(p)\le\tau_f(p^*)$. After possibly interchanging x,y,z, we may suppose that $\hat{\mathcal{O}}_{U,p}$ has regular parameters $x, y-\alpha, z-\beta$ for some $0\ne\alpha\in\mathbf{k}$, $0\ne\beta\in\mathbf{k}$. After interchanging u,v and y,z if necessary, we may assume that $a\ne 0$ and $ea-bd\ne 0$.

Define regular parameters $\overline{x},\overline{y},\overline{z}$ in $\hat{\mathcal{O}}_{X,p}$ by choosing $\lambda_{ij}\in\mathbf{Q}$, $\overline{\alpha}=\alpha^{\lambda_{21}}\beta^{\lambda_{22}}$, so that

$$x=\overline{x}y^{\lambda_{11}}z^{\lambda_{12}},\overline{y}+\overline{\alpha}=y^{\lambda_{21}}z^{\lambda_{22}},\overline{z}+\beta=z$$

satisfy

$$u=\overline{x}^a, v=\overline{x}^d(\overline{y}+\overline{\alpha}).$$

We may do this by setting

$$\lambda_{11}=-\frac{b}{a},\lambda_{12}=-\frac{c}{a},$$
$$\lambda_{21}=d\lambda_{11}+e=\frac{ea-bd}{a}\ne 0,\lambda_{22}=d\lambda_{12}=f.$$

We thus have an expression

$$\begin{array}{ll} u &=\overline{x}^a\\ v&=\overline{x}^d(\overline{y}+\overline{\alpha})\\ w&=P(\overline{x},\overline{y})+\overline{x}^m\Omega\end{array}\tag{9.40}$$

where $P(\overline{x},\overline{y})$ is a series and $\frac{\partial\Omega}{\partial\overline{z}}(p)\ne 0$.

We compute the Jacobian

$$\mathrm{Det}\left(\frac{\partial(u,v,w)}{\partial(\overline{x},\overline{y},\overline{z})}\right)=a\overline{x}^{a+d+m-1}\frac{\partial\Omega}{\partial\overline{z}}.$$

Comparing with (9.33), we see that $m=g$.

We have

$$\begin{array}{ll} x&=\overline{x}(\overline{y}+\overline{\alpha})^{\frac{\lambda_{11}}{\lambda_{21}}}(\overline{z}+\beta)^{\lambda_{12}-\frac{\lambda_{11}\lambda_{22}}{\lambda_{21}}}\\ y&=(\overline{y}+\overline{\alpha})^{\frac{1}{\lambda_{21}}}(\overline{z}+\beta)^{-\frac{\lambda_{22}}{\lambda_{21}}}\\ z&=\overline{z}+\beta,\end{array}$$

from which we see that

$$\begin{array}{ll} \frac{\partial}{\partial\overline{x}}&=y^{\lambda_{11}}z^{\lambda_{12}}\frac{\partial}{\partial x}\\ \frac{\partial}{\partial\overline{y}}&=\frac{\lambda_{11}}{\lambda_{21}}xy^{-\lambda_{21}}z^{-\lambda_{22}}\frac{\partial}{\partial x}+\frac{1}{\lambda_{21}}y^{1-\lambda_{21}}z^{-\lambda_{22}}\frac{\partial}{\partial y}\\ \frac{\partial}{\partial\overline{z}}&=\left(\frac{\lambda_{12}\lambda_{21}-\lambda_{11}\lambda_{22}}{\lambda_{21}}\right)xz^{-1}\frac{\partial}{\partial x}-\frac{\lambda_{22}}{\lambda_{21}}yz^{-1}\frac{\partial}{\partial y}+\frac{\partial}{\partial z}.\end{array}$$

Thus

$$\frac{\partial^m w}{\partial\overline{x}^m}=y^{m\lambda_{11}}z^{m\lambda_{12}}\frac{\partial^m w}{\partial x^m}.$$

From (9.32) we see that $\frac{\partial^m w}{\partial x^m} \in (x)$ if $m \neq a_i$ for some i and $m \not\geq g$. There is an expansion

$$P = \sum_{j,k} \beta_{j,k} \overline{x}^j \overline{y}^k$$

with

$$\beta_{j,k} = \frac{1}{k!j!} \frac{\partial^k}{\partial \overline{y}^k} \left(\frac{\partial^j w}{\partial \overline{x}^j} \right) (p).$$

Thus $\beta_{j,k} = 0$ if $j \neq a_i$ and $j \not\geq g$.

We may make a formal change of variables in (9.40), setting $\overline{\beta} = \Omega(p)$ and $z^* = \Omega - \overline{\beta}$, to get

$$\begin{aligned} u &= \overline{x}^a \\ v &= \overline{x}^d(\overline{y} + \overline{\alpha}) \\ w &= \sum \beta_{j,k} \overline{x}^j \overline{y}^k + \overline{x}^g (z^* + \overline{\beta}). \end{aligned} \tag{9.41}$$

We now compare $\tau_f(p)$ to $\tau_f(p^*)$. Let $q = f(p)$.

Suppose that q^* is a 3-point. If q is a 2-point then we have that $a_0 = 0$, and by (9.9), we have

$$\tau_f(p) = |(a\mathbf{Z} + d\mathbf{Z} + \sum_{\beta_{jk} \neq 0, j < g} j\mathbf{Z})/(a\mathbf{Z} + d\mathbf{Z})| \leq \tau_f(p^*).$$

If q^* is a 3-point and q is a 3-point, then by (9.9), we have

$$\tau_f(p) = |(a\mathbf{Z} + d\mathbf{Z} + \sum_{\beta_{jk} \neq 0, j < g} j\mathbf{Z})/(a\mathbf{Z} + d\mathbf{Z} + a_0\mathbf{Z})| \leq \tau_f(p^*).$$

Suppose that q^* is a 2-point. Then q is a 2-point, and by a similar calculation, $\tau_f(p) \leq \tau_f(p^*)$.

This completes the analysis that $\tau_f(p) \leq \tau_f(p^*)$ if p^* is a 3-point and $p \in U$.

The analysis when p^* is a 2-point or a 1-point is simpler, and we leave it to the reader.

We conclude that τ_f is upper semi-continuous.

□

DEFINITION 9.4. Suppose that $f : X \to Y$ is a prepared, proper morphism, and $\tau \in \mathbf{N}$ is such that $\tau_f(X) \leq \tau$. Let $G_X(f, \tau) = \{p \in D_X \mid \tau_f(X) = \tau\}$, $G_Y(f, \tau) = f(G_X(f, \tau))$. We will say that f is τ-prepared if $G_Y(f, \tau)$ contains no 2-curves and no 3-points.

By Lemma 9.3, $G_X(f, \tau)$ is a closed subset of X, and since f is proper, $G_Y(f, \tau)$ is a closed subset of Y.

CHAPTER 10

Super parameters

Throughout this chapter, we assume that $f : X \to Y$ is a dominant, proper morphism of nonsingular 3-folds.

DEFINITION 10.1. Assume that $f : X \to Y$ is prepared.

Suppose that $q \in Y$ is a 2-point. Permissible parameters u, v, w at q are **super parameters** for f at q if at all $p \in f^{-1}(q)$, there exist permissible parameters x, y, z for u, v, w at p such that we have one of the forms:

1. p is a 1-point

$$\begin{aligned} u &= x^a \\ v &= x^b(\alpha + y) \\ w &= x^c\gamma(x, y) + x^d(z + \beta) \end{aligned} \tag{10.1}$$

   where γ is a unit series (or zero), $0 \neq \alpha \in \mathbf{k}$ and $\beta \in \mathbf{k}$,

2. p is a 2-point of the form of (4.2) of Definition 4.1

$$\begin{aligned} u &= x^a y^b \\ v &= x^c y^d \\ w &= x^e y^f \gamma(x, y) + x^g y^h (z + \beta) \end{aligned} \tag{10.2}$$

   where $ad - bc \neq 0$, γ is a unit series (or zero), and $\beta \in \mathbf{k}$.

3. p is a 2-point of the form of (4.3) of Definition 4.1

$$\begin{aligned} u &= (x^a y^b)^k \\ v &= (x^a y^b)^t(\alpha + z) \\ w &= (x^a y^b)^l \gamma(x^a y^b, z) + x^c y^d \end{aligned} \tag{10.3}$$

   where $0 \neq \alpha \in \mathbf{k}$, $\gcd(a, b) = 1$, $ad - bc \neq 0$ and γ is a unit series (or zero).

4. p is a 3-point

$$\begin{aligned} u &= x^a y^b z^c \\ v &= x^d y^e z^f \\ w &= x^g y^h z^i \gamma + x^j y^k z^l \end{aligned} \tag{10.4}$$

   where $\text{rank}(u, v, x^j y^k z^l) = 3$, $\text{rank}(u, v, x^g y^h z^i) = 2$ and γ is a unit series in monomials M such that $\text{rank}(u, v, M) = 2$ (or γ is zero).

Suppose that $q \in Y$ is a 1-point. Permissible parameters u, v, w at q are **super parameters** for f at q if at all $p \in f^{-1}(q)$, there exist permissible parameters x, y, z for u, v, w at p such that we have one of the forms:

5. p is a 1-point

$$\begin{aligned} u &= x^a \\ v &= y \\ w &= x^c\gamma(x, y) + x^d(z + \beta) \end{aligned} \tag{10.5}$$

where γ is a unit series (or zero) and $\beta \in \mathbf{k}$,

6. p is a 2-point

$$
\begin{aligned}
u &= (x^a y^b)^k \\
v &= z \\
w &= (x^a y^b)^l \gamma(x^a y^b, z) + x^c y^d
\end{aligned}
\tag{10.6}
$$

where $\gcd(a,b) = 1$, $ad - bc \neq 0$ and γ is a unit series (or zero).

LEMMA 10.2. *Suppose that $f : X \to Y$ is prepared, and $\Phi : X_1 \to X$ is the blow up of a 2-curve or a 3-point. Let $f_1 = f \circ \Phi : X_1 \to Y$. Then f_1 is prepared, and $\tau_{f_1}(p) \le \tau_f(\Phi(p))$ for all $p \in D_{X_1}$.*

If $q \in Y$ and u, v, w are super parameters for f at q, then u, v, w are super parameters for f_1 at q.

PROOF. We prove this in the case when $\Phi : X_1 \to X$ is the blow up of a 2-curve C. The case when Φ is the blow up of a 3-point is similar.

Suppose that $p \in C$. Let $q = f(p)$. Then there are permissible parameters u, v, w at q and x, y, z for u, v, w at p such that either u, v are toroidal forms at p, or a form 2 (c) of Definition 4.6 holds at q. Further, $x = y = 0$ are local equations of C at p.

The most difficult case is when p is a 3-point, $q = f(p)$ is a 3-point and $\tau_f(p) \ge 1$. The other cases are similar. Assume that this case holds.

We have an expansion of the form of (9.1)

$$
\begin{aligned}
u &= x^a y^b z^c \\
v &= x^d y^e z^f \\
w &= \textstyle\sum_{(a_i,b_i,c_i) \not\geq (g,h,i)} \alpha_i x^{a_i} y^{b_i} z^{c_i} + x^g y^h z^i
\end{aligned}
\tag{10.7}
$$

with $\alpha_i \neq 0$ for all i. There are $\phi_i, \psi_i \in \mathbf{Q}$ such that

$$(a_i, b_i, c_i) = \phi_i(a, b, c) + \psi_i(d, e, f)$$

for all i.

Suppose that $p_1 \in \Phi^{-1}(p)$. Then (after possibly interchanging x and y) $\hat{\mathcal{O}}_{X_1,p_1}$ has regular parameters x_1, y_1, z where

$$x = x_1, y = x_1(y_1 + \alpha)$$

for some $\alpha \in \mathbf{k}$. We have

$$
\begin{aligned}
u &= x_1^{a+b}(y_1 + \alpha)^b z^c \\
v &= x_1^{d+e}(y_1 + \alpha)^e z^f \\
w &= \textstyle\sum_{(a_i,b_i,c_i) \not\geq (g,h,i)} \alpha_i x_1^{a_i+b_i}(y_1 + \alpha)^{b_i} z^{c_i} + x_1^{g+h}(y_1 + \alpha)^h z^i.
\end{aligned}
$$

We may assume that $\tau_{f_1}(p_1) \ge 1$.

Case 1. Assume that $0 \neq \alpha$ and $(a+b)f - c(d+e) \neq 0$. There exist regular parameters $\overline{x}_1, \overline{y}_1, \overline{z}_1$ in $\hat{\mathcal{O}}_{X_1,p_1}$ and $\overline{\beta} \in \mathbf{k}$, $0 \neq \overline{\alpha}_i \in \mathbf{k}$, such that

$$
\begin{aligned}
u &= \overline{x}_1^{a+b} \overline{z}_1^c \\
v &= \overline{x}_1^{d+e} \overline{z}_1^f \\
w &= \textstyle\sum_{(a_i+b_i,c_i) \not\geq (g+h,i)} \overline{\alpha}_i \overline{x}_1^{a_i+b_i} \overline{z}_1^{c_i} + \overline{x}_1^{g+h} \overline{z}_1^i (\overline{y}_1 + \overline{\beta})
\end{aligned}
$$

of the form of (9.3). The homomorphism $\Lambda : \mathbf{Z}^3 \to \mathbf{Z}^2$ defined by $(x,y,z) \mapsto (x+y,z)$ induces a surjection

$$\begin{aligned}&((a,b,c)\mathbf{Z}+(d,e,f)\mathbf{Z}+\textstyle\sum_{i\ge 0}(a_i,b_i,c_i)\mathbf{Z})/((a,b,c)\mathbf{Z}+(d,e,f)\mathbf{Z}+(a_0,b_0,c_0)\mathbf{Z})\\ &\to ((a+b,c)\mathbf{Z}+(d+e,f)\mathbf{Z}+\textstyle\sum_{i\ge 0}(a_i+b_i,c_i)\mathbf{Z})/\\ &\quad ((a+b,c)\mathbf{Z}+(d+e,f)\mathbf{Z}+(a_0+b_0,c_0)\mathbf{Z}).\end{aligned}$$

Thus

$$\begin{aligned}\tau_f(p) &\ge |((a+b,c)\mathbf{Z}+(d+e,f)\mathbf{Z}+\textstyle\sum_{i\ge 0}(a_i+b_i,c_i)\mathbf{Z})\\ &\quad /((a+b,c)\mathbf{Z}+(d+e,f)\mathbf{Z}+(a_0+b_0,c_0)\mathbf{Z})|\\ &\ge |((a+b,c)\mathbf{Z}+(d+e,f)\mathbf{Z}+\textstyle\sum_{(a_i+b_i,c_i)\not\ge(g+h,i)}(a_i+b_i,c_i)\mathbf{Z})\\ &\quad /((a+b,c)\mathbf{Z}+(d+e,f)\mathbf{Z}+(a_0+b_0,c_0)\mathbf{Z}))|\\ &= \tau_{f_1}(p_1).\end{aligned}$$

Case 2. Assume $0 \ne \alpha$, and $(a+b)f - c(d+e) = 0$. Then there exist $\overline{a},\overline{b} \in \mathbf{N}$ such that $(a+b,c) = k(\overline{a},\overline{b})$, $(d+e,f) = t(\overline{a},\overline{b})$ with $k,t \ne 0$, $\gcd(\overline{a},\overline{b}) = 1$.

There exist regular parameters $\overline{x}_1,\overline{y}_1,\overline{z}_1$ in $\hat{\mathcal{O}}_{X_1,p_1}$ and $0 \ne \overline{\alpha} \in \mathbf{k}$ such that

$$\begin{aligned}u &= \overline{x}_1^{a+b}\overline{z}_1^{c} = (\overline{x}_1^{\overline{a}}\overline{z}_1^{\overline{b}})^k\\ v &= \overline{x}_1^{d+e}\overline{z}_1^{f}(\overline{y}_1+\overline{\alpha}) = (\overline{x}_1^{\overline{a}}\overline{z}_1^{\overline{b}})^t(\overline{y}_1+\overline{\alpha})\\ w &= \textstyle\sum_{(a_i+b_i,c_i)\not\ge(g+h,i)}\alpha_i(\overline{x}_1^{a}\overline{z}_1^{b})^{\phi_i k+\psi_i t}(\overline{y}_1+\overline{\alpha})^{\psi_i} + \overline{x}_1^{g+h}\overline{z}_1^{i}\end{aligned}$$

of the form of (9.4). We have

$$\tau_{f_1}(p_1) \le |k\mathbf{Z}+t\mathbf{Z}+\sum_{(a_i+b_i,c_i)\not\ge(g+h,i)}(\phi_i k+\psi_i t)\mathbf{Z})/(k\mathbf{Z}+t\mathbf{Z}+(\phi_0 k+\psi_0 t)\mathbf{Z})|.$$

As in the argument of Case 1, we see that

$$\begin{aligned}\tau_f(p) &\ge |((a+b,c)\mathbf{Z}+(d+e,f)\mathbf{Z}+\textstyle\sum_{i\ge 0}(a_i+b_i,c_i)\mathbf{Z})\\ &\quad /((a+b,c)\mathbf{Z}+(d+e,f)\mathbf{Z}+(a_0+b_0,c_0)\mathbf{Z})|\\ &= |(k(\overline{a},\overline{b})\mathbf{Z}+t(\overline{a},\overline{b})\mathbf{Z}+\textstyle\sum_{i\ge 0}(\phi_i k(\overline{a},\overline{b})+\psi_i t(\overline{a},\overline{b})\mathbf{Z})\\ &\quad /(k(\overline{a},\overline{b})\mathbf{Z}+t(\overline{a},\overline{b})\mathbf{Z}+(\phi_0 k(\overline{a},\overline{b})+\psi_0 t(\overline{a},\overline{b}))\mathbf{Z}))|\\ &= |(k\mathbf{Z}+t\mathbf{Z}+\textstyle\sum_{i\ge 0}(\phi_i k+\psi_i t)\mathbf{Z})\\ &\quad /(k\mathbf{Z}+t\mathbf{Z}+(\phi_0 k+\psi_0 t)\mathbf{Z})|\\ &\ge \tau_{f_1}(p_1).\end{aligned}$$

Case 3. $0 = \alpha$. There exist regular parameters $\overline{x}_1,\overline{y}_1,\overline{z}_1$ in $\hat{\mathcal{O}}_{X_1,p_1}$ such that
We have

$$\begin{aligned}u &= \overline{x}_1^{a+b}\overline{y}_1^{b}\overline{z}^{c}\\ v &= \overline{x}_1^{d+e}\overline{y}_1^{e}\overline{z}^{f}\\ w &= \textstyle\sum_{(a_i+b_i,b_i,c_i)\not\ge(g+h,h,i)}\alpha_i\overline{x}_1^{a_i+b_i}\overline{y}_1^{b_i}\overline{z}^{c_i} + \overline{x}_1^{g+h}\overline{y}_1^{h}\overline{z}_1^{i}\end{aligned}$$

of the form of (9.1). Thus

$$\begin{aligned}\tau_{f_1}(p_1) &= |((a+b,b,c)\mathbf{Z}+(d+e,e,f)\mathbf{Z}+\textstyle\sum_{(a_i+b_i,b_i,c_i)\not\ge(g+h,h,i)}(a_i+b_i,b_i,c_i)\mathbf{Z})\\ &\quad /((a+b,b,c)\mathbf{Z}+(d+e,e,f)\mathbf{Z}+(a_0+b_0,b_0,c_0)\mathbf{Z})|\\ &\le \tau_f(p).\end{aligned}$$

In all of the above cases, we see that if u,v,w are super parameters at q, then u,v,w satisfy one of the cases (10.2) - (10.4) of Definition 10.1.

$\square$

LEMMA 10.3. *Suppose that $f : X \to Y$ is prepared, $q \in Y$ and u, v, w are super parameters at q. Then there exists a sequence of blow ups of 2-curves $\Phi : X_1 \to X$ such that $f_1 = f \circ \Phi : X_1 \to Y$ is prepared, $\tau_{f_1}(p) \le \tau_f(\Phi(p))$ for $p \in D_{X_1}$, u, v, w are super parameters at q for f_1 and if $p \in f_1^{-1}(q)$ with $\tau_{f_1}(p) > 1$, then $w = 0$ is a divisor supported on D_X at p.*

PROOF. The fact that for a sequence of blow ups of 2-curves $\Phi : X_1 \to Y$, $f_1 = f \circ \Phi : X_1 \to Y$ is prepared and $\tau_{f_1}(p) \le \tau_f(\Phi(p))$ for $p \in D_{X_1}$ follows from Lemma 10.2. Further, the condition that u, v, w are super parameters at q is preserved by blowup of 2-curves above X.

If $p \in f_1^{-1}(q)$ is a 1-point, then 1 or 5 of Definition 10.1 hold, and if $\tau_{f_1}(p) > 1$, then $w = 0$ is a divisor supported on D_X at p.

By Lemma 5.1, we may construct Φ so that if $p \in f_1^{-1}(q)$, then $(x^e y^f, x^g y^h)\hat{\mathcal{O}}_{X_1,p}$ is principal if 2 of Definition 10.1 holds, $((x^a y^b)^l, x^c y^d)\hat{\mathcal{O}}_{X,p}$ is principal if 3 holds,

$$(x^g y^h z^i, x^j y^k z^l)\hat{\mathcal{O}}_{X_1,p}$$

is principal if 4 holds, $((x^a y^b)^l, x^c y^d)\hat{\mathcal{O}}_{X_1,p}$ is principal if 6 holds.

We see that in all these cases that $w = 0$ is a divisor supported on D_{X_1} at p if $\tau_{f_1}(p) > 1$, so that the conclusions of the lemma hold. □

LEMMA 10.4. *Suppose that $f : X \to Y$ is prepared and $C \subset Y$ is a 2-curve. Then there exists a commutative diagram*

$$\begin{array}{ccc} X_1 & \xrightarrow{f_1} & Y_1 \\ \Phi_1 \downarrow & & \downarrow \Psi_1 \\ X & \xrightarrow{f} & Y \end{array}$$

where $\Psi_1 : Y_1 \to Y$ is the blow up of C, $\Phi_1 : X_1 \to X$ is a product of blow ups of 2-curves, Φ_1 is an isomorphism above $f^{-1}(Y - C)$ and f_1 is prepared. If $p_1 \in X_1$ and $p = \Phi_1(p_1)$, then

$$\tau_{f_1}(p_1) \le \tau_f(p).$$

If f is τ-prepared then f_1 is τ-prepared.

PROOF. Let $\Psi_1 : Y_1 \to Y$ be the blow up of C. By Lemma 5.2, there exists a commutative diagram

$$\begin{array}{ccc} X_1 & \xrightarrow{f_1} & Y_1 \\ \Phi_1 \downarrow & & \downarrow \Psi_1 \\ X & \xrightarrow{f} & Y \end{array}$$

such that $\Phi_1 : X_1 \to X$ is a sequence of blow ups of 2-curves, Φ_1 is an isomorphism above $f^{-1}(Y - C)$ and f_1 is a morphism. Further, by Lemma 10.2, $f \circ \Phi_1$ is prepared, and $\tau_{f \circ \Phi_1}(p) \le \tau_f(\Phi_1(p))$ for $p \in D_{X_1}$.

We will show that $f_1 : X_1 \to Y_1$ is prepared, and $\tau_{f_1}(p) \le \tau_f(\Phi_1(p))$ for $p \in D_{X_1}$.

Suppose that $q \in C \subset Y$ and $p \in (f \circ \Phi_1)^{-1}(q)$. Let $q_1 = f_1(p)$, $p' = \Phi_1(p)$.

Suppose that q is a 3-point. Let u, v, w be permissible parameters at q. Then after possibly permuting u, v and w, we have that u, v, w have one of the forms (4.1) - (4.4) of Definition 4.1 at p', and u, v, w also have one of the forms (4.1) - (4.4) of Definition 4.1 at p.

First assume q_1 is a 3-point. Further assume that p is a 3-point. Then u, v, w have the form of (9.1) at p, and have a form (4.4) of Definition 4.1 at p. Furthermore, p' is a 3-point. If $\tau_f(p') = -\infty$, then $\tau_{f\circ\Phi_1}(p) = -\infty$, and if $\tau_f(p) \geq 1$, then $\tau_{f\circ\Phi_1}(p) \geq 1$ and $H_{f,p'} = H_{f\circ\Phi_1,p}$, $A_{f,p'} = A_{f\circ\Phi_1,p}$ and $\tau_f(p') = \tau_{f\circ\Phi_1}(p)$. After possibly interchanging u and v, q_1 has permissible parameters u_1, v_1, w_1 such that one of the following equations (10.8), (10.9) or (10.10) hold:

$$u = u_1, v = u_1v_1, w = w_1 \tag{10.8}$$

or

$$u = u_1, v = v_1, w = u_1w_1 \tag{10.9}$$

or

$$u = u_1w_1, v = v_1, w = w_1. \tag{10.10}$$

Assume that (10.8) or (10.9) holds. Then u_1, v_1, w_1 have a form (4.4) of Definition 4.1 and (9.1) at p. If $\tau_{f\circ\Phi_1}(p) = -\infty$, then $\tau_{f_1}(p) = -\infty$, and if $\tau_{f\circ\Phi_1}(p) \geq 1$, then

$$H_{f\circ\Phi_1,p} = H_{f_1,p},$$

$A_{f\circ\Phi_1,p} = A_{f_1,p}$ and $\tau_{f_1}(p) = \tau_{f\circ\Phi_1}(p) = \tau_f(p')$.

Assume that (10.10) holds. If $\tau_{f\circ\Phi_1}(p) = -\infty$, then we can interchange u, v and w to obtain the case (10.8), which we have already analyzed, so we may assume that $\tau_{f\circ\Phi_1}(p) \geq 1$. There exist permissible parameters x, y, z for u, v, w at p such that u, v, w has an expression (9.1).

If $\operatorname{rank}(v, M_0) = 2$, then w, v is a toroidal form at p, so we have, after a change of variables in x, y, z at p, an expression of w, v, u of the form of (9.1), and (by Lemma 9.2) we obtain the same calculation of $H_{f\circ\Phi_1,p}$ and $A_{f\circ\Phi_1,p}$ for these new parameters. Thus (10.10) has been transformed into (10.9), from which it follows that w_1, v_1, u_1 have a form (4.4) of Definition 4.1 at p and $\tau_{f_1}(p) = \tau_{f\circ\Phi_1}(p) = \tau_f(p')$.

If $\operatorname{rank}(v, M_0) = 1$, then $\operatorname{rank}(w, u) = 2$ and w, u is a toroidal form at p. As in the above paragraph, we change variables to obtain a form (10.8), from which it follows that w_1, u_1, v_1 have a form (4.4) of Definition 4.1 at p and $\tau_{f_1}(p) = \tau_{f\circ\Phi_1}(p) = \tau_f(p')$.

Suppose that q_1 is a 3-point and p is a 2-point. Then u, v, w are such that u, v are toroidal forms of type (4.2) or type (4.3) of Definition 4.1 at p, and $w = M\gamma$ where M is a monomial in x, y, and γ is a unit series in $\hat{\mathcal{O}}_{X_1,p}$.

First assume that u, v are of type (4.2) of Definition 4.1 at p. After possibly interchanging u and v, q_1 has permissible parameters u_1, v_1, w_1 of one of the forms (10.8) (10.9) or (10.10). In any of these cases, we have an expression

$$u_1 = M_1\gamma_1, v_1 = M_2\gamma_2, w_1 = M_3\gamma_3$$

where $\gamma_1, \gamma_2, \gamma_3$ are unit series and M_1, M_2, M_3 are monomials in x, y with

$$\operatorname{rank}(M_1, M_2, M_3) = 2.$$

Thus (after possibly interchanging u_1, v_1, w_1) u_1, v_1, w_1 have a form (4.2) of Definition 4.1 at p.

Now assume that u, v is of type (4.3) of Definition 4.1 at p (and there does not exist a permutation of u, v, w such that u, v are of type (4.2) at p. We continue to assume that q_1 is a 3-point. Then there are permissible parameters x, y, z at p such that

$$\begin{aligned} u &= (x^a y^b)^k \\ v &= (x^a y^b)^t(\alpha + z) \\ w &= (x^a y^b)^l[\gamma(x^a y^b, z) + x^c y^d] \end{aligned}$$

where γ is a unit series and $ad - bc \neq 0$.

If $u = v = 0$ are local equations for C at q then after possibly permuting u and v, we may assume that q_1 has permissible parameters u_1, v_1, w_1 defined by

$$u = u_1, v = u_1 v_1, w = w_1.$$

Thus u_1, v_1, w_1 have a form (4.3) of Definition 4.1 at p. Otherwise, we can assume, after possibly interchanging u and v that $u = w = 0$ are local equations for C. If permissible parameters are defined at q_1 by

$$u = u_1, v = v_1, w = u_1 w_1,$$

then u_1, v_1, w_1 have a form (4.3) of Definition 4.1 at p.

The remaining possibility is that u_1, v_1, w_1 are permissible parameters at q_1, where

$$u = u_1 w_1, v = v_1, w = w_1.$$

Then

$$\begin{aligned} u_1 &= (x^a y^b)^{k-l}[\gamma + x^c y^d]^{-1} \\ v_1 &= (x^a y^b)^t(\alpha + z) \\ w_1 &= (x^a y^b)^l[\gamma + x^c y^d]. \end{aligned}$$

Define new regular parameters $\overline{x}, \overline{y}, z$ at p by $x = \overline{x}\lambda_x, y = \overline{y}\lambda_y$, where λ_x, λ_y are unit series in $\hat{\mathcal{O}}_{X_1,p}$ such that

$$x^a y^b = (\gamma + x^c y^d)^{\frac{-1}{l}} \overline{x}^a \overline{y}^b.$$

Then

$$\begin{aligned} u_1 &= (\overline{x}^a \overline{y}^b)^{k-l}(\gamma + x^c y^d)^{-\frac{k}{l}} \\ v_1 &= (\overline{x}^a \overline{y}^b)^t[\gamma + x^c y^d]^{\frac{-t}{l}}(\alpha + z) \\ w_1 &= (\overline{x}^a \overline{y}^b)^l. \end{aligned}$$

If

$$\frac{\partial \gamma}{\partial z}(0,0,0) \neq 0,$$

then w_1, u_1 have a form (4.3) of Definition 4.1 at p. If

$$\frac{\partial \gamma}{\partial z}(0,0,0) = 0,$$

then w_1, v_1 have a form (4.3) of Definition 4.1 at p.

In all of these cases (q_1 a 3-point and p a 2-point) we calculate that $\tau_{f_1}(p) \leq \tau_{f \circ \Phi_1}(p)$.

There is a similar analysis (to the case when (4.3) holds at p) if p is a 1-point (and q_1 is a 3-point). After possibly permuting u, v, w, we find permissible parameters u_1, v_1, w_1 at q_1 such that one of the forms (10.8) – (10.10) hold. As in the above paragraph, we see that (after possibly interchanging u_1, v_1, w_1) u_1, v_1, w_1 have a form (4.1) of Definition 4.1 at p. We have that $\tau_{f_1}(p) \leq \tau_{f \circ \Phi_1}(p)$.

Still assuming that q is a 3-point, assume that q_1 is a 2-point. If p is a 3-point then there are permissible parameters x, y, z for u, v, w at p such that

$$\begin{aligned} u &= x^a y^b z^c \gamma_1 \\ v &= x^d y^e z^f \gamma_2 \\ w &= x^l y^m z^n \gamma_3 \end{aligned}$$

where $\gamma_1, \gamma_2, \gamma_3$ are unit series, and

$$\operatorname{rank}\begin{pmatrix} a & b & c \\ d & e & f \\ l & m & n \end{pmatrix} \geq 2.$$

After possibly permuting u, v and w, we may assume that q_1 has permissible parameters u_1, v_1, w_1 defined by

$$u = u_1, v = v_1, w = u_1(w_1 + \alpha) \tag{10.11}$$

with $0 \neq \alpha \in \mathbf{k}$. Thus $(a, b, c) = (l, m, n)$, so that

$$\operatorname{rank}\begin{pmatrix} a & b & c \\ d & e & f \end{pmatrix} = 2$$

and $\tau_{f_1}(p) = \tau_{f\circ\Phi_1}(p) = \tau_f(p') \geq 1$. We thus have that u_1, v_1 are toroidal forms at p, of type (4.4) of Definition 4.1 and $\tau_{f_1}(p) = \tau_f(p')$.

We have a similar analysis if q_1 is a 2-point, p is a 2-point (q is a 3-point), and u, v satisfy (4.2) of Definition 4.1 at p. Then q_1 has permissible parameters u_1, v_1, w_1 satisfying (10.11), and u_1, v_1 are toroidal forms of type (4.2) of Definition 4.1 at p.

Now assume that p is a 2-point, q_1 is a 2-point, (q is a 3-point) and u, v satisfy (4.3) of Definition 4.1 at p and (4.2) of Definition 4.1 does not hold at p for any permutation of u, v, w. Then we have permissible parameters x, y, z at p such that

$$\begin{aligned} u &= (x^a y^b)^l \\ v &= (x^a y^b)^t (\beta + z) \\ w &= (x^a y^b)^m (\gamma(x^a y^b, z) + x^c y^d) \end{aligned}$$

where γ is a unit series and $\beta \neq 0$. If $u = w = 0$ are local equations of C at q, then q_1 has regular parameters u_1, v_1, w_1 defined by

$$u = u_1, v = v_1, w = u_1(w_1 + \alpha) \tag{10.12}$$

with $\alpha \neq 0$. Thus u_1, v_1 are toroidal forms of type (4.3) of Definition 4.1 at p.

We have a similar analysis if $v = w = 0$ are local equations of C at q.

If $u = v = 0$ are local equations of C at q, then q_1 has permissible parameters u_1, w_1, v_1 defined by

$$u = u_1, v = u_1(v_1 + \beta), w = w_1 \tag{10.13}$$

with $0 \neq \beta \in \mathbf{k}$ and we thus have $t = l$. We have

$$\begin{aligned} u_1 &= (x^a y^b)^l \\ w_1 &= (x^a y^b)^m (\gamma + x^c y^d) \\ v_1 &= z \end{aligned}$$

and u_1, w_1, v_1 have the form 2 (c) of Definition 4.6 at p.

In all these cases, we have $\tau_{f_1}(p) \leq \tau_{f\circ\Phi_1}(p)$.

There is a similar analysis in the case when (q is a 3-point) q_1 is a 2-point and p is a 1-point. Then (after possibly permuting u, v, w) u, v satisfy (4.1) of Definition

4.1 at p. If $u = w = 0$ are local equations of C at q, then q_1 has permissible parameters u_1, v_1, w_1 defined by (10.12), and u_1, v_1 are toroidal forms of type (4.1) of Definition 4.1 at p. If $u = v = 0$ are local equations of C at q, then q_1 has permissible parameters u_1, w_1, v_1 defined by (10.13), and u_1, w_1, v_1 have the form 2 (b) of Definition 4.6 at p. We further check that $\tau_{f_1}(p) \le \tau_{f\circ\Phi_1}(p)$ in all these cases (when q is a 3-point, q_1 is a 2-point, p is a 1-point).

Suppose that q is a 2-point

Let u, v, w be permissible parameters at q. Suppose that p is a 3-point (satisfying 1 of Definition 4.6) or p is a 3-point which satisfies 2 (a) of Definition 4.6. q_1 is a 2-point, which has permissible parameters u_1, v_1, w_1 defined by

$$u = u_1, v = u_1v_1, w = w_1,$$

or

$$u = u_1v_1, v = v_1, w = w_1,$$

or q_1 is a 1-point, which has permissible parameters u_1, v_1, w_1 defined by

$$u = u_1, v = u_1(v_1 + \alpha), w = w_1,$$

with $0 \ne \alpha$. By a similar calculation to the case when q is a 3-point, we see that u_1, v_1 are toroidal forms at p, f_1 is prepared at p and $\tau_{f_1}(p) \le \tau_{f\circ\Phi_1}(p)$.

We further check that $\tau_{f_1}(p) \le \tau_f(p')$ if p is a 1-point or a 2-point (and p' satisfies 2 (a) of Definition 4.6).

Now suppose that u, v, w satisfy 2 (b) of Definition 4.6 at p. If permissible parameters at q_1 are u_1, v_1, w with $u = u_1, v = u_1v_1, w = w_1$, then u_1, v_1, w_1 have the form 2 (b) at p also. If $u = u_1v_1, v = v_1$ then we can make an appropriate change of permissible parameters at p to get a form 2 (b) for v, u, w at p. Thus we obtain a form 2 (b) for u_1, v_1, w_1 at p. If $u = u_1, v = u_1(v_1 + \alpha)$, with $0 \ne \alpha \in \mathbf{k}$, then q_1 is a 1-point, and u_1, w_1 are toroidal forms at p. We further check that $\tau_{f_1}(p) \le \tau_{f\circ\Phi_1}(p)$.

Suppose that u, v, w satisfy 2 (c) of Definition 4.6 at p. If q_1 is a 2-point, which has regular parameters u_1, v_1, w_1 defined by

$$u = u_1v_1, v = v_1, w = w_1 \tag{10.14}$$

or

$$u = u_1, v = u_1v_1, w = w_1 \tag{10.15}$$

then at p there is an expression of u_1, v_1, w_1 of the form 2 (c) of Definition 4.6. If q_1 is a 1-point, which has regular parameters u_1, v_1, w_1 defined by

$$u = u_1, v = u_1(v_1 + \alpha), w = w_1, \tag{10.16}$$

with $0 \ne \alpha \in \mathbf{k}$, then u_1, w_1 are toroidal forms at p. We further check that $\tau_{f_1}(p) \le \tau_{f\circ\Phi_1}(p)$.

Comparing the above expressions, we see that f_1 is prepared, and that f_1 satisfies the conclusions of the lemma.

□

LEMMA 10.5. *Suppose that $f : X \to Y$ is prepared and $q \in Y$ is a 2-point. Then there exists a commutative diagram*

$$\begin{array}{ccc} X_1 & \xrightarrow{f_1} & Y_1 \\ \Phi \downarrow & & \downarrow \Psi \\ X & \xrightarrow{f} & Y \end{array} \tag{10.17}$$

where Φ and Ψ are products of blow ups of 2-curves, such that f_1 is prepared, $\tau_{f_1}(p) \le \tau_f(\Phi(p))$ for $p \in D_{X_1}$, and there exist no points of form 2 (b) or 2 (c) of Definition 4.6 for f_1 above points of $\Psi^{-1}(q)$.

PROOF. There exist sequences of blow ups of 2-curves $Y_1 \to Y$ such that the rational map $X \to Y_1$ is defined at all points $p \in f^{-1}(q)$ such that f has at p an expression of form 2 (b) or 2 (c) of Definition 4.6, and p maps to 1-point. By Lemma 10.4, by blowing up 2-curves above X, we can construct f_1 which has the desired property. No new points of the form 2 (b) or 2 (c) of Definition 4.6 are created by this construction. □

LEMMA 10.6. *Suppose that $f : X \to Y$ is prepared, Ω is a set of 2-points of Y, and we have assigned to each $q_0 \in \Omega$ permissible parameters $u = u_{q_0}, v = v_{q_0}, w = w_{q_0}$ at q_0. Then there exists a commutative diagram*

$$\begin{array}{ccc} X_1 & \xrightarrow{f_1} & Y_1 \\ \Phi \downarrow & & \downarrow \Psi \\ X & \xrightarrow{f} & Y \end{array} \tag{10.18}$$

such that

1. *f_1 is prepared.*
2. *Φ, Ψ are products of blow ups of 2-curves.*
3. *Let*

$$\Omega_1 = \left\{ \begin{array}{l} q_1 \in Y_1 \text{ such that } q_1 \text{ is a 2-point and } q_1 = f_1(p_1) \\ \text{for some 3-point } p_1 \in (f \circ \Phi)^{-1}(\Omega). \end{array} \right.$$

Suppose that $q_1 \in \Omega_1$ with $q_0 = \Psi(q_1) \in \Omega$. Then there exist permissible parameters u_1, v_1, w_1 at q_1 such that

$$\begin{aligned} u_{q_0} &= u_1^{\overline{a}} v_1^{\overline{b}} \\ v_{q_0} &= u_1^{\overline{c}} v_1^{\overline{d}} \\ w_{q_0} &= w_1 \end{aligned}$$

for some $\overline{a}, \overline{b}, \overline{c}, \overline{d} \in \mathbf{N}$ with $\overline{a}\overline{d} - \overline{b}\overline{c} = \pm 1$, and if $p_1 \in f_1^{-1}(q_1)$ is a 3-point, then there exist permissible parameters x, y, z at p_1 for u_1, v_1, w_1 such that

$$\begin{aligned} u_1 &= x_1^{a_1} y_1^{b_1} z_1^{c_1} \\ v_1 &= x_1^{d_1} y_1^{e_1} z_1^{f_1} \\ w_1 &= \gamma_1 + N_1 \end{aligned} \tag{10.19}$$

where $N_1 = x_1^{g_1} y_1^{h_1} z_1^{i_1}$, with rank$(u_1, v_1, N_1) = 3$, $\gamma_1 = \sum_i \alpha_i M_i$ where $\alpha_i \in \mathbf{k}$ and each M_i is a monomial in x_1, y_1, z_1 such that there are expressions

$$M_i^{e_i} = u_1^{a_i} v_1^{b_i} \tag{10.20}$$

with $a_i, b_i, e_i \in \mathbf{N}$ and gcd$(a_i, b_i, e_i) = 1$ for all i. Further, there is a bound $r \in \mathbf{N}$ such that $e_i \le r$ for all M_i in expressions (10.20).

4. *Φ is an isomorphism above $f^{-1}(Y-\Sigma(Y))$.*

PROOF. Suppose that $q_0\in\Omega$. Let the 3-points in $f^{-1}(q_0)$ be $\{p_1,\ldots,p_t\}$. Each p_i has permissible parameters x,y,z such that there is an expression of the form (9.1),

$$\begin{aligned} u_{q_0} &= x^ay^bz^c\\ v_{q_0} &= x^dy^ez^f\\ w_{q_0} &= \gamma+N \end{aligned}\tag{10.21}$$

where $N=x^gy^hz^i$, $\gamma=\sum\alpha_iM_i$ with relations

$$M_i^{e_i}=u^{a_i}v^{b_i}\tag{10.22}$$

with $a_i,b_i,e_i\in\mathbf{Z}$, $e_i>0$.

We construct an infinite commutative diagram of morphisms

$$\begin{array}{ccc} \vdots & & \vdots\\ \downarrow & & \downarrow\\ X_n & \overset{f_n}{\rightarrow} & Y_n\\ \Phi_n\downarrow & & \downarrow\Psi_n\\ \vdots & & \vdots\\ \Phi_2\downarrow & & \downarrow\Psi_2\\ X_1 & \overset{f_1}{\rightarrow} & Y_1\\ \Phi_1\downarrow & & \downarrow\Psi_1\\ X & \overset{f}{\rightarrow} & Y \end{array}\tag{10.23}$$

as follows. Order the 2-curves of Y, and let $\Psi_1:Y_1\rightarrow Y$ be the blow up of the 2-curve C of smallest order. Then construct (by Lemma 10.4) a commutative diagram

$$\begin{array}{ccc} X_1 & \overset{f_1}{\rightarrow} & Y_1\\ \Phi_1\downarrow & & \downarrow\Psi_1\\ X & \overset{f}{\rightarrow} & Y \end{array}\tag{10.24}$$

where f_1 is prepared, Φ_1 is a product of blow up of 2-curves and Φ_1 is an isomorphism above $f^{-1}(Y-C)$. Order the 2-curves of Y_1 so that the 2-curves contained in the exceptional divisor of Ψ_1 have larger order than the order of the (strict transforms of the) 2-curves of Y.

Let $\Psi_2:Y_2\rightarrow Y_1$ be the blow up of the 2-curve C_1 on Y_1 of smallest order, and construct a commutative diagram

$$\begin{array}{ccc} X_2 & \overset{f_2}{\rightarrow} & Y_2\\ \Phi_2\downarrow & & \downarrow\Psi_2\\ X_1 & \overset{f_1}{\rightarrow} & Y_1 \end{array}$$

as in (10.24). We now iterate to construct (10.23). Let $\overline{\Psi}_n=\Psi_1\circ\cdot\circ\Psi_n:Y_n\rightarrow Y$, $\overline{\Phi}_n=\Phi_1\circ\cdots\circ\Phi_n:X_n\rightarrow X$.

Let ν be a 0-dimensional valuation of $\mathbf{k}(X)$. Let p_n be the center of ν on X_n, $q_n=f_n(p_n)$. We will say that ν is resolved on X_n if one of the following holds:

1. $\overline{\Psi}_n(q_n)\not\in\Omega$ or
2. $\overline{\Psi}_n(q_n)=q_0\in\Omega$ and

(a) p_n is not a 3-point or
(b) p_n is a 3-point such that a form (10.19) holds for p_n and $q_n = f_n(p_n)$ so that (10.20) holds.

Observe that if ν is resolved on X_n, then there exists a neighborhood U of the center of ν in X_n such that if ω is a 0-dimensional valuation of $\mathbf{k}(X)$ whose center is in U, then ω is resolved on X_n, and if $n' > n$, then ν is resolved on $X_{n'}$.

We will now show that for every 0-dimensional valuation ν of $\mathbf{k}(X)$, there exists $n \in \mathbf{N}$ such that ν is resolved on X_n.

If the center of ν on Y is not in Ω or if the center of ν on X is not a 3-point, then ν is resolved on X, so we may assume that the center of ν on Y is $q_0 \in \Omega$ and the center of ν on X is a 3-point p.

Suppose that $\nu(u_{q_0})$ and $\nu(v_{q_0})$ are rationally dependent. Then there exists n such that the center of ν on Y_n is a 1-point. Thus ν is resolved on X_n.

Suppose that $\nu(u_{q_0})$ and $\nu(v_{q_0})$ are rationally independent. At the center p of ν on X,

$$u = u_{q_0}, v = v_{q_0}, w = w_{q_0}$$

have an expression (9.1). We may identify ν with an extension of ν to the quotient field of $\hat{\mathcal{O}}_{X,p}$ which dominates $\hat{\mathcal{O}}_{X,p}$. We have

$$u = x^a y^b z^c, v = x^d y^e z^f,$$

and

$$M_i^{e_i} = u^{k_i} v^{l_i} = x^{a_i} y^{b_i} z^{c_i}$$

with $k_i, l_i \in \mathbf{Z}$, $e_i > 0$, $a_i, b_i, c_i \in \mathbf{N}$. Thus, for all i, $(k_i, l_i) \in \sigma$, where

$$\sigma = \{(k,l) \in \mathbf{Q}^2 \mid ka + ld \geq 0, kb + le \geq 0, kc + lf \geq 0\}.$$

Since

$$\operatorname{rank} \begin{pmatrix} a & b & c \\ d & e & f \end{pmatrix} = 2,$$

σ is a rational polyhedral cone which contains no nonzero linear subspaces, and is contained in the (irrational) half space

$$\{(k,l) \mid k\nu(u) + l\nu(v) \geq 0\}.$$

Let $\lambda_1 = (m_1, m_2)$, $\lambda_2 = (n_1, n_2)$ be integral vectors such that $\sigma = \mathbf{Q}_+\lambda_1 + \mathbf{Q}_+\lambda_2$. Since λ_1, λ_2 are rational points in σ, we have $\nu(u^{m_1}v^{m_2}) > 0$ and $\nu(u^{n_1}v^{n_2}) > 0$.

Since $\nu(u)$ and $\nu(v)$ are rationally independent, there exists (by Theorem 2.7 [**C2**]) a sequence of quadratic transforms $\mathbf{k}[u,v] \to \mathbf{k}[u_1, v_1]$ such that the center of ν on $\mathbf{k}[u_1, v_1]$ is (u_1, v_1), there is an expression

$$u = u_1^{\overline{a}} v_1^{\overline{b}}, v = u_1^{\overline{c}} v_1^{\overline{d}}$$

for some $\overline{a}, \overline{b}, \overline{c}, \overline{d} \in \mathbf{N}$ with $\overline{a}\overline{d} - \overline{b}\overline{c} = \pm 1$, and $u^{m_1}v^{m_2}, u^{n_1}v^{n_2} \in \mathbf{k}[u_1, v_1]$. Thus there exists a rational polyhedral cone $\sigma_1 \subset \mathbf{Q}^2$ containing λ_1 and λ_2 such that $\mathbf{k}[u_1, v_1] = \mathbf{k}[\sigma_1 \cap \mathbf{Z}^2]$. We thus have $M_i^{e_i} \in \mathbf{k}[u_1, v_1]$ for all i, so that for all i, $M_i^{e_i}$ is a monomial in u_1 and v_1.

There exists an n such that the center of ν on Y_n has permissible parameters u_1, v_1, w_1 where

$$u = u_1^{\overline{a}} v_1^{\overline{b}}, v = u_1^{\overline{c}} v_1^{\overline{d}}, w = w_1. \tag{10.25}$$

We thus have $\tilde{a}_i, \tilde{b}_i \in \mathbf{N}$ such that

$$M_i^{e_i} = u_1^{\tilde{a}_i} v_1^{\tilde{b}_i} \tag{10.26}$$

for all i. Let p_n be the center of ν on X_n. If p_n is not a 3-point, then ν is resolved on X_n.

If p_n is a 3-point, then u_1, v_1, w_1 (defined by (10.25)) are permissible parameters at the 2-point $f_n(p_n)$. There exist permissible parameters x_1, y_1, z_1 at p_n for u_1, v_1, w_1 defined by

$$\begin{aligned} x &= x_1^{a_{11}} y_1^{a_{12}} z_1^{a_{13}} \\ y &= x_1^{a_{21}} y_1^{a_{22}} z_1^{a_{23}} \\ z &= x_1^{a_{31}} y_1^{a_{132}} z_1^{a_{33}} \end{aligned}$$

where $a_{ij} \in \mathbf{N}$, and $\mathrm{Det}(a_{ij}) = \pm 1$. Substituting into the expression (9.1) of u, v, w at p, we have expressions

$$u_1 = x_1^{\tilde{a}} y_1^{\tilde{b}} z_1^{\tilde{c}}, v_1 = x_1^{\tilde{d}} y_1^{\tilde{e}} z_1^{\tilde{f}}, N = x_1^{\tilde{g}} y_1^{\tilde{h}} z_1^{\tilde{i}},$$

where

$$\mathrm{Det}\begin{pmatrix} \tilde{a} & \tilde{b} & \tilde{c} \\ \tilde{d} & \tilde{e} & \tilde{f} \\ \tilde{g} & \tilde{h} & \tilde{i} \end{pmatrix} \neq 0.$$

Let $\vec{v}_1 = (\tilde{a}, \tilde{b}, \tilde{c})$, $\vec{v}_2 = (\tilde{d}, \tilde{e}, \tilde{f})$,

$$\sigma_2 = \mathbf{Q}_+ \vec{v}_1 + \mathbf{Q}_+ \vec{v}_2 \subset \mathbf{Q}^2.$$

By Gordan's Lemma, (Proposition 1 [**F**]) $\sigma_2 \cap \mathbf{Z}^3$ is a finitely generated semi group. Let $\vec{w}_1, \dots, \vec{w}_n \in \sigma_2 \cap \mathbf{Z}^3$ be generators. There exists $0 \neq r \in \mathbf{N}$ and $\delta_j, \epsilon_j \in \mathbf{N}$ such that

$$\vec{w}_j = \frac{\delta_j}{r} \vec{v}_1 + \frac{\epsilon_j}{r} \vec{v}_2$$

for $1 \leq j \leq n$. Since the exponents of

$$M_i = u_1^{\frac{\tilde{a}_i}{e_i}} v_1^{\frac{\tilde{b}_i}{e_i}} = (x_1^{\tilde{a}} y_1^{\tilde{b}} z_1^{\tilde{c}})^{\frac{\tilde{a}_i}{e_i}} (x_1^{\tilde{d}} y_1^{\tilde{e}} z_1^{\tilde{f}})^{\frac{\tilde{b}_i}{e_i}}$$

are in $\sigma_2 \cap \mathbf{Z}^3$ for all i, we have an expression

$$M_i = u_1^{\frac{\overline{a}_i}{r}} v_1^{\frac{\overline{b}_i}{r}}$$

with $\overline{a}_i, \overline{b}_i \in \mathbf{N}$ for all M_i appearing in the expansion (9.1) of w. Thus ν is resolved on X_n.

By compactness of the Zariski-Riemann manifold of X ([**Z**]), there exist finitely many X_i, $1 \leq i \leq t$, such that the center of any 0-dimensional valuation ν of $\mathbf{k}(X)$ is resolved on some X_i. Thus $X_t \to Y_t$ satisfies the conclusions of the Lemma.

□

LEMMA 10.7. *Suppose that $f : X \to Y$ is prepared, $q \in Y$ is a 2-point contained in a 2-curve E of Y and u, v, w are permissible parameters at q. Then there exists a commutative diagram*

$$\begin{array}{ccc} X_1 & \overset{f_1}{\to} & Y_1 \\ \Phi_1 \downarrow & & \downarrow \Psi_1 \\ X & \overset{f}{\to} & Y \end{array} \tag{10.27}$$

such that

1. *f_1 is prepared.*
2. *Φ_1 is a product of blow ups of 2-curves and 3-points such that Φ_1 is an isomorphism above $f^{-1}(Y-E)$. Ψ_1 is a product of blow ups of 2-curves.*
3. *Suppose that $q_1 \in \Psi_1^{-1}(q)$ is a 2-point, so that q_1 has permissible parameters u_1, v_1, w_1 defined by*

$$
\begin{aligned}
u &= u_1^a v_1^b \\
v &= u_1^c v_1^d \\
w &= w_1
\end{aligned}
\tag{10.28}
$$

for some $a, b, c, d \in \mathbf{N}$ with $ad - bc \neq 0$. Then u_1, v_1, w_1 are super parameters at q_1.

PROOF. By Lemma 10.4 and Lemma 10.2, any diagram (10.27) satisfying 2 satisfies 1. Further, if $p \in f^{-1}(q)$, u, v, w are super parameters at p and $p_1 \in \Phi_1^{-1}(p)$ is such that $f_1(p_1) = q_1$ is a 2-point, then the permissible parameters u_1, v_1, w_1 of (10.28) at q_1 are super parameters at p_1.

Step 1. We will show that there exists a sequence of blow ups of 2-curves and 3-points $\Phi_1 : X_1 \to X$ such that Φ_1 is an isomorphism over $f^{-1}(Y-E)$ and u, v, w are super parameters at all 3-points $p \in (f \circ \Phi_1)^{-1}(q)$.

Suppose that $p \in f^{-1}(q)$ is a 3-point, so that there exist permissible parameters $x, y, z \in \hat{\mathcal{O}}_{X,p}$ at p for u, v, w such that

$$
\begin{aligned}
u &= x^a y^b z^c \\
v &= x^d y^e z^f \\
w &= g(x,y,z) + N
\end{aligned}
\tag{10.29}
$$

of the form of (4.4) of Definition 4.1 and (4.10) of Lemma 4.2. There exist regular parameters $\overline{x}, \overline{y}, \overline{z}$ in $\mathcal{O}_{X,p}$, and unit series $\lambda_1, \lambda_2, \lambda_3 \in \hat{\mathcal{O}}_{X,p}$ such that

$$x = \overline{x}\lambda_1, y = \overline{y}\lambda_2, z = \overline{z}\lambda_3.$$

$\overline{xyz} = 0$ is a local equation of D_X at p.

There is an expression $g = \sum \alpha_i M_i$ where $0 \neq \alpha_i \in \mathbf{k}$ and $M_i = x^{a_i} y^{b_i} z^{c_i}$ are monomials in x, y, z such that $\text{rank}(u, v, M_i) = 2$. Let I^p be the ideal in $\mathcal{O}_{X,p}$ generated by the $\overline{x}^{a_i}\overline{y}^{b_i}\overline{z}^{c_i}$ for a_i, b_i, c_i appearing in some M_i. There exists an r such that

$$I^p = (\overline{x}^{a_0}\overline{y}^{b_0}\overline{z}^{c_0}, \overline{x}^{a_1}\overline{y}^{b_1}\overline{z}^{c_1}, \ldots, \overline{x}^{a_r}\overline{y}^{b_r}\overline{z}^{c_r}).$$

We have relations $M_i^{e_i} = u^{\alpha_i} v^{\beta_i}$ with $e_i, \alpha_i, \beta_i \in \mathbf{Z}$ and $e_i > 0$ for all i. Thus for $a \in \text{spec}(\mathcal{O}_{X,p})$, $(I^p)_a$ is principal if u or v is not in a.

By Lemma 5.3, there exists a sequence of blow ups of 2-curves and 3-points $\Phi_1 : X_1 \to X$ such that Φ_1 is an isomorphism over $f^{-1}(X-E)$ and $I^p \mathcal{O}_{X_1,p_1}$ is invertible for all 3-points $p \in X$ such that $f(p) = q$ and $p_1 \in \Phi_1^{-1}(p)$. $f \circ \Phi_1 : X_1 \to Y$ is prepared by Lemma 10.2.

Suppose that $p_1 \in (f \circ \Phi_1)^{-1}(q)$ is a 3-point. Then $p = \Phi_1(p_1)$ is also a 3-point, and $\tau_{f\circ\Phi_1}(p_1) = \tau_f(p)$ (by Lemma 10.2). Let notation be as in (10.29). There exist permissible parameters x_1, y_1, z_1 for the permissible parameters u, v, w at p_1 such that x_1, y_1, z_1 are defined by

$$
\begin{aligned}
x &= x_1^{a_{11}} y_1^{a_{12}} z_1^{a_{13}} \\
y &= x_1^{a_{21}} y_1^{a_{22}} z_1^{a_{23}} \\
z &= x_1^{a_{31}} y_1^{a_{32}} z_1^{a_{33}},
\end{aligned}
\tag{10.30}
$$

where $a_{ij} \in \mathbf{N}$, $\mathrm{Det}(a_{ij}) = \pm 1$. Thus all of the M_i and N are distinct monomials when expanded in the variables x_1, y_1, z_1. So that since $I_p\mathcal{O}_{X_1,p_1}$ is principal, $g(x,y,z)$ is a monomial in x_1, y_1, z_1 times a unit series (in x_1, y_1, z_1). Thus u, v, w are super parameters at p_1.

Step 2. We will show that there exists a sequence of blow ups of 2-curves $\Phi_2 : X_2 \to X_1$, where $\Phi_1 : X_1 \to X$ is the map constructed in Step 1, such that Φ_2 is an isomorphism over $(f \circ \Phi_1)^{-1}(Y - E)$ and u, v, w are super parameters at all $p \in (f \circ \Phi_1 \circ \Phi_2)^{-1}(q)$ for which p is a 3-point or p is a 2-point of type (4.2) of Definition 4.1 for u, v, w.

Let $f_1 = f \circ \Phi_1$. Let ν be a 0-dimensional valuation of $\mathbf{k}(X)$. Let p be the center of ν on X_1. Say that ν is resolved on X_1 if

1. $f_1(p) \neq q$, or
2. p is not a 2-point of type (4.2) of Definition 4.1, or
3. p is a 2-point of type (4.2) and u, v, w are super parameters at p.

The condition of being resolved is an open condition on the Zariski Riemann manifold of X.

We construct an infinite sequence of morphisms

$$\cdots \to X_n \overset{\Phi_n}{\to} \cdots \overset{\Phi_3}{\to} X_2 \overset{\Phi_2}{\to} X_1 \tag{10.31}$$

as follows. Order the 2-curves C of X_1 such that $q \in (f \circ \Phi_1)(C) \subset E$. Let $\Phi_2 : X_2 \to X_1$ be the blow up of the 2-curve C_1 on X_1 of smallest order. Order the 2-curves C' of X_2 such that $q \in (f \circ \Phi_1 \circ \Phi_2)(C') \subset E$ so that the 2-curves contained in the exceptional divisor of Φ_2 have order larger than the order of the (strict transform of the) 2-curves C of X_1 such that $q \in f_1(C) \subset E$. Let $\Phi_3 : Y_3 \to Y_2$ be the blow ups of the 2-curve C_2 on Y_3 of smallest order, and repeat to inductively construct the morphisms $\Phi_n : X_n \to X_{n-1}$. Let $\overline{\Phi}_n = \Phi_2 \circ \cdots \Phi_n : X_n \to X_1$. The morphisms $f \circ \overline{\Phi}_n$ are prepared by Lemma 10.2.

Let ν be a 0-dimensional valuation of $\mathbf{k}(X)$ that is not resolved on X. Let p be the center of ν on X. Then $p \in f_1^{-1}(q)$ is a 2-point satisfying (4.2) of Definition 4.1, so there exist permissible parameters x, y, z at p for u, v, w such that

$$\begin{aligned} u &= x^a y^b \\ v &= x^c y^d \\ w &= g(x,y) + x^e y^f z \end{aligned}$$

with $ad - bc \neq 0$. There exist regular parameters $\overline{x}, \overline{y}, \overline{z}$ in $\mathcal{O}_{X_1,p}$ and unit series $\lambda_1, \lambda_2 \in \hat{\mathcal{O}}_{X_1,p}$ such that

$$x = \overline{x}\lambda_1, y = \overline{y}\lambda_2.$$

$\overline{xy} = 0$ is a local equation of D_{X_1} at p.

If $\nu(\overline{x}), \nu(\overline{y})$ are rationally independent, then the center of ν on X_n is a 2-point for all n, there exists an n such that the center of ν on X_n is a 2-point such that u, v, w satisfy (4.2) of Definition 4.1 and u, v, w are super parameters at p_1, by embedded resolution of plane curve singularities (cf. Section 3.4 and Exercise 3.3 [**C6**]) applied to $g(x,y) = 0$. If $\nu(\overline{x}), \nu(\overline{y})$ are rationally dependent, then there exists an n such that the center p_1 of ν on X_n is a 1-point.

By compactness of the Zariski-Riemann manifold of $\mathbf{k}(X)$ [**Z**], there exists an n such that all points of X_n are resolved. Thus there exists a sequence of blow ups of 2-curves $\Phi_2 : X_2 \to X_1$ such that the conclusions of Step 2 hold.

Step 3. We will show that there exists a diagram (10.27) satisfying the conclusions of the lemma.

After replacing f with $f \circ \Phi_1 \circ \Phi_2$, we can assume that f satisfies the conclusions of Step 2. We construct a sequence of diagrams (10.27) satisfying 1, 2 and 3 of the conclusions of the lemma as follows. Let $\Psi_1 : Y_1 \to Y$ be the blow up of the 2-curve C containing q. We order the two 2-curves in Y_1 which dominate C. Let $\Psi_2 : Y_2 \to Y_1$ be the blow up of the 2-curve of smallest order. Now extend the ordering to the 2-curves of Y_2 which dominate C, by requiring that the two 2-curves on the exceptional divisor of Ψ_2 which dominate C have larger order than the order of the (strict transform of the) 2-curve on Y_1 dominating C (which was not blown up by Ψ_2). Now let $\Psi_3 : Y_3 \to Y_2$ be the blow up of the 2-curve of smallest order. We continue this process to construct a sequence of blow ups of 2-curves

$$\cdots \to Y_n \overset{\Psi_n}{\to} Y_{n-1} \to \cdots \to Y_1 \overset{\Psi_1}{\to} Y.$$

Let $\overline{\Psi}_n = \Psi_1 \circ \cdots \Psi_{n-1} \circ \Psi_n$. Let

$$U = \left\{ \begin{array}{l} p \in f^{-1}(q) \text{ such that } u, v, w \text{ have a form (4.1) or (4.3) of} \\ \text{Definition 4.1 or of 2 (b) or 2 (c) of Definition 4.6 at } p \end{array} \right\}.$$

U is an open subset of $f^{-1}(q)$. For each $p \in U$, there exists $n(p)$ such that $n \geq n(p)$ implies the rational map $\overline{\Psi}_n^{-1} \circ f$ is defined at p, and $(\overline{\Psi}_n^{-1} \circ f)(p_1)$ is a 1-point, for p_1 in some neighborhood U_p of p in U. $\{U_p\}$ is an open cover of U, so there exists a finite subcover $\{U_{p_1}, \ldots, U_{p_m}\}$ of U. Let $n = \max\{n(p_1), \ldots, n(p_m)\}$. We have that $\overline{\Psi}_n^{-1} \circ f(p)$ is a 1-point if $p \in U$.

By Lemma 10.4, we can now construct a commutative diagram

$$\begin{array}{ccc} X_n & \overset{f_n}{\to} & Y_n \\ \overline{\Phi}_n \downarrow & & \downarrow \overline{\Psi}_n \\ X & \overset{f}{\to} & Y \end{array}$$

such that 1 and 2 of the conclusions of the lemma hold.

If $p_1 \in X_n$ is such that $f_n(p_1) = q_1 \in \overline{\Psi}_n^{-1}(q)$ is a 2-point, then $p = \overline{\Phi}_n(p_1) \in X - U$, and thus u, v, w have a form 2 or 4 of Definition 10.1 at p. As observed at the beginning of the proof, p_1 must then have one of the forms 1 – 4 of Definition 10.1 with respect to the permissible parameters u_1, v_1, w_1 at q_1 defined by 3 of the statement of Lemma 10.7. Thus f_n satisfies 3 of the statement of Lemma 10.7. □

CHAPTER 11

Good and perfect points

Throughout this section, we suppose that $f : X \to Y$ is a dominant, proper morphism of nonsingular 3-folds.

DEFINITION 11.1. Suppose that $f : X \to Y$ is prepared, $p \in D_X$ and $q = f(p)$. Let u, v, w be permissible parameters at q. We say that w is good at p for f if one of the following expressions holds:

p a 1-point, q a 1-point

$$u = x^a, v = y, w = \sum_{a \not| i} a_{ij} x^i y^j + x^n(z + \alpha) \tag{11.1}$$

p a 2-point, q a 1-point

$$u = (x^a y^b)^k, v = z, w = \sum_{k \not| i} a_{ij} (x^a y^b)^i z^j + x^c y^d \tag{11.2}$$

p a 1-point, q a 2-point

$$u = x^a, v = x^b(\alpha + y), w = \sum_{d \not| i} a_{ij} x^i y^j + x^n(z + \beta) \tag{11.3}$$

where $d = \gcd(a, b)$.

p a 2-point, q a 2-point

$$u = x^a y^b, v = x^c y^d, w = \sum_{(i,j) \notin \mathbf{Z}(a,b) + \mathbf{Z}(c,d)} a_{ij} x^i y^j + x^e y^f z \tag{11.4}$$

p a 2-point, q a 2-point

$$u = (x^a y^b)^k, v = (x^a y^b)^t(\alpha + z), w = \sum_{d \not| i} a_{ij} (x^a y^b)^i z^j + x^c y^d \tag{11.5}$$

where $d = \gcd(k, t)$.

p a 3-point, q a 2-point

$$u = x^a y^b z^c, v = x^d y^e z^f, w = \sum a_{ijk} x^i y^j z^k + x^g y^h z^i \tag{11.6}$$

where the sum is over i, j, k such that

$$\operatorname{Det} \begin{pmatrix} a & b & c \\ d & e & f \\ i & j & k \end{pmatrix} = 0, (i, j, k) \notin \mathbf{Z}(a, b, c) + \mathbf{Z}(d, e, f).$$

DEFINITION 11.2. Suppose that $f : X \to Y$ is prepared, $p \in D_X$ and $q = f(p)$ is a 1-point. Let u, v, w be permissible parameters at q. We say that w is weakly good at p for f if one of the following forms hold:

1. p is a 1-point,
$$u = x^a, v = y, w = \sum_{j=0}^{m} a_{\sigma_j}(y)x^{\sigma_j} + x^n(z+\alpha)$$
where $\alpha \in \mathbf{k}$, $\sigma_0 < \sigma_1 < \cdots < \sigma_m < n$, σ_i are all nonzero, and $a \not| \sigma_0$.
2. p is a 2-point,
$$u = (x^a y^b)^k, v = z, w = \sum_{j=0}^{m} a_{\sigma_j}(z)(x^a y^b)^{\sigma_j} + x^c y^d$$
where $\gcd(a,b) = 1$, $ad - bc \neq 0$, $\sigma_m(a,b) \not> (c,d)$, $\sigma_0 < \sigma_1 < \cdots < \sigma_m$, σ_i are all nonzero, and $k \not| \sigma_0$.

REMARK 11.3. *Suppose that $f : X \to Y$ is prepared and $\tau_f(p) = 1$, u, v, w are permissible parameters at $q = f(p)$, and w is good (weakly good) at p for f. Then u, v, w are monomial forms (Definition 4.4) at p.*

REMARK 11.4. *Suppose that $f; X \to Y$ is prepared. If q is a 1-point, and u, v, w are permissible parameters at q satisfying the conclusions of Lemma 4.9, then for all $p \in f^{-1}(q)$, there exists a series $\phi_p(u,v)$ such that $w - \phi_p(u,v)$ is good (weakly good) at p for f.*

LEMMA 11.5. *Suppose that $f : X \to Y$ is prepared, $q \in Y$ and $u, v, w \in \mathcal{O}_{Y,q}$ are permissible parameters at q (such that the conclusions of Lemma 4.9 hold if q is a 1-point). Suppose that $p \in f^{-1}(q)$ and there exists $\phi(u,v) \in \hat{\mathcal{O}}_{Y,q}$ such that $w - \phi(u,v)$ is good (weakly good) at p for f.*

Suppose that p is an n-point. Then there exists an affine neighborhood $V =$ Spec(S) of p such that $w - \phi(u,v)$ is good (weakly good) at p' for f for all n-points $p' \in f^{-1}(q) \cap V$.

PROOF. We will prove this in the case that p and q are 1-points, and a form (11.1) of Definition 11.1 holds for $u, v, w - \phi$ in $\hat{\mathcal{O}}_{X,p}$. The proof in the other cases is similar.

There exists an affine neighborhood $V = \text{spec}(S)$ of p, regular parameters $x, y, z \in \hat{\mathcal{O}}_{X,p}$ and a finite etale morphism $\pi : V_1 = \text{spec}(S_1) \to S$ such that x, y, z are uniformizing parameters on V_1, and regular parameters in $\mathcal{O}_{V_1,p'}$ for $p' \in \pi^{-1}(p)$ such that
$$\begin{aligned} u &= x^a \\ v &= y \\ w &= \textstyle\sum_{i<n} a_{ij}x^i y^j + x^n(\gamma(x,y,z)z + \Omega(x,y)) \end{aligned}$$
in $\hat{\mathcal{O}}_{V_1,p'} = \hat{\mathcal{O}}_{X,p}$ where γ is a unit series, $\Omega(x,y)$ is a series.

Let $U = \text{Spec}(R)$ be an affine neighborhood of q such that $f(V) \subset U$.

In $\hat{\mathcal{O}}_{X,p}$,
$$w - \phi(u,v) = \sum_{a \not| i, i<n} a_{ij}x^i y^j + x^n(\gamma z + \tilde{\Omega})$$
where $\tilde{\Omega}(x,y)$ is a series. We see that
$$x^n \text{ divides } \frac{\partial(w-\phi)}{\partial z} \text{ in } \hat{\mathcal{O}}_{X,p},$$
so that

$$x^n \text{ divides } \frac{\partial(w-\phi)}{\partial z} \text{ in } \mathcal{O}_{S_1,p'} \otimes_{R_q} \hat{R}_q$$

at all points $p' \in \pi^{-1}(p)$.

We also have that

$$\frac{\partial^{n+1}(w-\phi)}{\partial z \partial x^n}(p) \neq 0$$

which implies

$$\frac{\partial^{n+1}(w-\phi)}{\partial z \partial x^n}(p') \neq 0$$

at all $p' \in \pi^{-1}(p)$.

Finally, we see that

$$x \text{ divides } \frac{\partial^i(w-\phi)}{\partial x^i}$$

in $\mathcal{O}_{S_1,p'} \otimes_{R_q} \hat{R}_q$ at all points p' of $\pi^{-1}(p)$.

Thus there exists a Zariski closed subset C of V_1 which is disjoint from $\pi^{-1}(p)$ such that if $\overline{p} \in (f \circ \pi)^{-1}(q) \cap (V_1 - C)$, then

$$x^n \text{ divides } \frac{\partial(w-\phi)}{\partial z} \text{ in } \mathcal{O}_{S_1,\overline{p}} \otimes_{R_q} \hat{R}_q \tag{11.7}$$

$$\frac{\partial^{n+1}(w-\phi)}{\partial z \partial x^n} \text{ is a unit in } \mathcal{O}_{S_1,\overline{p}} \otimes_{R_q} \hat{R}_q \tag{11.8}$$

and

$$x \text{ divides } \frac{\partial^i(w-\phi)}{\partial x^i} \text{ in } \mathcal{O}_{S_1,\overline{p}} \otimes_{R_q} \hat{R}_q \tag{11.9}$$

if $i < n$ and a divides i.

Let $\overline{C} = \pi(C)$. $\pi : \mathrm{Spec}(S_1) - \pi^{-1}(\overline{C}) \rightarrow V - \overline{C}$ is finite etale. Let $\overline{V} = \mathrm{spec}(\overline{S})$ be an affine neighborhood of p in $V - \overline{C}$, and let $\pi^{-1}(\overline{V}) = \mathrm{Spec}(\overline{S}_1)$. After replacing V with $\overline{V}$, S with $\overline{S}$, V_1 with $\pi^{-1}(\overline{V})$ and S_1 with $\overline{S}_1$, we have that (11.7), (11.8) and (11.9) hold at all $\overline{p} \in (f \circ \pi)^{-1}(q)$.

Now consider the expression of $u, v, w - \phi(u,v)$ at $\overline{p} \in (f \circ \pi)^{-1})(q)$. There exists $\alpha \in \mathbf{k}$ such that $x, y, z - \alpha$ are regular parameters at $\overline{p}$. We have

$$\begin{aligned} u &= x^a \\ v &= y \\ w - \phi &= \textstyle\sum \frac{1}{i!j!k!} \frac{\partial^{i+j+k}(w-\phi)}{\partial x^i \partial y^j \partial z^k}(0,0,\alpha) x^i y^j (z-\alpha)^k. \end{aligned}$$

(11.7) implies

$$\frac{\partial^{i+j+k}(w-\phi)}{\partial x^i \partial y^j \partial z^k}(0,0,\alpha) = 0$$

if $i < n$ and $k \geq 1$, (11.9) implies

$$\frac{\partial^{i+j}(w-\phi)}{\partial x^i \partial y^j}(0,0,\alpha) = 0$$

if a divides i and $i < n$, and (11.8) implies

$$\frac{\partial^{n+1}(w-\phi)}{\partial x^n \partial z}(0,0,\alpha) \neq 0.$$

Thus

$$w-\phi=\sum_{i<n,a\not| i}\frac{1}{i!j!}\frac{\partial^{i+j}(w-\phi)}{\partial x^i\partial y^j}(0,0,\alpha)x^iy^j+x^n(\gamma'(z-\alpha)+\Omega'(x,y))$$

where γ' is a unit series, so that $w-\phi$ is good at $\overline{p}$. □

LEMMA 11.6. *Suppose that $f:X\rightarrow Y$ is prepared, and $C\subset Y$ is an irreducible curve in the nonfinite locus of f such that C contains a 1-point.*

Suppose that $U\subset Y$ is an affine open subset, with uniformizing parameters u,v,w such that u,v,w are permissible parameters in $\mathcal{O}_{Y,q}$ for a 1-point $q\in C\cap U$ such that $u=w=0$ are local equations of C. Then for a general point $\overline{q}$ of $C\cap U$, and appropriate (general) $\alpha\in\mathbf{k}$, $u,\overline{v}=v-\alpha,w$ are permissible parameters at $\overline{q}$ and for $p\in f^{-1}(\overline{q})$, either p is a 1-point and we have a form at p

$$\begin{aligned} u&=x^a\\ \overline{v}&=y\\ w&=\textstyle\sum_{i<n}\phi_i(y)x^i+x^n(z+\delta)\end{aligned}\tag{11.10}$$

with $\delta\in\mathbf{k}$ and $\phi_i(0)\neq 0$ whenever $\phi_i\neq 0$, or p is a 2-point with a form at p

$$\begin{aligned} u&=(x^ay^b)^t\\ \overline{v}&=z\\ w&=\textstyle\sum\phi_i(z)(x^ay^b)^i+x^cy^d\end{aligned}\tag{11.11}$$

with $gcd(a,b)=1$, $ad-bc\neq 0$ and $\phi_i(0)\neq 0$ whenever $\phi_i\neq 0$.

PROOF. To prove this theorem, we may replace q with a point of $C\cap U$ such that C is the only component of the nonfinite locus of f through q, and q is a nonsingular point of C. Suppose $p\in f^{-1}(q)$. Then there exists a Zariski open neighborhood $V=V^p=\text{Spec}(S)$ of p in X, and an etale neighborhood $W=W^p=\text{Spec}(S_1)$ of V^p with uniformizing parameters x,y,z in S_1, with induced morphism

$$\pi:\text{Spec}(S_1)\rightarrow\text{Spec}(S),$$

such that x,y,z are regular parameters in $\mathcal{O}_{W^p,p_1}$ for $p_1\in\pi^{-1}(p)$, and by Lemma 4.9 and its proof, we have one of the following cases:

Case 1. Suppose that p is a 1-point. Then we have in $\hat{\mathcal{O}}_{X,p}=\hat{S}_p$:

$$u=x^a,v=y,w=\sum_{i<n}\phi_i(y)x^i+x^n(\gamma z+\psi(x,y))\tag{11.12}$$

where γ is a unit series. In (11.12), for $i<n$ we have

$$\frac{1}{i!}\frac{\partial^i w}{\partial x^i}=\phi_i(y)+x\Omega\in\hat{S}_p\tag{11.13}$$

for some $\Omega\in\hat{S}_p$ and

$$\frac{\partial^{n+1}w}{\partial x^n\partial z}(0,0,0)\neq 0.$$

We can choose V^p so that for $p_1\in W^p\cap D_X$, with regular parameters

$$x,\overline{y}=y-\alpha,\overline{z}=z-\beta$$

in $\hat{\mathcal{O}}_{W^p,p_1}$, for $i<n$ we have

$$\frac{\partial^{i+1}w}{\partial z\partial x^i}(0,\alpha,\beta)=0\tag{11.14}$$

and

$$\frac{\partial^{n+1} w}{\partial z \partial x^n}(0, \alpha, \beta) \neq 0. \tag{11.15}$$

We can choose V^p so that for $i < n$, all irreducible components of $x = \frac{\partial^i w}{\partial x^i} = 0$ in W^p contain (a preimage of) p.

Case 2. Suppose that p is a 2-point. Then we have in $\hat{\mathcal{O}}_{X,p} = \hat{S}_p$:

$$u = (x^a y^b)^t, v = z, w = \sum \phi_i(z)(x^a y^b)^i + x^c y^d \gamma \tag{11.16}$$

with $\gcd(a, b) = 1$, $ad - bc \neq 0$ and $c > ai$ or $d > bi$ for all i in the series. Further, γ is a unit series.

We have $\Omega_1, \Omega_2 \in \hat{S}_p = \hat{\mathcal{O}}_{X,p}$ such that

(11.17)

$$\frac{1}{j!k!}\frac{\partial^{j+k} w}{\partial x^j \partial y^k} = \begin{cases} x\Omega_1 + y\Omega_2 & \text{if } jb - ka \neq 0 \text{ and } j < c \text{ or } k < d \\ \phi_i(z) + x\Omega_1 + y\Omega_2 & \text{if there exists } i \text{ such that } (j,k) = i(a,b) \\ & \text{and } j < c \text{ or } k < d \end{cases}$$

There exists $\Omega_1 \in \hat{\mathcal{O}}_{X,p}$ such that

(11.18)

$$\frac{1}{j!}\frac{\partial^j w}{\partial x^j} = \begin{cases} x\Omega_1 & \text{if } j < c \text{ and there do not exist } k, i \text{ such that} \\ & (j,k) = i(a,b) \\ y^{ib}\phi_i(z) + x\Omega_1 & \text{if } j < c \text{ and there exist } k, i \text{ such that } (j,k) = i(a,b). \end{cases}$$

There exists $\Omega_1 \in \hat{\mathcal{O}}_{X,p}$ such that

$$\frac{1}{k!}\frac{\partial^k w}{\partial y^k} = \begin{cases} y\Omega_1 & \text{if } k < d \text{ and there do not exist } j, i \text{ such that} \\ & (j,k) = i(a,b) \\ x^{ia}\phi_i(z) + y\Omega_1 & \text{if } k < d \text{ and there exist } j, i \text{ such that } (j,k) = i(a,b). \end{cases}$$

Furthermore,

$$\frac{\partial^{c+d} w}{\partial x^c \partial y^d}(0, 0, 0) \neq 0.$$

We can choose V^p so that

1. for $j < c$ or $k < d$, all irreducible components of
$$x = y = \frac{\partial^{j+k} w}{\partial x^j \partial y^k} = 0$$
in W^p contain (a preimage of) p,
2. for $j < c$, all irreducible components of
$$x = \frac{\partial^j w}{\partial x^j} = 0$$
in W^p contain (a preimage of) p, and
3. for $k < d$, all irreducible components of
$$y = \frac{\partial^k w}{\partial y^k} = 0$$
in W^p contain (a preimage of) p.

There exist $V_1 = V^{p_1}, \ldots, V_n = V^{p_n}$ such that $\{V_1, \ldots, V_n\}$ is an affine cover of $f^{-1}(q)$. We may assume that $V_1, \ldots, V_n$ is an affine cover of $f^{-1}(U)$.

Suppose that $\overline{q} \in C \cap U$ is a general point. Then $\mathcal{O}_{Y,q}$ has regular parameters $u, \overline{v} = v - \alpha, w$ for a general $\alpha \in \mathbf{k}$. $u, \overline{v}, w$ are permissible parameters at $\overline{q}$. Suppose that $\overline{p} \in f^{-1}(\overline{q})$. $\overline{p} \in V^p = V_i$ for some i. We identify p and $\overline{p}$ with points in $\pi^{-1}(p)$, $\pi^{-1}(\overline{p})$.

Suppose that p is a 1-point. There exists $\beta \in \mathbf{k}$ such that $x, \overline{y} = y - \alpha, z - \beta$ are permissible parameters in $\hat{\mathcal{O}}_{X,\overline{p}}$. We have an expression

$$
\begin{aligned}
u &= x^a \\
\overline{v} &= \overline{y} = y - \alpha \qquad (11.19)\\
w &= \sum \tfrac{1}{i!j!k!} \tfrac{\partial^{i+j+k} w}{\partial x^i \partial y^j \partial z^k}(0, \alpha, \beta) x^i (y - \alpha)^j (z - \beta)^k
\end{aligned}
$$

in $\hat{\mathcal{O}}_{X,\overline{p}}$.

By (11.14) and (11.15), in (11.19), we have

$$w = \sum_{i<n} \overline{\phi}_i(\overline{y}) x^i + x^n(\overline{z\gamma} + \overline{\Psi}(x, \overline{y}))$$

where $\overline{\gamma}$ is a unit series, and

$$\overline{\phi}_i(\overline{y}) = \sum_{j=0}^{\infty} \frac{1}{i!j!} \frac{\partial^{i+j} w}{\partial x^i \partial y^j}(0, \alpha, \beta) \overline{y}^j.$$

If $\phi_i(y) = 0$ we have $\overline{\phi}_i(\overline{y}) = 0$.

Suppose that $\phi_i(y) \neq 0$. We will show that $\overline{\phi}_i(0) \neq 0$. If $\phi_i(0) \neq 0$, then $\frac{\partial^i w}{\partial x^i}$ does not vanish on $W^p \cap D_X$, so that

$$\overline{\phi}_i(0) = \frac{\partial^i w}{\partial x^i}(\overline{p}) \neq 0.$$

Suppose that $\phi_i(0) = 0$. Further suppose that

$$\overline{\phi}_i(0) = \frac{\partial^i w}{\partial x^i}(\overline{p}) = 0.$$

Then there exists an irreducible curve Λ which is a component of $x = \frac{\partial^i w}{\partial x^i} = 0$ in W^p containing $\overline{p}$.

By our construction of W^p, we may assume that our choice of preimage of p in W^p (which we have identified with p) satisfies $p \in \Lambda$. Let I_Λ be the prime ideal of Λ in S_1. Let $\hat{S}_1$ be the completion of S_1 at the maximal ideal m_p of p in S_1. $\frac{\partial^i w}{\partial x^i}, x \in I_\Lambda \hat{S}_1$ implies $\phi_i(y) \in I_\Lambda \hat{S}_1$. Since $I_\Lambda \hat{S}_1$ is reduced, we have $y \in I_\Lambda \hat{S}_1$. As $(S_1)_{m_p} \to \hat{S}_1 = \hat{\mathcal{O}}_{X,p}$ is faithfully flat, we have $y \in I_\Lambda (S_1)_{m_p}$.

Since I_Λ is a prime ideal and $(I_\Lambda)_{m_p} \neq (S_1)_{m_p}$, we have that $y \in I_\Lambda$.

Let $m_{\overline{p}}$ be the ideal of $\overline{p}$ in S_1. The maximal ideal of $(S_1)_{m_{\overline{p}}}$ is $(x, y - \alpha, z - \beta)$.

$$I_\Lambda (S_1)_{m_{\overline{p}}} \subset (x, y - \alpha, z - \beta)$$

implies $y \in (x, y - \alpha, z - \beta)$ which is impossible since $0 \neq \alpha$. Thus we have $\overline{\phi}_i(0) \neq 0$.

Suppose that p is a 2-point and $\overline{p}$ is a 2-point. Then there is $\alpha \in \mathbf{k}$ such that $x, y, z - \alpha$ are regular parameters in $\hat{\mathcal{O}}_{X,\overline{p}}$, and we have an expression

$$
\begin{aligned}
u &= (x^a y^b)^t \\
\overline{v} &= \overline{z} = z - \alpha \qquad (11.20)\\
w &= \sum \tfrac{1}{i!j!k!} \tfrac{\partial^{i+j+k} w}{\partial x^i \partial y^j \partial z^k}(0, 0, \alpha) x^i y^j (z - \alpha)^k
\end{aligned}
$$

in $\hat{\mathcal{O}}_{X,\overline{p}}$.

We have

$$\frac{\partial^{c+d}w}{\partial x^c \partial y^d}(0,0,\alpha)\neq 0.$$

Further if $j<c$ or $k<d$, $jb-ka\neq 0$ and $l\geq 0$, we have

$$\frac{\partial^{j+k+l}w}{\partial x^j\partial y^k\partial z^l}(0,0,\alpha)=0.$$

Thus in (11.20) we have

$$w=\sum \overline{\phi}_i(z)(x^ay^b)^i+x^cy^d\overline{\gamma}, and$$

where $\overline{\gamma}$ is a unit series, the sum is over i such that $ai<c$ or $bi<d$, and

$$\overline{\phi}_i(\overline{z})=\sum_{k=0}^{\infty}\frac{1}{(ia)!(jb)!k!}\frac{\partial^{ia+jb+k}w}{\partial x^{ia}\partial y^{ib}\partial z^k}(0,0,\alpha)\overline{z}^k.$$

By a similar analysis as for Case 1, we see that if $\phi_i(z)=0$ then $\overline{\phi}_i(\overline{z})=0$ and if $\phi_i(z)\neq 0$, we have $\overline{\phi}_i(0)\neq 0$.

Suppose that p is a 2-point and $\overline{p}$ is a 1-point. Then after possibly interchanging x and y, there exist $\alpha,\beta\in\mathbf{k}$ and regular parameters $x, y-\beta, z-\alpha$ in $\hat{\mathcal{O}}_{X,\overline{p}}$ with $0\neq\alpha,\beta$.

Set

$$\overline{x}=xy^{\frac{b}{a}}, \overline{y}=y-\beta, \overline{z}=z-\alpha$$

Then $\overline{x},\overline{y},\overline{z}$ are regular parameters in $\hat{\mathcal{O}}_{X,\overline{p}}$. From the Jacobian of $u,\overline{v},w$ we see that we have an expression

$$\begin{aligned}u&=\overline{x}^{at}\\ \overline{v}&=\overline{z}\\ w&=P(\overline{x},\overline{z})+\overline{x}^c\Omega\end{aligned}$$

where P and Ω are series, $\frac{\partial\Omega}{\partial y}(\overline{p})\neq 0$, and P has degree $<c$ in $\overline{x}$.

We have

$$\begin{aligned}x&=\overline{x}(\overline{y}+\beta)^{-\frac{b}{a}}\\ y&=\overline{y}+\beta\\ z&=\overline{z}+\alpha.\end{aligned}$$

Further,

$$\begin{aligned}\frac{\partial}{\partial\overline{x}}&=y^{-\frac{b}{a}}\frac{\partial}{\partial x}\\ \frac{\partial}{\partial\overline{y}}&=-\tfrac{b}{a}xy^{-1}\frac{\partial}{\partial x}+\frac{\partial}{\partial y}\\ \frac{\partial}{\partial\overline{z}}&=\frac{\partial}{\partial z}.\end{aligned}$$

We have an expansion

$$P=\sum_{j<c}\overline{\phi}_j(\overline{z})\overline{x}^j=\sum_{j<c}\left(\sum_{k=0}^{\infty}\frac{\partial^{j+k}w}{\partial\overline{z}^k\partial\overline{x}^j}(\overline{p})\overline{z}^k\right)\overline{x}^j$$

where

$$\frac{\partial^{j+k}w}{\partial\overline{z}^k\overline{x}^j}(\overline{p})=\beta^{-j\frac{b}{a}}\frac{\partial^{j+k}w}{\partial z^k\partial x^j}(\overline{p}).$$

By (11.18) we have that $\overline{\phi}_j(\overline{z})=0$ if there do not exist k,i such that $(j,k)=i(a,b)$.

Suppose that $j < c$ and there exists k, i such that $(j,k) = i(a,b)$ and $\phi_j(z) \neq 0$. Suppose that $\overline{\phi}_j(0) = 0$. Then

$$0 = \overline{\phi}_j(0) = \frac{\partial^j w}{\partial \overline{x}^j}(\overline{p}) = \beta^{-j\frac{b}{a}} \frac{\partial^j w}{\partial x^j}(\overline{p})$$

implies there exists an irreducible curve Λ which is a component of

$$x = \frac{\partial^j w}{\partial x^j} = 0$$

in W^p which contains p and $\overline{p}$. Let I_Λ be the prime ideal of Λ in S_1. Let $\hat{S}_1$ be the completion of S_1 at the maximal ideal m_p of p in S_1. $\frac{\partial^j w}{\partial x^j}, x \in I_\Lambda \hat{S}_1$ implies

$$y^{ib}\phi_i(z) \in I_\Lambda \hat{S}_1.$$

Since $y \notin I_\Lambda \hat{S}_1$, as Λ is not a 2-curve, we have $z \in I_\Lambda \hat{S}_1$. Thus $z \in I_\Lambda (S_1)_{m_p}$. As $(I_\Lambda)_{m_p} \neq (S_1)_{m_p}$, we have that $z \in I_\Lambda$, which is a contradiction, since $\alpha \neq 0$. Thus $\overline{\phi}_i(0) \neq 0$.

□

LEMMA 11.7. *Suppose that $f : X \to Y$ is prepared. Then there exists an open subset V of Y such that $V \cap C \neq \emptyset$ for every integral curve $C \subset D_Y$ contained in the nonfinite locus of f which contains a 1-point, and if $U = f^{-1}(V)$, $\overline{f} = f \mid U$, then there exists a commutative diagram*

$$\begin{array}{ccc} U_1 & \xrightarrow{\overline{f}_1} & V_1 \\ \overline{\Phi}_1 \downarrow & & \downarrow \overline{\Psi}_1 \\ U & \xrightarrow{\overline{f}} & V \end{array}$$

such that $\overline{\Phi}_1$ and $\overline{\Psi}_1$ are products of blow ups of curves which dominate a curve C contained in the nonfinite locus of $\overline{f}$ which are possible centers (for the preimage of $D_V = D_Y \cap V$) and $\overline{f}_1$ is toroidal.

PROOF. By Lemma 11.6, there exists an open set $V \subset Y$ such that $V \cap C = \emptyset$ if C is a 2-curve or an isolated point contained in the nonfinite locus of f, $V \cap C \neq \emptyset$ for all curves C contained in the nonfinite locus of f which contain a 1-point, and if $\overline{q} \in C \cap V$ is a 1-point, then there exist permissible parameters u, v, w at $\overline{q}$ such that $u = w = 0$ are local equations of C, and if $p_1 \in f^{-1}(\overline{q})$, then we have permissible parameters x, y, z in $\hat{\mathcal{O}}_{X,p_1}$ such that p_1 is a 1-point:

$$\begin{array}{ll} u & = x^a \\ v & = y \\ w & = \sum_{j=0}^m \phi_{i_j}(y) x^{i_j} + x^n(\overline{\alpha} + z) \end{array} \tag{11.21}$$

where $i_0 < i_1 < \cdots < i_m < n$, $\overline{\alpha} \in \mathbf{k}$, all $\phi_{i_j}(y)$ are nonzero and $\phi_{i_j}(0) \neq 0$ for all j, or p_1 is a 2-point:

$$\begin{array}{ll} u & = (x^a y^b)^t \\ v & = z \\ w & = \sum_{j=0}^m \phi_{i_j}(z)(x^a y^b)^{i_j} + x^c y^d \end{array} \tag{11.22}$$

where $ad - bc \neq 0$, all $\phi_{i_j}(y)$ are nonzero, $\phi_{i_j}(0) \neq 0$ for all j, $i_0 < i_1 < \cdots < i_m$ and $i_j a < c$ or $i_j b < d$ for all j.

Let C be the nonfinite locus of $\overline{f}: U \rightarrow V$. There exists $\Phi_1': U_1' \rightarrow U$ which is a product of blow ups of 2-curves (which dominate an irreducible component of C) such that all local forms (11.22) at points $p \in (\overline{f} \circ \Phi_1')^{-1}(\overline{q})$ for $\overline{q} \in C \cap V$ are such that either $m = -1$ (so that $\sum \phi_{i_j}(z)(x^a y^b)^{i_j} = 0$), or $(x^a y^b)^{i_0}$ divides $x^c y^d$.

Suppose that $\overline{q} \in C$. The set of points $p \in (\overline{f} \circ \Phi_1')^{-1}(\overline{q})$ such that $\mathcal{I}_C \mathcal{O}_{U_1'}$ is not invertible is a union of points p such that p has permissible parameters x, y, z of the form

$$u = x^a, v = y, w = x^n z \tag{11.23}$$

with $n < a$ or

$$u = (x^a y^b)^t, v = z, w = x^c y^d \tag{11.24}$$

with $(at - c)(bt - d) < 0$.

Let $\overline{\Psi}_1 : V_1 \rightarrow V$ be the blow up of C (which has local equations $u = w = 0$ at $\overline{q} \in C$).

We can blow up curves above U_1' which dominate a component of C to obtain $\overline{\Phi}_1 : U_1 \rightarrow U_1'$ such that there exists a factorization $\overline{f}_1 : U_1 \rightarrow V_1$, and if $\overline{q} \in C$, $p_1 \in (\overline{f} \circ \Phi_1' \circ \overline{\Phi}_1)^{-1}(\overline{q})$, then an expression (11.21) or (11.22) holds.

Suppose that $\overline{q} \in C$, with permissible parameters u, v, w as above. Let $q_1 \in \overline{\Psi}_1^{-1}(\overline{q})$. q_1 has permissible parameters u_1, v_1, w_1 with q_1 a 1-point

$$u = u_1, v = v_1, w = u_1(w_1 + \alpha) \tag{11.25}$$

or q_1 a 2-point

$$u = u_1 v_1, v = w_1, w = u_1. \tag{11.26}$$

Suppose that $p_1 \in \overline{f}_1^{-1}(q_1)$.

Case 1 Suppose that $0 \neq \alpha$ in (11.25). Then $a = i_0$ and $\phi_{i_0}(0) = \alpha$ (or $m = -1$, $a = n$ and $\overline{\alpha} = \alpha$) if p_1 satisfies (11.21), $t = i_0$ and $\phi_{i_0}(0) = \alpha$ if p_1 satisfies (11.22).

We thus have that at p_1,

$$u_1 = x^a, v = y, w_1 = (\phi_{i_0}(y) - \alpha) + \sum_{j=1}^{m} \phi_{i_j}(y) x^{i_j - i_0} + x^{n - i_0}(\overline{\alpha} + z) \tag{11.27}$$

of the form (11.21) or

(11.28)

$$u_1 = (x^a y^b)^t, v = z, w_1 = (\phi_{i_0}(z) - \alpha) + \sum_{j=1}^{m} \phi_{i_j}(z)(x^a y^b)^{i_j - i_0} + x^{c - i_0 a} y^{d - i_0 b}.$$

of the form of (11.22).

If $\overline{f}_1$ is not toroidal at p_1, we have that $u_1 = w_1 - (\phi_{i_0}(v) - \alpha) = 0$ are (formal) local equations of a branch of the nonfinite locus of $\overline{f}_1$ through q_1. After possibly replacing V with an open subset of V, for $\overline{q} \in C$, $\overline{q}$ is a general point of a component of C, so the nonfinite locus of $\overline{f}_1$ through q_1 must be the germ of a nonsingular algebraic curve.

Case 2 Suppose that $0 = \alpha$ in (11.25). Then $i_0 > a$ (or $m = -1$ and $n > a$ or $m = -1$, $n = a$ and $\overline{\alpha} = 0$) in (11.21), or $i_0 > t$ (or $m = -1$ and $(c, d) > t(a, b)$

in (11.22)). We have that u_1, v, w_1 are permissible parameters at q_1 for $\overline{f}_1$ of the form (11.21) or (11.22).

Case 3 (11.26) holds. Then $a > i_0$ (or $m = -1$, $0 \neq \overline{\alpha}$ and $n < a$) in (11.21) or $t > i_0$ (or $m = -1$ and $(c,d) \leq (at, bt)$) in (11.22).

Suppose that (11.21) holds and $m \geq 0$. We change variables at p_1 to get an expression

$$u = \sum_{i=0}^{\tilde{m}} \tilde{\phi}_{\tilde{i}_j}(y)\tilde{x}^{\tilde{i}_j} + \tilde{x}^{n-i_0+a}(\tilde{z} + \cdots), v = y, w = \tilde{x}^{i_0}$$

with $a = \tilde{i}_0 < \cdots < \tilde{i}_m < n - i_0 + a$. As in the proof of Lemma 11.6, $\tilde{\phi}_{\tilde{i}_j}(0) \neq 0$ for all $\tilde{i}_j$ since $\overline{q}$ is a general point of a component of C. q_1 is a 2-point, and we have:

$$(11.29) \qquad u_1 = \tilde{x}^{i_0}, v_1 = \sum_{j=0}^{\tilde{m}} \tilde{\phi}_{\tilde{i}_j}(y)\tilde{x}^{\tilde{i}_j - i_0} + \tilde{x}^{n-2i_0+a}(\tilde{z} + \cdots), w_1 = y.$$

Note that $u_1 = v_1 = 0$ are local equations of the nonfinite locus of $\overline{f}_1$ at q_1 if $\overline{f}_1$ is not toroidal at p_1. This is the germ of a 2-curve which dominates a component of C.

Suppose that (11.22) and (11.26) hold and $m \geq 0$. We change variables at p_1 to have an expression

$$u = \sum_{j=0}^{\tilde{m}} \tilde{\phi}_{\tilde{i}_j}(z)(\tilde{x}^a\tilde{y}^b)^{\tilde{i}_j} + \tilde{x}^{c-i_0a+ta}\tilde{y}^{d-i_0b+tb}, v = z, w = (\tilde{x}^a\tilde{y}^b)^{i_0}$$

with $\tilde{i}_j a < c - i_0 a + ta$ or $\tilde{i}_j b < d - i_0 b + tb$ for all $\tilde{i}_j$.

As in the proof of Lemma 11.6, $\tilde{\phi}_{\tilde{i}_j}(0) \neq 0$ for all $\tilde{i}_j$, since $\overline{q}$ is a general point of a component of C. q_1 is a 2-point and we have

$$(11.30) \quad u_1 = (\tilde{x}^a\tilde{y}^b)^{i_0}, v_1 = \sum_{j=0}^{\tilde{m}} \tilde{\phi}_{\tilde{i}_j}(z)(\tilde{x}^a\tilde{y}^b)^{\tilde{i}_j - i_0} + \tilde{x}^{c-2i_0a+ta}\tilde{y}^{d-2i_0b+tb}, w_1 = z.$$

$u_1 = v_1 = 0$ are local equations of the nonfinite locus of $\overline{f}_1$ at q_1 if $\overline{f}_1$ is not toroidal at p_1. This is the germ of a 2-curve which dominates a component of C.

The nonfinite locus C_1 of $f_1 : U_1 \rightarrow V_1$ is a (disjoint) union of nonsingular curves which dominate components of C. If γ_1 is a component of C_1 then γ_1 consists of 1-points or γ_1 consists of 2-points. We will construct a commutative diagram

$$(11.31) \qquad \begin{array}{ccc} U_2 & \overset{\overline{f}_2}{\rightarrow} & V_2 \\ \overline{\Phi}_2 \downarrow & & \downarrow \overline{\Psi}_2 \\ U_1 & \overset{\overline{f}_1}{\rightarrow} & V_1 \end{array}$$

where $\overline{\Psi}_2 : V_2 \rightarrow V_1$ is the blow up of C_1.

Suppose that γ_1 is a component of C_1 and $q_1 \in \gamma_1$.

First suppose that γ_1 consists of 1-points. Then (11.21) or (11.22) holds at all points $p \in \overline{f}_1^{-1}(q_1)$. The construction of (11.31) above points of γ_1 is as in the construction of $\overline{f}_1$ above.

Suppose that γ_1 consists of 2-points. Then there exist permissible parameters u_1, v_1, w_1 at q_1 such that (11.29) or (11.30) holds at all $p \in \overline{f}_1^{-1}(q_1)$, and $u_1 = v_1 = 0$ are local equations of γ_1 at q_1.

If $p \in \overline{f}_1^{-1}(q_1)$ is such that $\mathcal{I}_{C_1}\mathcal{O}_{U_2,p}$ is not invertible, then we have permissible parameters x, y, z at p such that

$$u_1 = x^a, v_1 = x^n z, w_1 = y$$

with $n < a$ or

$$u_1 = (x^a y^b)^t, v_1 = x^c y^d, w_1 = z.$$

In particular, $\overline{f}_1$ is toroidal at p.

We now blow up curves 2-curves (above U_1) which dominate γ_1 and are supported in the locus where $U_1 \rightarrow V_1$ is torodial to obtain the construction of $\overline{\Phi}_2 : U_2 \rightarrow U_1$ above γ_1. $\overline{f}_2$ is toroidal above the torodial locus of f. Let $q_2 \in \overline{\Psi}_2^{-1}(q_1)$. q_2 has permissible parameters u_2, v_2, w_2 defined by one of the following 3 cases.

q_2 is a 1-point

$$u_1 = u_2, v_1 = u_2(w_2 + \alpha), w_1 = v_2 \tag{11.32}$$

with $\alpha \neq 0$ or q_2 is a 2-point

$$u_1 = u_2, v_1 = u_2 v_2, w_1 = w_2 \tag{11.33}$$

or q_2 is a 2-point

$$u_1 = u_2 v_2, v_1 = v_2, w_1 = w_2. \tag{11.34}$$

As in the case when q_2 is a 1-point, we see that if (11.32) holds, then all points above q_2 have the form (11.21) or (11.22), and that if (11.33) or (11.34) holds, then all points p_2 above q_2 have the form (11.29) or (11.30).

We iterate to construct a commutative diagram

$$\begin{array}{ccc}
\vdots & & \vdots \\
\downarrow & & \downarrow \\
U_n & \overset{\overline{f}_n}{\rightarrow} & V_n \\
\overline{\Phi}_n \downarrow & & \downarrow \overline{\Psi}_n \\
U_{n-1} & \overset{\overline{f}_{n-1}}{\rightarrow} & V_{n-1} \\
\downarrow & & \downarrow \\
\vdots & & \vdots \\
\downarrow & & \downarrow \\
U_1 & \overset{\overline{f}_1}{\rightarrow} & V_1 \\
\downarrow & & \downarrow \\
U & \overset{\overline{f}}{\rightarrow} & V
\end{array} \tag{11.35}$$

where each $V_r \rightarrow V_{r-1}$ is the blow up of the nonfinite locus C_{r-1} of $\overline{f}_{r-1}$, which is a disjoint union of nonsingular curves which dominate components of C.

All points of U_n have a form (11.21), (11.22), (11.29) or (11.30). We continue the construction as long as $\overline{f}_n$ is not toroidal.

Suppose that (11.35) does not converge to a toroidal morphism in a finite number of steps. Then there exists a 0-dimensional valuation ν of $\mathbf{k}(X)$ with center on U such that $\overline{f}_n$ is not toroidal at the center p_n of ν on U_n for all n. Let q_n be the center of ν on V_n.

Suppose that (11.21) holds for $p = p_0$. There exists $r(1)$ such that $q_{r(1)}$ has (formal) permissible parameters $u_{r(1)}, v_{r(1)}, \overline{w}_{r(1)}$ defined by

$$u = u_{r(1)}^e, w = u_{r(1)}^f(\overline{w}_{r(1)} + \phi_{i_0}(0))$$

where $\gcd(e, f) = 1$ and $\frac{e}{f} = \frac{a}{i_0}$.

The germ of $\overline{f}$ at p factors through $\overline{\Psi}_{r(1)}$, so we have

$$u_{r(1)} = x^{\frac{a}{e}}, \overline{w}_{r(1)} = (\phi_{i_0}(y) - \phi_{i_0}(0)) + \sum_{j=1}^{m} \phi_{i_j}(y)x^{i_j - i_0} + x^{n-i_0}(\overline{\alpha} + z).$$

Set $w_{r(1)} = \overline{w}_{r(1)} - [\phi_{i_0}(v) - \phi_{i_0}(0)]$.

$u_{r(1)}, v, w_{r(1)}$ are (formal) regular parameters at $q_{r(1)}$, and $u_{r(1)} = w_{r(1)} = 0$ are equations of the nonfinite locus at $q_{r(1)}$. We see that at $p_{r(1)}$,

$$u_{r(1)} = x^{a(1)}, v = y, w_{r(1)} = \sum_{j=0}^{m(1)} \phi(1)_{i(1)_j}(y)x^{i(1)_j} + x^{n(1)}(\overline{\alpha} + z)$$

where $a(1) = \gcd(a, i_0)$, $m(1) = m - 1$, $n(1) = n - i_0$, $i(1)_j = i_{j+1} - i_0$ for $0 \le j \le m(1)$, $\phi(1)_{i(1)_j}(y) = \phi_{i_{j+1}}(y)$.

We iterate to get for $k \le m+1$, $r(k)$ such that $q_{r(k)}$ has permissible parameters $u_{r(k)}, v, \overline{w}_{r(k)}$ defined by

$$u_{r(k-1)} = u_{r(k)}^{e_k}, w_{r(k-1)} = u_{r(k)}^{f_k}(\overline{w}_{r(k)} + \phi(k-1)_{i(k-1)_0}(0))$$

where $\gcd(e_k, f_k) = 1$.

The rational map $U \to V_{r(k)}$ is a morphism at $p = p_{r(k)}$. Set

$$w_{r(k)} = \overline{w}_{r(k)} - [\phi(k-1)_{i(k-1)_0}(v) - \phi(k-1)_{i(k-1)_0}(0)].$$

We have an expression

$$u_{r(k)} = x^{a(k)}, v = y, w_{r(k)} = \sum_{j=0}^{m(k)} \phi(k)_{i(k)_j}(y)x^{i(k)_j} + x^{n(k)}(\overline{\alpha} + z).$$

We have (for $k \le m + 1$) $a(k) = \gcd(a, i_0, i_1, \ldots, i_{k-1})$, $n(k) = n - i_{k-1}$, $m(k) = m - k$, $i(k)_j = i_{j+k} - i_{k-1}$ for $0 \le j \le m(k)$.

We further have

$$\frac{e_k}{f_k} = \frac{a(k-1)}{i(k-1)_0} = \frac{\gcd(a, i_0, \ldots, i_{k-2})}{i_{k-1} - i_{k-2}}. \tag{11.36}$$

$u_{r(k)} = w_{r(k)} = 0$ are (formal) equations of the nonfinite locus at $q_{r(k)}$.

$q_{r(m+1)}$ has permissible parameters $u_{r(m+1)}, v, w_{r(m+1)}$ defined by

$$u_{r(m+1)} = x^{a(m+1)}, v = y, w_{r(m+1)} = x^{n(m+1)}(\overline{\alpha} + z).$$

The rational map $U \to V_{r(m+1)}$ is a morphism at p. We have

$$a(m+1) = \gcd(a, i_0, i_1, \ldots, i_m)$$

and $n(m+1) = n - i_m$.

Finally, we see that there exists $r(m+2)$ such that $\overline{f}_{r(m+2)}$ is toroidal at $p_{r(m+2)}$, a contradiction.

A similar argument holds if (11.22) holds at $p = p_0$.

We conclude that (11.35) converges after a finite number of iterations in a diagram which satisfies the conclusions of Lemma 11.7.

$\square$

REMARK 11.8. *It follows from a simple variation of the proof of Lemma 11.7 that if $f: X \to Y$ is prepared, and $C \subset Y$ is a 2-curve, then there exists a Zariski open subset V of Y such that $C \cap V = \emptyset$ and there exists a commutative diagram*

$$\begin{array}{ccc} U_1 & \overset{\overline{f}_1}{\to} & V_1 \\ \overline{\Phi}_1 \downarrow & & \downarrow \overline{\Psi}_1 \\ U & \overset{f}{\to} & V \end{array}$$

as in the conclusions of Lemma 11.7 such that $\overline{f}_1$ is toroidal.

DEFINITION 11.9. Suppose that $f: X \to Y$ is prepared and $q \in G_Y(f, \tau)$ is a 1-point. Then q is **perfect** for f if the nonfinite locus of f through q is a germ of a nonsingular curve γ and if u, v, w are algebraic permissible parameters at q such that $u = w = 0$ are local equations of γ at q then there exist finitely many series $\phi_i(u, v) \in \mathbf{k}[[u, v]]$ such that

1. $u, v, w_i = w - \phi_i(u, v)$ are super parameters at q for all i.
2. For all $p \in f^{-1}(q)$, some w_i is weakly good for f at p.

LEMMA 11.10. *Suppose that $f: X \to Y$ is prepared. Let $V \subset G_Y(f, \tau)$ be the set of perfect 1-points. Then $G_Y(f, \tau) - V$ is a finite set.*

PROOF. Suppose that $\overline{q}$ is a general point of a curve $C \subset G_Y(f, \tau)$ (so that $\overline{q}$ is a 1-point). Let u, v, w be algebraic permissible parameters at $\overline{q}$ such that $u = w = 0$ are local equations of C.

In a neighborhood of $\overline{q}$, we construct a diagram (11.35). (11.35) is finite by the conclusions of Lemma 11.7.

Suppose that $p \in f^{-1}(\overline{q})$. At p we have permissible parameters $x, y, z \in \hat{\mathcal{O}}_{X,p}$ such that if p is a 1-point:

$$(11.37) \qquad \begin{aligned} u &= x^a \\ v &= y \\ w &= \textstyle\sum_{j=0}^m \phi_{i_j}(y) x^{i_j} + x^n(\beta + z) \end{aligned}$$

where $\beta \in \mathbf{k}$, $i_0 < i_1 < \cdots < i_m < n$, all $\phi_{i_j}(y)$ are non zero and $\phi_{i_j}(0) \neq 0$ for all j, or if p is a 2-point:

$$(11.38) \qquad \begin{aligned} u &= (x^b y^c)^a \\ v &= z \\ w &= \textstyle\sum_{j=0}^m \phi_{i_j}(z)(x^b y^c)^{i_j} + x^d y^e \end{aligned}$$

where $\gcd(b, c) = 1$, $be - cd \neq 0$, all $\phi_{i_j}(z)$ are non zero, $i_0 < i_1 < \cdots < i_m$, $(b, c) \not\geq (d, e)$, and $\phi_{i_j}(0) \neq 0$ for all j. In either case, there exists a largest $l \leq m$ such that $a \mid i_j$ if $j \leq l$.

If at p there is a form (11.37), set

$$\phi_p(u,v) = \sum_{j\le l} \phi_{i_j}(y)x^{i_j} = \sum_{j\le l}\phi_{i_j}(v)u^{\frac{i_j}{a}}.$$

If at p there is a form (11.38), set

$$\phi_p(u,v) = \sum_{j\le l} \phi_{i_j}(z)(x^by^c)^{i_j} = \sum_{j\le l}\phi_{i_j}(v)u^{\frac{i_j}{a}}.$$

In both cases $w - \phi_p(u,v)$ is weakly good for f at p.

By Lemma 11.5, there exist finitely many points $p_1,\ldots,p_n \in X$ such that if we set $\phi_i(u,v) = \phi_{p_i}(u,v)$, then for all $p \in f^{-1}(\overline{q})$, some $w_i = w - \phi_i(u,v)$ is weakly good for f at p.

Since the ϕ_{i_j} are units in $\hat{\mathcal{O}}_{X,p_i}$, we see that u,v,w_i satisfy 5 or 6 of Definition 10.1 of super parameters at p_i.

Let $p = p_i$ for some i, with the notation of (11.37) or (11.38).

Suppose that $\overline{p} \in f^{-1}(\overline{q})$. We must show that u,v,w_i are super parameters at $\overline{p}$.

At $\overline{p}$ we have permissible parameters $\overline{x},\overline{y},\overline{z} \in \hat{\mathcal{O}}_{X,\overline{p}}$ such that if $\overline{p}$ is a 1-point:

$$(11.39)\qquad \begin{aligned} u &= \overline{x}^{\overline{a}}\\ v &= \overline{y}\\ w &= \textstyle\sum_{j=0}^{\overline{m}} \overline{\phi}_{\overline{i}_j}(\overline{y})\overline{x}^{\overline{i}_j} + \overline{x}^{\overline{n}}(\overline{\beta}+\overline{z})\end{aligned}$$

where $\overline{\beta}\in\mathbf{k}$, $\overline{i}_{\overline{m}} < \overline{n}$, all $\overline{\phi}_{\overline{i}_j}(\overline{y})$ are non zero,

or if $\overline{p}$ a 2-point:

$$(11.40)\qquad \begin{aligned} u &= (\overline{x}^{\overline{b}}\overline{y}^{\overline{c}})^{\overline{a}}\\ v &= \overline{z}\\ w &= \textstyle\sum_{j=0}^{\overline{m}} \overline{\phi}_{\overline{i}_j}(\overline{z})(\overline{x}^{b}\overline{y}^{c})^{\overline{i}_j} + \overline{x}^{\overline{d}}\overline{y}^{\overline{e}}\end{aligned}$$

where $\gcd(\overline{b},\overline{c}) = 1$, $\overline{i}_{\overline{m}}(\overline{b},\overline{c}) \not\geq (\overline{d},\overline{e})$, all $\overline{\phi}_{\overline{i}_j}(\overline{z})$ are non zero.

We know from Lemma 11.6 that all ϕ_{i_j} and $\overline{\phi}_{\overline{i}_j}$ are units in $\hat{\mathcal{O}}_{X,p}$ and $\hat{\mathcal{O}}_{X,\overline{p}}$ respectively.

It will follow that w_i are super parameters at $\overline{p}$ after we have proven that if there exists t with $t \le \min\{l,\overline{m}\}$ and

$$\frac{i_j}{a} = \frac{\overline{i}_j}{\overline{a}} \text{ and } \phi_{i_j}(0) = \overline{\phi}_{\overline{i}_j}(0)$$

for $0 \le j \le t$, then we have equality of power series in u,

$$\overline{\phi}_{\overline{i}_j}(v) = \phi_{i_j}(v)$$

for $0 \le j \le t$, and thus equality of series

$$\sum_{j=0}^{t}\phi_{i_j}(v)u^{\frac{i_j}{a}} = \sum_{j=0}^{t}\overline{\phi}_{\overline{i}_j}(v)u^{\frac{\overline{i}_j}{\overline{a}}}.$$

We will prove this in the case when $p = p_i$ satisfies (11.37) and $\overline{p}$ satisfies (11.39).

The proof of the remaining three cases is similar.

We prove this by induction. First assume that $l \ge 0$ and

$$\frac{i_0}{a} = \frac{\overline{i}_0}{\overline{a}} \tag{11.41}$$

and $\phi_{i_0}(0) = \overline{\phi}_{\overline{i}_0}(0)$.

Let ν be a valuation of $\mathbf{k}(X)$ such that the center of ν on X is p, and identifying ν with an extension of ν to the quotient field of $\hat{\mathcal{O}}_{X,p}$ which dominates $\hat{\mathcal{O}}_{X,p}$, we have

$$\nu(w - \sum_{j=0}^{k} \phi_{i_j}(y)x^{i_j}) > \nu(w - \sum_{j=0}^{k-1} \phi_{i_j}(y)x^{i_j})$$

for $0 \le k \le m$.

Let p_n be the center of ν on U_n, q_n be the center of ν on V_n in the commutative diagram (11.35).

With the notation of the proof of Lemma 11.7, we see that the rational map $U \to V_{r(1)}$ is a morphism at $p = p_{r(1)}$, and $\overline{f}_{r(1)}(p) = q_{r(1)}$ has (formal) permissible parameters $u_{r(1)}, v, w_{r(1)}$ defined by

$$u = u_{r(1)}^e, w = u_{r(1)}^f(\overline{w}_{r(1)} + \phi_{i_0}(0)), w_{r(1)} = \overline{w}_{r(1)} - [\phi_{i_0}(v) - \phi_{i_0}(0)] \tag{11.42}$$

with $\gcd(e,f) = 1$ and $\frac{e}{f} = \frac{a}{i_0}$.

$$u_{r(1)} = w_{r(1)} = 0$$

are local equations of (a branch of) the nonfinite locus of $\overline{f}_{r(1)} : U \to V_{r(1)}$ at $q_{r(1)}$. We see from (11.42), (11.39) and (11.41) that $U \to V_{r(1)}$ is a morphism at $\overline{p} = \overline{p}_{r(1)}$, that $r(1) = \overline{r}(1)$, and $\overline{f}_{r(1)}(\overline{p}) = q_{r(1)}$. Further,

$$u_{r(1)} = \overline{w}_{r(1)} - [\overline{\phi}_{\overline{i}_0}(v) - \overline{\phi}_{\overline{i}_0}(0)] = 0$$

are also (formal) local equations of (a branch of) the nonfinite locus of $U_{r(1)} \to V_{r(1)}$ at $q_{r(1)}$. Since $\overline{q}$ is a general point of C, the nonfinite locus of $U_{r(1)} \to V_{r(1)}$ is a nonsingular curve. Thus

$$\phi_{i_0}(v) = \overline{\phi}_{\overline{i}_0}(v).$$

(11.41) implies that

$$\gcd(a, i_0) = \frac{a}{e}$$

and

$$\gcd(\overline{a}, \overline{i}_0) = \frac{\overline{a}}{e}.$$

Suppose that we further have that $l \ge 1$, $\frac{i_1}{a} = \frac{\overline{i}_1}{\overline{a}}$ and $\phi_{i_1}(0) = \overline{\phi}_{\overline{i}_1}(0)$. Then from (11.41) we have

$$\frac{i_1 - i_0}{a} = \frac{\overline{i}_1 - \overline{i}_0}{\overline{a}}.$$

Thus

$$\frac{\gcd(a, i_0)}{i_1 - i_0} = \frac{\gcd(\overline{a}, \overline{i}_0)}{\overline{i}_1 - \overline{i}_0}.$$

We have (with the notation of (11.36) of the proof of Lemma 11.7) that

$$\frac{e_2}{f_2} = \frac{\gcd(a, i_0)}{i_1 - i_0},$$

the rational map $U \to X_{r(2)}$ is a morphism at $p = p_{r(2)}$, and $q_{r(2)} = \overline{f}_{r(2)}(p)$ has permissible parameters $u_{r(2)}, v, w_{r(2)}$ defined by

$$u_{r(1)} = u_{r(2)}^{e_2}, v_{r(1)} = v_{r(2)}, w_{r(1)} = u_{r(2)}^{f_2}(\overline{w}_{r(2)} + \phi_{i_1}(0)),$$
$$w_{r(2)} = \overline{w}_{r(2)} - [\phi_{i_1}(v) - \phi_{i_1}(0)].$$

We see that $\overline{f}_{r(2)}$ is a morphism at $\overline{p}$, and $\phi_{i_1}(0) = \overline{\phi}_{i_1}(0)$ implies that $\overline{f}_{r(2)}(\overline{p}) = q_{r(2)}$. Let

$$w_{r(2)} = \overline{w}_{r(2)} - [\phi_{i_1}(v) - \phi_{i_1}(0)].$$

We have

$$u_{r(2)} = w_{r(2)} = 0$$

and

$$u_{r(2)} = \overline{w}_{r(2)} - [\overline{\phi}_{i_1}(v) - \overline{\phi}_{i_1}(0)] = 0$$

are local equations of (branches of) the nonfinite locus of $\overline{f}_{r(1)}$ at $q_{r(1)}$. Thus, since the nonfinite locus of $\overline{f}_{r(1)}$ is nonsingular,

$$\phi_{i_1}(v) = \overline{\phi}_{\overline{i}_1}(v).$$

Assume that $l \geq 2$,

$$\frac{i_0}{a} = \frac{\overline{i}_0}{\overline{a}}, \frac{i_1}{a} = \frac{\overline{i}_1}{\overline{a}} \text{ and } \frac{i_2}{a} = \frac{\overline{i}_2}{\overline{a}}$$

and $\phi_{i_2}(0) = \overline{\phi}_{\overline{i}_2}(0)$, as well as $\phi_{i_0}(0) = \overline{\phi}_{i_0}(0)$ and $\phi_{i_1}(0) = \overline{\phi}_{i_1}(0)$.

Then

$$\frac{\gcd(a, i_0, i_1)}{a} = \frac{\gcd(\overline{a}, \overline{i}_0, \overline{i}_1)}{\overline{a}}.$$

Now

$$\frac{i_2 - i_1}{a} = \frac{\overline{i}_2 - \overline{i}_1}{\overline{a}}$$

implies

$$\frac{\gcd(a, i_0, i_1)}{i_2 - i_1} = \frac{\gcd(\overline{a}, \overline{i}_0, \overline{i}_1)}{\overline{i}_2 - \overline{i}_1}.$$

The rational map $U \to V_{r(3)}$ is a morphism at p and $\overline{p}$, $\overline{f}_{r(3)}(p) = \overline{f}_{r(3)}(\overline{p}) = q_{r(3)}$, and

$$\phi_{i_2}(v) = \overline{\phi}_{i_2}(v)$$

since the fundamental locus of $\overline{f}_{r(3)}$ is nonsingular.

Iterating, we see that if $j \leq t$,

$$\frac{i_j}{a} = \frac{\overline{i}_j}{\overline{a}}$$

and

$$\phi_{i_j}(0) = \overline{\phi}_{i_j}(0)$$

for $j \leq t$, then

$$\frac{\gcd(a, i_0, \ldots, i_{j-1})}{i_j - i_{j-1}} = \frac{\gcd(\overline{a}, \overline{i}_0, \ldots, \overline{i}_{j-1})}{\overline{i}_j - \overline{i}_{j-1}}$$

for $j \leq t$ and

$$\phi_{i_j}(v) = \overline{\phi}_{\overline{i}_j}(v)$$

for $j \leq t$.

We have verified that a general point of every one dimensional component of $G_Y(f,\tau)$ is perfect. Thus the conclusions of the lemma hold.

□

DEFINITION 11.11. Suppose that $f : X \to Y$ is prepared. Let V be the largest open subset of Y on which the conclusions of Lemma 11.7 hold. Let $\Theta(Y) = \Theta(f,Y)$ be the set of perfect 1-points in $V \cap G_Y(f,\tau)$.

REMARK 11.12. *Suppose that f is τ-prepared. Then $G_Y(f,\tau) - \Theta(f,Y)$ is a finite set by Lemma 11.10, Lemma 11.7 and Lemma 11.5.*

CHAPTER 12

Relations

In this chapter, we suppose that Y is a nonsingular projective 3-fold with toroidal structure D_Y, and $f : X \to Y$ is a dominant proper morphism of nonsingular 3-folds, with toroidal structures D_Y and $D_X = f^{-1}(D_Y)$, such that D_X contains the singular locus of f.

DEFINITION 12.1. A quasi-pre-relation R on Y is an association U from a locally closed subset $U(R) \subset D_Y$, such that $U(R)$ contains no non trivial open subsets of 2-curves, $U(R)$ contains no 3-points and dim $U(R) \le 1$.

If $q \in U(R)$ is a 2-point,

$$R(q) = (S_R(q), E_{R,1}(q), E_{R,2}(q), w_{R(q)}, u_{R(q)}, v_{R(q)}, e_R(q), a_R(q), b_R(q), \lambda_R(q))$$

with $a_R(q), b_R(q), e_R(q) \in \mathbf{Z}$, $\gcd(a_R(q), b_R(q), e_R(q)) = 1$, $e_R(q) > 1$,

$$u_{R(q)}, v_{R(q)}, w_{R(q)}$$

are (possibly formal) permissible parameters at q with $u_{R(q)}, v_{R(q)} \in \mathcal{O}_{Y,q}$, $0 \ne \lambda_R(q) \in \mathbf{k}$. $S_R(q) = \mathrm{Spec}(\mathcal{O}_{Y,q}/(w_{R(q)}))$, $E_{R,1}(q)$ is the divisor $u_R(q) = 0$ on $S_R(q)$ and $E_{R,2}(q)$ is the divisor $v_{R(q)} = 0$ on $S_R(q)$.

We will also allow quasi-pre-relations with $a_R(q) = b_R(q) = \infty$, $e_R(q) = 1$ and $\lambda_R(q) = 1$.

If $q \in U(R)$ is a 1-point,

$$R(q) = (S_R(q), E_R(q), w_{R(q)}, u_{R(q)}, v_{R(q)}, e_R(q), a_R(q), \lambda_R(q))$$

with $a_R(q), e_R(q) \in \mathbf{Z}$, $\gcd(a_R(q), e_R(q)) = 1$, $e_R(q) > 1$, $u_{R(q)}, v_{R(q)}, w_{R(q)}$ are (possibly formal) permissible parameters at q with $u_{R(q)}, v_{R(q)} \in \mathcal{O}_{Y,q}$, $0 \ne \lambda_R(q) \in \mathbf{k}$. $S_R(q) = \mathrm{Spec}(\mathcal{O}_{Y,q}/(w_{R(q)}))$, $E_R(q)$ is the divisor $u_{R(q)} = 0$ on $S_R(q)$.

We will also allow quasi-pre-relations with $a_R(q) = \infty$, $e_R(q) = 1$ and $\lambda_R(q) = 1$.

A restriction R' of a quasi-pre-relation R is the association from a locally closed subset $U(R')$ of $U(R)$ such that $R'(q) = R(q)$ for $q \in U(R')$.

Suppose that R is a quasi-pre-relation and $q \in U(R)$ is a 2-point. Let

$$u = u_{R(q)}, v = v_{R(q)}, w = w_{R(q)}, a = a_R(q), b = b_R(q), c = e_R(q), \lambda = \lambda_R(q).$$

If $a, b \ne \infty$, then $R(q)$ is determined by the expression

$$w^e - \lambda u^a v^b. \tag{12.1}$$

Depending on the signs of a and b, this expression determines a (formal) germ at q of an (irreducible) surface singularity

$$F = F_{R(q)} = 0 \tag{12.2}$$

of one of the following forms:

$$F = w^e - \lambda u^a v^b = 0$$

if $a, b \geq 0$ and $a + b > 0$,

$$F = w^e u^{-a} - \lambda v^b = 0$$

if $a < 0$, $b > 0$,

$$F = w^e v^{-b} - \lambda u^a = 0$$

if $b < 0$, $a > 0$.

In the remaining case, $a, b \leq 0$,

$$F = w^e u^{-a} v^{-b} - \lambda$$

is a unit in $\hat{\mathcal{O}}_{Y,q}$ and $F(q) \neq 0$.

If $a, b = \infty$, and $q \in U(R)$ is a 2-point, then $R(q)$ is determined by the expression

$$F = F_{R(q)} = w_{R(q)} = 0. \tag{12.3}$$

Suppose that R is a quasi-pre-relation, and $q \in U(R)$ is a 1-point. Let

$$u = u_{R(q)}, v = v_{R(q)}, w = w_{R(q)}, a = a_R(q), e = e_R(q), \lambda = \lambda_R(q).$$

If $a \neq \infty$, then $R(q)$ is determined by the expression

$$w^e - \lambda u^a. \tag{12.4}$$

This expression determines a (formal) germ at q of an (irreducible) surface singularity

$$F = F_{R(q)} = 0 \tag{12.5}$$

of the form

$$F = w^e - \lambda u^a = 0.$$

if $a > 0$. In the remaining case, $a \leq 0$, so that

$$F = w^e u^{-a} - \lambda$$

is a unit in $\hat{\mathcal{O}}_{Y,q}$ and $F(q) \neq 0$.

If $a = \infty$, and $q \in U(R)$ is a 1-point, then $R(q)$ is determined by the expression

$$F = F_{R(q)} = w_{R(q)} = 0. \tag{12.6}$$

A quasi-pre-relation R is resolved if $F_{R(q)}$ is a unit in $\hat{\mathcal{O}}_{Y,q}$ for all $q \in U(R)$ (This includes the case $U(R) = \emptyset$).

Definition 12.2. A subvariety G of Y is an admissible center for a quasi-pre-relation R on Y if one of the following holds:

1. G is a 2-point.
2. G is a 1-point.
3. G is a 2-curve of Y.
4. $G \subset D_Y$ is a nonsingular curve which contains a 1-point and makes SNCs with D_Y. If $q \in U(R) \cap G$ then the (formal) germ of G at q is contained in the germ $w_{R(q)} = 0$.

Observe that admissible centers are possible centers.

Suppose that R is a quasi-pre-relation on Y, G is an admissible center for R, and $\Psi : Y_1 \to Y$ is the blow up of G.

Let W be the union over $q \in U(R)$ of points q_1 in $\Psi^{-1}(q)$ such that q_1 is on the strict transform of $w_{R(q)} = 0$. Assume that this is a locally closed subset of D_{Y_1} of dimension ≤ 1 which contains no 2-curves or 3-points (This condition will always be satisfied when R is a pre-relation, Definition 12.3). The transform R^1 of R on Y_1 is then the quasi-pre-relation on Y_1 defined by the condition that $U(R^1) = W$. For such q_1, $R^1(q_1)$ is determined by the following rules:

If $q \in U(R) \cap G$, and

$$u = u_{R(q)}, v = v_{R(q)}, w = w_{R(q)},$$

then G has local equations of one of the following forms at q (corresponding to the cases of Definition 12.2):

1.,2. $u = v = w = 0$,

3. $u = v = 0$,

4. a) q a 2-point, $u = w = 0$ or $v = w = 0$

4. b) q a 1-point, $u = w = 0$.

If $q_1 \in U(R_1) \cap \Psi^{-1}(q)$, then

$$u_1 = u_{R^1(q_1)}, v_1 = v_{R^1(q_1)}, w_1 = w_{R^1(q_1)}$$

are defined, respectively, by

1., 2. $u = u_1, v = u_1(v_1+\alpha), w = u_1w_1$ for some $\alpha \in \mathbf{k}$, or $u = u_1v_1, v = v_1, w = v_1w_1$,

3. $u = u_1, v = u_1(v_1 + \alpha), w = w_1$ for some $\alpha \in \mathbf{k}$, or $u = u_1v_1, v = v_1, w = w_1$,

4. a) $u = u_1, w = u_1w_1$ or $v = v_1, w = v_1w_1$

4. b) $u = u_1, v = v_1, w = u_1w_1$.

Suppose that $q_1 \in U(R^1) \cap \Psi^{-1}(q)$ and case 1 holds. Suppose that

$$R(q) = (S, E, w, u, v, e, a, b, \lambda).$$

If $0 \ne \alpha$ and $a, b \ne \infty$, we define

$$a_{R^1}(q_1) = a + b - e, e_{R^1}(q_1) = e, \lambda_{R^1}(q_1) = \lambda\alpha^b.$$

$R^1(q_1)$ is thus determined by

$$w_1^e - \lambda\alpha^b u_1^{a+b-e}.$$

If $a, b \ne \infty$ and

$$u = u_1, v = u_1v_1, w = u_1w_1,$$

we define

$$a_{R^1}(q_1) = a + b - e, b_{R^1}(q_1) = b, e_{R^1}(q_1) = e, \lambda_{R^1}(q_1) = \lambda.$$

$R^1(q_1)$ is determined by $w_1^e - \lambda u_1^{a+b-e}v_1^b$.

If $a, b \ne \infty$ and

$$u = u_1v_1, v = v_1, w = v_1w_1,$$

we define

$$a_{R^1}(q_1) = a, b_{R^1}(q_1) = a + b - e, e_{R^1}(q_1) = e, \lambda_{R^1}(q_1) = \lambda.$$

$R^1(q_1)$ is determined by $w_1^e - \lambda u_1^a v_1^{a+b-e}$.

If $a = b = \infty$, we define

$$a_{R^1}(q_1) = \infty, b_{R^1}(q_1) = \infty, e_{R^1}(q_1) = 1, \lambda_{R^1}(q_1) = 1,$$

and $R^1(q_1)$ is determined by $w_{R^1(q_1)}$.

In cases 2,3 and 4, we define $R^1(q_1)$ in an analogous way.

Suppose that $\Psi_1 : Y_1 \rightarrow Y$ is a sequence of blow ups of admissible centers for (the transforms of) R, R^1 is the transform of R on Y_1, $q \in U(R)$ and $q_1 \in \Psi_1^{-1}(q) \cap U(R^1)$. Let

$$u = u_{R(q)}, v = v_{R(q)}, w = w_{R(q)}.$$

$$u_1 = u_{R^1(q_1)}, v_1 = v_{R^1(q_1)}, w_1 = w_{R^1(q_1)}.$$

u and v are related to u_1, v_1 birationally. That is, $\mathbf{k}(u,v) = \mathbf{k}(u_1, v_1)$. We have one of the following expressions:

$$u = u_1^a v_1^b, v = u_1^c v_1^d, w = u_1^e v_1^f w_1 \tag{12.7}$$

with $ad - bc = \pm 1$,

$$u = u_1^a \gamma_1(u_1, v_1), v = u_1^b \gamma_2(u_1, v_1), w = u_1^c \gamma_3(u_1, v_1) w_1 \tag{12.8}$$

where $\gamma_1, \gamma_2, \gamma_3$ are unit series,

$$u = (u_1^a v_1^b)^t \gamma_1(u_1, v_1), v_1 = (u_1^a v_1^b)^k \gamma_2(u_1, v_1), w = u_1^e v_1^f \gamma_3(u_1, v_1) w_1 \tag{12.9}$$

where $\gcd(a,b) = 1$, and $\gamma_1, \gamma_2, \gamma_3$ are unit series.

Definition 12.3. A quasi-pre-relation R is a pre-relation if there exists a nonsingular 3-fold $\overline{Y}_R$ with toroidal structure $D_{\overline{Y}}$, a pre-relation R^0 on $\overline{Y}_R$ such that $U(R^0) = \{\overline{q}\}$ is a single point with

$$R^0(\overline{q}) = (\cdots, w_{R^0(\overline{q})}, u_{R^0(\overline{q})}, v_{R^0(\overline{q})}, \cdots)$$

and a sequence of possible blow ups

$$\overline{\Psi}_R : Y = Y_n \rightarrow \cdots \rightarrow Y_1 \rightarrow Y_0 = \overline{Y}_R$$

where each Y_i has a quasi-pre-relation R^i which is the restriction of the transform of R^{i-1}, and $Y_{i+1} \rightarrow Y_i$ is an admissible blow up for R^i, and $R = R^n$.

Definition 12.4. A pre-relation R on Y is algebraic if there exists an open subset V of Y and a nonsingular irreducible closed surface $\Omega(R) \subset V$ such that $\Omega(R)$ makes SNCs with D_Y, $U(R) \subset \Omega(R)$ and $S_R(q)$ is the (formal) germ of $\Omega(R)$ at q for all $q \in U(R)$. Further, if $q \in \Omega(R) \cap D_Y$, then there exist permissible parameters u_q, v_q, w_q at q such that $w_q = 0$ is a local equation of $\Omega(R)$.

Suppose that R is algebraic, and $\Psi : Y_1 \rightarrow Y$ is an admissible blow up. then after possibly replacing $\Omega(R)$ with an open subset of $\Omega(R)$ (containing $U(R)$), we have that the transform R^1 of R is algebraic, where $\Omega(R^1)$ is the strict transform of $\Omega(R)$ by Ψ.

Definition 12.5. Suppose that $f : X \rightarrow Y$ is prepared.

A primitive relation R for f is

1. A pre-relation $\overline{R}$ on Y.

2. A locally closed subset $T(R) \subset f^{-1}(U(\overline{R})) \cap G_X(f,\tau)$ such that if $p \in T(R)$ and $f(p)$ is a 2-point with
$$\overline{R}(f(p)) = (S, E_1, E_2, w, u, v, e, a, b, \lambda),$$
then u, v are toroidal forms at p. If $a, b \neq \infty$, then there exist permissible parameters x, y, z at p for u, v, w such that
$$w^e = u^a v^b \overline{\Lambda}(x, y, z) \tag{12.10}$$
where $\overline{\Lambda}(0, 0, 0) = \lambda$.

If $a = b = \infty$, then u, v, w have a monomial form (Definition 4.4) at p.

If $f(p)$ is a 1-point with
$$\overline{R}(f(p)) = (S, E, w, u, v, e, a, \lambda),$$
then u, v are toroidal forms at p. If $a \neq \infty$, then there exist permissible parameters x, y, z at p for u, v, w such that
$$w^e = u^a \overline{\Lambda}(x, y, z) \tag{12.11}$$
where $\overline{\Lambda}(0, 0, 0) = \lambda$.

If $a = \infty$, then u, v, w have a monomial form (Definition 4.4) at p.

In all cases, we define $R(p) = \overline{R}(f(p))$.

A relation R for f is a finite set of pre-relations $\{\overline{R}_i\}$ on Y with associated primitive relations R_i for f such that the sets $T(R_i)$ are pairwise disjoint.

We denote $U(R) = \cup_i U(\overline{R}_i)$ and $T(R) = \cup_i T(R_i)$, and define
$$R(p) = R_i(p)$$
if $p \in T(R_i)$.

If $U(\overline{R}_i) \cap U(\overline{R}_j) \neq \emptyset$, then we further require that $Y_{\overline{R}_i} = Y_{\overline{R}_j}$ (with the notation of Definition 12.3) and $u_{(\overline{R}_i)^0(\overline{q})} = u_{(\overline{R}_j)^0(\overline{q})}, v_{(\overline{R}_i)^0(\overline{q})} = v_{(\overline{R}_j)^0(\overline{q})}$. This implies that
$$u_{\overline{R}_i(q)} = u_{\overline{R}_j(q)}, v_{\overline{R}_i(q)} = v_{\overline{R}_j(q)}$$
if $q \in U(\overline{R}_i) \cap U(\overline{R}_j)$. We will call $\{\overline{R}_i\}$ the pre-relations associated to R. We will say that R is algebraic if each $\overline{R}_i$ is algebraic and
$$\Omega(\overline{R}_i) \cap U(R) = U(\overline{R}_i) \tag{12.12}$$
for all i. We will also denote $\Omega(R_i) = \Omega(\overline{R}_i)$. For $p \in T(R_i) \subset T(R)$, we denote
$$R(p) = \begin{pmatrix} S = S_R(p), E_1(p), E_2(p), w = w_{R(p)}, u = u_{R(p)}, \\ v = v_{R(p)}, e = e_R(p), a = a_R(p), b = b_R(p), \lambda = \lambda_R(p) \end{pmatrix}$$
if $f(p)$ is a 2-point,
$$R(p) = \begin{pmatrix} S = S_R(p), E(p), w = w_{R(p)}, u = u_{R(p)}, \\ v = v_{R(p)}, e = e_R(p), a = a_R(p), \lambda = \lambda_R(p) \end{pmatrix}$$
if $f(p)$ is a 1-point.

A relation R is resolved if $T(R) = \emptyset$.

DEFINITION 12.6. Suppose that $f : X \to Y$ is prepared, R is a relation for f and

$$\begin{array}{ccc} X_1 & \xrightarrow{f_1} & Y_1 \\ \Phi \downarrow & & \downarrow \Psi \\ X & \xrightarrow{f} & Y \end{array}$$

is a commutative diagram such that

1. Ψ is a product of blow ups which are admissible for all of the pre-relations $\overline{R}_i$ associated to R (and their transforms) and Φ is a product of blow ups of possible centers
2. f_1 is prepared.
3. Let $\overline{R}_i^1$ be the transforms of the $\overline{R}_i$ on Y_1 and let
$$T_i = \Phi^{-1}(T(R_i)) \cap f_1^{-1}(U(\overline{R}_i^1)) \cap G_{X_1}(f_1, \tau).$$
Suppose that if $p \in T_i$ then $u_{\overline{R}_i^1(f_1(p))}, v_{\overline{R}_i^1(f_1(p))}$ are toroidal forms at p.

Then the transform R^1 of R for f_1 is the relation for f_1 defined by

$$T(R^1) = \cup T_i,$$
$$R^1(p) = \overline{R}_i^1(f_1(p))$$

for $p \in T_i$.

It is straightforward to verify that R^1 satisfies the conditions of Definition 12.5, substituting from (12.10), (12.11) and (12.7) - (12.9) into (4.1).

CHAPTER 13

Well prepared morphisms

Suppose that $\tau \in \mathbf{N}$ and $f : X \to Y$ is a dominant, proper, prepared morphism of nonsingular 3-folds with toroidal structures D_Y and $D_X = f^{-1}(D_Y)$. Further suppose that the singular locus of f is contained in D_X. If R is a relation for f with associated pre-relations $\{\overline{R}_i\}$, then for $p \in T(R_i)$ such that $f(p) = q$ is a 2-point, we have that

(13.1)
$$R(p) = \left(\begin{array}{c} S_i = S_R(p), E_1 = E_{R,1}(p), E_2 = E_{R,2}(p), w_i = w_{R(p)}, u = u_{R(p)}, \\ v = v_{R(p)}, e_i = e_R(p), a_i = a_R(p), b_i = b_R(p), \overline{\lambda}_i = \lambda_R(p) \end{array} \right)$$

which we will abbreviate (as in (12.1)) as

$$R(p) = w_i^{e_i} - \overline{\lambda}_i u^{a_i} v^{b_i}, \tag{13.2}$$

with $e_i > 1$, if $a_i, b_i \neq \infty$, or (as in (12.3))

$$R(p) = w_i \tag{13.3}$$

if $a_i, b_i = \infty$. In this case, we require that u, v, w_i have a monomial form (Definition 4.4) at p.

For $p \in T(R_i)$ such that $f(p) = q$ is a 1-point, we have that

$$R(p) = \left(\begin{array}{c} S_i = S_R(p), E_1 = E_R(p), w_i = w_{R(p)}, u = u_{R(p)}, \\ v = v_{R(p)}, e_i = e_R(p), a_i = a_R(p), \overline{\lambda}_i = \lambda_R(p) \end{array} \right) \tag{13.4}$$

which we will abbreviate (as in (12.4)) as

$$R(p) = w_i^{e_i} - \overline{\lambda}_i u^{a_i}, \tag{13.5}$$

with $e_i > 1$, if $a_i \neq \infty$, or (as in (12.5))

$$R(p) = w_i \tag{13.6}$$

if $a_i = \infty$. In this case, we require that u, v, w_i have a monomial form (Definition 4.4) at p.

Recall that if $p' \in T(R)$ is such that $f(p') = f(p)$, then $u_{R(p')} = u_{R(p)} = u$ and $v_{R(p')} = v_{R(p)} = v$. Let I be an index set for the pre-relations $\{\overline{R}_i\}$ associated to R.

Definition 13.1. Suppose that $\tau \geq 1$. A prepared morphism $f : X \to Y$ is pre-τ-quasi-well prepared with relation R if:

1. $T(R) = G_X(f, \tau) \cap f^{-1}(U(R))$
2. Suppose that $p \in T(R)$. Then $\tau > 1$ implies $R(p)$ has a form (13.2) or (13.5), $\tau = 1$ implies $R(p)$ has a form (13.3) or (13.6).
3. Suppose that $\tau > 1$, $q \in U(\overline{R}_i)$ and $p \in f^{-1}(q) \cap T(R_i)$. Let
$$u = u_{\overline{R}_i(q)}, v = v_{\overline{R}_i(q)}, w = w_{\overline{R}_i(q)}, e = e_{\overline{R}_i}(q).$$

a) Suppose that q is a 2-point.
   (i) Suppose that u, v satisfy (4.1) at p, and $w = x^c\gamma$, where γ is a unit in $\hat{\mathcal{O}}_{X,p}$. Then
$$e = |(a\mathbf{Z} + b\mathbf{Z} + c\mathbf{Z})/(a\mathbf{Z} + b\mathbf{Z})|.$$
   (ii) Suppose that u, v satisfy (4.2) at p, and $w = x^f y^g\gamma$ where γ is a unit in $\hat{\mathcal{O}}_{X,p}$. Then
$$e = |[(a,b)\mathbf{Z} + (c,d)\mathbf{Z} + (f,g)\mathbf{Z}]/[(a,b)\mathbf{Z} + (c,d)\mathbf{Z}]|.$$
   (iii) Suppose that u, v satisfy (4.3) at p and $w = (x^a y^b)^l\gamma$, where γ is a unit in $\hat{\mathcal{O}}_{X,p}$. Then
$$e = |(k\mathbf{Z} + t\mathbf{Z} + l\mathbf{Z})/(k\mathbf{Z} + t\mathbf{Z})|.$$
   (iv) Suppose that u, v satisfy (4.4) at p and $w = x^g y^h z^i\gamma$ where γ is a unit in $\hat{\mathcal{O}}_{X,p}$ and
$$\text{Det}\begin{pmatrix} a & b & c \\ d & e & f \\ g & h & i \end{pmatrix} = 0.$$
   Then
$$e = |[(a,b,c)\mathbf{Z} + (d,e,f)\mathbf{Z} + (g,h,i)\mathbf{Z}]/[(a,b,c)\mathbf{Z} + (d,e,f)\mathbf{Z}]|$$

b) Suppose that q is a 1-point.
   (i) Suppose that u, v satisfy (4.5) at p, and $w = x^b\gamma$ where γ is a unit in $\hat{\mathcal{O}}_{X,p}$. Then
$$e = |(a\mathbf{Z} + b\mathbf{Z})/a\mathbf{Z}|.$$
   (ii) Suppose that u, v satisfy (4.6) at p, and $w = (x^a y^b)^l\gamma$ where γ is a unit in $\hat{\mathcal{O}}_{X,p}$. Then
$$e = |(k\mathbf{Z} + l\mathbf{Z})/k\mathbf{Z}|.$$

4. If $q \in U(\overline{R}_i) \cap U(\overline{R}_j)$, then there exists $\lambda_{ij}(u,v) \in \mathbf{k}[[u,v]]$, with
$$u = u_{\overline{R}_i(q)} = u_{\overline{R}_j(q)}, v = v_{\overline{R}_i(q)} = v_{\overline{R}_j(q)}, w_i = w_{\overline{R}_i(q)}, w_j = w_{\overline{R}_j(q)}$$
such that
$$w_j = w_i + \lambda_{ij}(u,v),$$
and with the notation of Definition 12.5, there exists a series
$$(\lambda_{ij})^0(u_{(\overline{R}_j)^0(\overline{q})}, v_{(\overline{R}_j)^0(\overline{q})})$$
such that
$$w_{(\overline{R}_j)^0(\overline{q})} - w_{(\overline{R}_i)^0(\overline{q})} = (\lambda_{ij})^0(u_{(\overline{R}_j)^0(\overline{q})}, v_{(\overline{R}_j)^0(\overline{q})}),$$
and $\lambda_{ij}(u,v)$ is obtained from $\lambda^0_{ij}(u_{(\overline{R}_j)^0(\overline{q})}, v_{(\overline{R}_j)^0(\overline{q})})$ from the appropriate expression (12.7) - (12.9).
5. Suppose that $q \in U(\overline{R}_i)$, where $\overline{R}_i$ is a relation associated to R. Then $u = u_{\overline{R}_i(q)}, v = v_{\overline{R}_i(q)}, w_i = w_{\overline{R}_i(q)}$ are super parameters at q (Definition 10.1).

f is τ-quasi-well prepared with relation R if f is pre-τ-quasi-well prepared with $T(R) = G_X(f,\tau)$.

REMARK 13.2. *All of the equations of 3 of Definition 13.1 follow from the condition that* $p \in T(R_i) \cap f^{-1}(q)$ *except for a) i where (4.1) holds, and a) iii where (4.3) holds.*

REMARK 13.3. *If* $f : X \to Y$ *is* τ*-quasi-well prepared, then* f *is* τ *prepared.*

We will allow a prepared morphism without relation ($U(R) = \emptyset$) as a type of pre-τ-quasi-well prepared morphism.

DEFINITION 13.4. $f : X \to Y$ is τ-quasi-well prepared with relation R and pre-algebraic structure if f is τ-quasi-well prepared with relation R and

$$u_{\overline{R}_i(q)}, v_{\overline{R}_i(q)}, w_{\overline{R}_i(q)} \in \mathcal{O}_{Y,q}$$

for all $\overline{R}_i$ associated to R, and $q \in U(\overline{R}_i)$.

DEFINITION 13.5. $f : X \to Y$ is τ-well prepared with relation R if

1. f is τ-quasi-well prepared with relation R and pre-algebraic structure.
2. The primitive pre-relations $\{\overline{R}_i\}$ associated to R are algebraic, and R is algebraic (Definition 12.5).
3. Suppose that $q \in U(\overline{R}_i) \cap U(\overline{R}_j)$. Let $w_i = w_{\overline{R}_i(q)}$ and $w_j = w_{\overline{R}_j(q)}$, $u = u_{\overline{R}_i(q)} = u_{\overline{R}_j(q)}$, $v = v_{\overline{R}_i(q)} = v_{\overline{R}_j(q)}$.

   Suppose that q is a 2-point. Then there exists a unit series $\phi_{ij} \in \mathbf{k}[[u,v]]$ and $a_{ij}, b_{ij} \in \mathbf{N}$ (or $\phi_{ij} = 0$ with $a_{ij} = b_{ij} = -\infty$) with

$$w_j = w_i + u^{a_{ij}} v^{b_{ij}} \phi_{ij}. \tag{13.7}$$

   Suppose that q is a 1-point. Then there exists a unit series $\phi_{ij} \in \mathbf{k}[[u,v]]$ and $c_{ij} \in \mathbf{N}$ (or $\phi_{ij} = 0$ with $c_{ij} = -\infty$) with

$$w_j = w_i + u^{c_{ij}} \phi_{ij}. \tag{13.8}$$

4. For $q \in U(R)$ a 2-point, set $I_q = \{i \in I \mid q \in U(\overline{R}_i)\}$. Then the set

$$\{(a_{ij}, b_{ij}) \mid i, j \in I_q\} \tag{13.9}$$

   from equation (13.7) is totally ordered.

DEFINITION 13.6. Suppose that $f : X \to Y$ is pre-τ-quasi-well prepared, $\overline{q} \in Y$ is a 1-point such that $\overline{q} \notin U(R)$. A curve $C \subset D_Y$ such that $\overline{q} \in C$ is called a **resolving curve** for f and R at $\overline{q}$ if

1. C makes SNCs with D_Y.
2. $C \cap G_Y(f, \tau) \subset \{\overline{q}\}$.
3. If $q \in C$ is a 2-point, then there exist super parameters u, v, w at q such that $u = w = 0$ are local equations of C at q.
4. If $q \in C$ is a 1-point, then there exist permissible parameters u, v, w at q such that u, v are toroidal forms at p for all $p \in f^{-1}(q)$, and $u = v = 0$ are local equations of C at q.

DEFINITION 13.7. Suppose that $f : X \to Y$ is pre-τ-quasi-well prepared with relation R.

1. A 2-point $q \in U(R)$ is prepared for R.
2. A 1-point or 2-point $q \in Y$ such that $q \notin U(R)$ is prepared for R if there exist super parameters u, v, w at q.
3. A 2-curve $C \subset Y$ is prepared for R.
4. A 1-point $q \in U(R)$ is prepared.

5. A resolving curve C for f and R at a 1-point $q \notin U(R)$ is prepared.

If E is a component of D_Y, $\overline{R}_i$ is pre-algebraic, and $q \in U(\overline{R}_i)$, we will denote $\overline{E \cdot S_{\overline{R}_i}(q)}$ for the Zariski closure in Y of the curve germ $u = w_{\overline{R}_i(q)} = 0$ at q, where $u = 0$ is a local equation of E.

DEFINITION 13.8. Suppose that $f : X \to Y$ is τ-well prepared with relation R for f. A nonsingular curve $C \subset D_Y$ which makes SNCs with D_Y is prepared for R of type 6 if

1. $C = \overline{E_\alpha \cdot S_{\overline{R}_i}(q_\beta)}$ for some component E_α of D_Y, pre-relation $\overline{R}_i$ associated to R and $q_\beta \in U(\overline{R}_i)$.
2. $\Omega(\overline{R}_i)$ contains C.
3. If $C' = \overline{E_\gamma \cdot S_{\overline{R}_j}(q_\delta)}$ is such that $C' \subset \Omega(\overline{R}_j)$, $C \neq C'$, and $q \in C \cap C'$, then $q \in U(\overline{R}_i) \cap U(\overline{R}_j)$ and $C' = \overline{E_\gamma \cdot S_{\overline{R}_j}(q)}$.
4. If $j \neq i$ and $C = \overline{E_\gamma \cdot S_{\overline{R}_j}(q_\delta)}$ then C satisfies 1 and 2 and 3 of this definition (for $\overline{R}_j$). (In this case we have by (12.12) that $U(\overline{R}_j) \cap C = U(\overline{R}_i) \cap C$).
5. Let
$$I_C = \{j \in I \mid C = \overline{E_\gamma \cdot S_{\overline{R}_j}(q_\delta)} \text{ for some } \overline{R}_j, E_\gamma, q_\delta \in U(\overline{R}_j)\}.$$
Suppose that $q \in C$ is a 1-point or a 2-point such that $q \notin U(R)$. Then there exist $u, v \in \mathcal{O}_{Y,q}$ such that for $j \in I_C$ there exists $\tilde{w}_j \in \mathcal{O}_{Y,q}$ such that
   (a) $\tilde{w}_j = 0$ is a local equation of $\Omega(\overline{R}_j)$ and $u, v, \tilde{w}_j$ are permissible parameters at q such that $u = \tilde{w}_j = 0$ are local equations of C at q.
   (b) $u, v, \tilde{w}_j$ are super parameters at q.
   (c) If $i, j \in I_C$ and q is a 1-point, there exist relations
   $$\tilde{w}_j = \tilde{w}_i + u^{c_{ij}}\phi_{ij}(u, v)$$
   where ϕ_{ij} is a unit series (or $\phi_{ij} = 0$ and $c_{ij} = -\infty$).
   (d) If $i, j \in I_C$ and q is a 2-point (with $q \notin U(R)$) then there exist relations
   $$\tilde{w}_j = \tilde{w}_i + u^{a_{ij}}v^{b_{ij}}\phi_{ij}(u, v)$$
   where ϕ_{ij} is a unit series (or $\phi_{ij} = 0$ and $a_{ij} = b_{ij} = -\infty$), and the set $\{(a_{ij}, b_{ij})\}$ is totally ordered.

If $f : X \to Y$ is τ-well prepared with relation R, and $\overline{R}_i$ is a pre-relation associated to R, we will feel free to replace $\Omega(\overline{R}_i)$ with an open subset of $\Omega(\overline{R}_i)$ containing $U(\overline{R}_i)$, and all curves $C = \overline{E \cdot S_{\overline{R}_i}(q)}$ such that E is a component of D_Y, $q \in U(\overline{R}_i)$ and C is prepared for R of type 6. This convention will allow some simplification of the statements of the theorems and proofs.

DEFINITION 13.9. $f : X \to Y$ is τ-very-well prepared with relation R if

1. f is τ-well prepared with relation R.
2. If E is a component of D_Y and $q \in U(\overline{R}_i) \cap E$, then $C = \overline{E \cdot S_{\overline{R}_i}(q)}$ is prepared for R of type 6 (Definition 13.8).
3. For all $\overline{R}_i$ associated to R, let
$$V_i(Y) = \left\{\gamma = \overline{E_\alpha \cdot S_{\overline{R}_i}(q_\gamma)} \mid q_\gamma \in U(\overline{R}_i), E_\alpha \text{ is a component of } D_Y\right\}.$$

Then

$$F_i = \sum_{\gamma \in V_i(Y)} \gamma$$

is a SNC divisor on $\Omega(\overline{R}_i)$ whose intersection graph is a forest (its connected components are trees).

If $f : X \to Y$ is τ-very-well prepared, we will feel free to replace $\Omega(\overline{R}_i)$ with an open neighborhood of F_i in $\Omega(\overline{R}_i)$. This will allow some simplification of the proofs.

REMARK 13.10. *Suppose that $f : X \to Y$ is τ-very well prepared. Then it follows from Definition 13.9 and (12.12) that $F_i \cap U(R) = U(\overline{R}_i)$ for all $\overline{R}_i$ associated to R.*

DEFINITION 13.11. Suppose that $f : X \to Y$ is pre-τ-quasi-well prepared (or τ-well prepared or τ-very-well prepared) with relation R. Let $\{\overline{R}_i\}$ be the pre-relations associated to R. Suppose that G is a point or nonsingular curve in Y which is an admissible center for all of the $\overline{R}_i$. Then G is called a permissible center for R if there exists a commutative diagram

$$\begin{array}{ccc} X_1 & \overset{f_1}{\to} & Y_1 \\ \Phi \downarrow & & \downarrow \Psi \\ X & \overset{f}{\to} & Y \end{array} \tag{13.10}$$

where Ψ is the blow up of G and Φ is a sequence of blow ups

$$X_1 = \overline{X}_n \to \cdots \to \overline{X}_1 \to X$$

of nonsingular curves and 3-points γ_i which are possible centers such that

1. f_1 is prepared and the assumptions of Definition 12.6 hold so that the transform R^1 of R for f_1 is defined.
2. $$\tau_{f_1}(p) \le \tau_f(\phi(p))$$
   for $p \in D_{X_1}$.
3. $f_1 : X_1 \to Y_1$ is pre-τ-quasi-well prepared, (or τ-well prepared or τ-very-well prepared) with relation R^1.

(13.10) is called a pre-τ-quasi-well prepared (or τ-well prepared or τ-very-well prepared) diagram of R and Ψ.

DEFINITION 13.12. Suppose that $f : X \to Y$ is τ-well prepared (or τ-very-well prepared) with relation R and $C \subset Y$ is prepared for R of type 6. Then C is a $*$-permissible center for R if there exists a commutative diagram

$$\begin{array}{ccc} X_1 & \overset{f_1}{\to} & Y_1 \\ \Phi \downarrow & & \downarrow \Psi \\ X & \overset{f}{\to} & Y \end{array} \tag{13.11}$$

such that

1. f_1 is τ-prepared and the assumptions of Definition 12.6 hold so that the transform R^1 of R for f_1 is defined.

2.
$$\tau_{f_1}(p) \le \tau_f(\phi(p))$$
for $p \in D_{X_1}$.

3. $f_1 : X_1 \to Y_1$ is τ-well prepared (or τ-very-well prepared).
4. (13.11) has a factorization

$$\begin{array}{ccc} X_1 = \overline{X}_m & \overset{f_1=\overline{f}_m}{\to} & \overline{Y}_m = Y_1\\ \downarrow & & \downarrow\\ \vdots & & \vdots\\ \downarrow & & \downarrow\\ \overline{X}_2 & \overset{\overline{f}_2}{\to} & \overline{Y}_2\\ \overline{\Phi}_2 \downarrow & & \downarrow \overline{\Psi}_2\\ \overline{X}_1 & \overset{\overline{f}_1}{\to} & \overline{Y}_1\\ \overline{\Phi}_1 \downarrow & & \downarrow \overline{\Psi}_1\\ X & \overset{f}{\to} & Y \end{array} \tag{13.12}$$

where $\overline{\Psi}_1$ is the blow up of C,

$$\begin{array}{ccc} \overline{X}_1 & \overset{\overline{f}_1}{\to} & \overline{Y}_1\\ \overline{\Phi}_1 \downarrow & & \downarrow \overline{\Psi}_1\\ X & \overset{f}{\to} & Y \end{array} \tag{13.13}$$

is a τ-well prepared diagram of R and $\overline{\Psi}_1$ of the form (13.10), each $\overline{\Psi}_{i+1} : \overline{Y}_{i+1} \to \overline{Y}_i$ for $i \ge 1$ is the blow up of a 2-point $q \in \overline{Y}_i$ which is prepared for the transform R^i of R on $\overline{X}_i$ of type 2 of Definition 13.7, and

$$\begin{array}{ccc} \overline{X}_{i+1} & \overset{\overline{f}_{i+1}}{\to} & \overline{Y}_{i+1}\\ \overline{\Phi}_{i+1} \downarrow & & \downarrow \overline{\Psi}_{i+1}\\ \overline{X}_i & \overset{\overline{f}_i}{\to} & \overline{Y}_i \end{array}$$

is a τ-well prepared diagram of R^i and $\overline{\Psi}_{i+1}$ of the form of (13.10).

5. Suppose that E is the strict transform of $\overline{\Psi}_1^{-1}(C)$ on Y_1. then $\overline{E \cdot \overline{R}_i^1(q)}$ is prepared for R^1 of type 6 for all primitive relations $\overline{R}_i^1$ associated to R^1, and $q \in U(R_i^1) \cap E$.
6. Suppose that $\gamma \subset Y$ is a curve which is prepared for R of type 6. Then the strict transform of γ on Y_1 is prepared for R^1 of type 6.

Definition 13.13. Suppose that $f : X \to Y$ is pre-τ-quasi-well prepared (or τ-well prepared or τ-very-well prepared) with relation R. Suppose that

$$\begin{array}{ccc} X_1 & \overset{f_1}{\to} & Y_1\\ \Phi \downarrow & & \downarrow \Psi\\ X & \overset{f}{\to} & Y \end{array} \tag{13.14}$$

is a commutative diagram such that there is a factorization

$$
(13.15)\qquad
\begin{array}{ccc}
X_1 = \overline{X}_m & \overset{f_1=\overline{f}_m}{\rightarrow} & \overline{Y}_m = Y_1 \\
\downarrow & & \downarrow \\
\vdots & & \vdots \\
\downarrow & & \downarrow \\
\overline{X}_2 & \overset{\overline{f}_2}{\rightarrow} & \overline{Y}_2 \\
\overline{\Phi}_2 \downarrow & & \downarrow \overline{\Psi}_2 \\
\overline{X}_1 & \overset{\overline{f}_1}{\rightarrow} & \overline{Y}_1 \\
\overline{\Phi}_1 \downarrow & & \downarrow \overline{\Psi}_1 \\
X & \overset{f}{\rightarrow} & Y
\end{array}
$$

where each commutative diagram

$$
\begin{array}{ccc}
\overline{X}_{i+1} & \rightarrow & \overline{Y}_{i+1} \\
\overline{\Phi}_{i+1} \downarrow & & \downarrow \overline{\Psi}_{i+1} \\
\overline{X}_i & \rightarrow & \overline{Y}_i
\end{array}
$$

is either of the form (13.10) or of the form (13.11). Then (13.14) is called a pre-τ-quasi-well prepared (or τ-well prepared or τ-very-well prepared) diagram of R and Ψ.

CHAPTER 14

Construction of τ-well prepared diagrams

LEMMA 14.1. *Suppose that $f : X \to Y$ is pre-τ-quasi-well prepared (or τ-well prepared or τ-very-well prepared) and $C \subset Y$ is a 2-curve. Then C is a permissible center for R, and there exists a pre-τ-quasi-well-prepared (or τ-well prepared or τ-very-well prepared) diagram (13.10) of R and the blow up $\Psi : Y_1 \to Y$ of C such that Φ is a product of blow ups of 2-curves. Furthermore,*

1. *Φ is an isomorphism over $f^{-1}(Y - C)$ and $\tau_{f_1}(p_1) \le \tau_f(\Phi(p_1))$ if $p_1 \in D_{X_1}$.*
2. *Suppose that f is τ-well prepared. Then*
   (a) *Let E be the exceptional divisor for Ψ. Suppose that $q \in U(\overline{R}_i^1) \cap E$ for some $\overline{R}_i$ associated to R. Let $\gamma_i = \overline{S_{\overline{R}_i^1}(q) \cdot E}$. Then $\gamma_i = \Psi^{-1}(\Psi(q))$ is a prepared curve for R^1 of type 6.*
   (b) *If γ is a prepared curve for R, then the strict transform of γ on Y_1 is a prepared curve for R^1.*
3. *Suppose that $\overline{q} \in C$, $p \in f^{-1}(\overline{q})$, $p' \in \Phi^{-1}(p)$, u, v, w are permissible parameters at $\overline{q}$ such that $u = v = 0$ are local equations of C, and w is good at p. If w is not good at p', then $\tau_{f_1}(p') < \tau_f(p)$.*
4. *If f is τ prepared then f_1 is τ prepared.*

PROOF. By Lemma 10.4, there exists a commutative diagram

$$\begin{array}{ccc} X_1 & \xrightarrow{f_1} & Y_1 \\ \Phi \downarrow & & \downarrow \Psi \\ X & \xrightarrow{f} & Y \end{array}$$

where Φ is a product of blow ups of 2-curves and f_1 is prepared, with the property that $\tau_{f_1}(p_1) \le \tau_f(\Phi(p_1))$ if $p_1 \in D_{X_1}$. We further have that Φ is an isomorphism over $f^{-1}(Y - C)$, and f_1 is τ-prepared if f is τ-prepared.

Let $\{\overline{R}_i\}$ be the pre-relations on Y associated to R. C is an admissible center for the $\{\overline{R}_i\}$ (Definition 12.2). Let $\{\overline{R}_i^1\}$ be the transforms of the $\{\overline{R}_i\}$ on Y_1.

By consideration of the local forms of the map f_1 given in the proof of Lemma 10.4, we see that the conditions of Definition 12.6 hold so that we can define the transform R^1 of R for f_1.

We consider one explicit case in detail. Suppose that $q_1 \in U(\overline{R}_i^1)$, and $p_1 \in f_1^{-1}(q_1) \cap \Phi^{-1}(T(R_i))$ is a 3-point. Let $q = \Psi(q_1)$, $p = \Phi(p_1)$. We have that q and q_1 are 2-points, and p is a 3-point (since Lemma 5.2 is used to construct Φ). There exist permissible parameters $u = u_{R_i(p)}, v = v_{R_i(p)}, w_i = w_{R_i(p)}$ at q such that $R_i(p) = \overline{R}_i(q)$ is determined by

$$w_i^e - \lambda u^a v^b$$

with $e = e_{R_i}(p)$, $\lambda = \lambda_{R_i}(p)$ if $a = a_{R_i}(p), b = b_{R_i}(p) \neq \infty$, and by

$$w_i = 0$$

if $a = a_{R_i}(p) = b = b_{R_i}(p) = \infty$. There exist permissible parameters x, y, z for u, v, w_i at p such that an expression (9.1) of Definition 9.1 holds for u, v, w_i and we have a relation

$$w_i^e = u^a v^b \Lambda(x, y, z)$$

where $\Lambda(x, y, z)$ is a unit series in $\hat{\mathcal{O}}_{X,p}$ with $\Lambda(0,0,0) = \lambda$ if $a, b \neq \infty$ and u, v, w_i have a monomial form in x, y, z if $a = b = \infty$. After possibly interchanging u and v, we may assume (since q_1 is a 2-point) that q_1 has permissible parameters $\overline{u}, \overline{v}, w_i$ such that

$$u = \overline{u}, v = \overline{uv}.$$

Since p_1 is a 3-point, $\hat{\mathcal{O}}_{X_1,p_1}$ has regular parameters x_1, y_1, z_1 such that

$$\begin{aligned} x &= x_1^{a_{11}} y_1^{a_{12}} z_1^{a_{13}} \\ y &= x_1^{a_{21}} y_1^{a_{22}} z_1^{a_{23}} \\ z &= x_1^{a_{31}} y_1^{a_{32}} z_1^{a_{33}} \end{aligned}$$

with $\mathrm{Det}(a_{ij}) = \pm 1$. Thus $\overline{u}, \overline{v}, w_i$ has an expression of the form of (9.1) in x_1, y_1, z_1. If $a, b \neq \infty$ we have the relation

$$w_i^e = \overline{u}^{a+b} \overline{v}^b \Lambda,$$

and $\overline{R}_i^1(q_1)$ is determined by

$$w_i^e - \overline{u}^{a+b} \overline{v}^b \lambda. \tag{14.1}$$

If $a = b = \infty$, $\overline{u}, \overline{v}, w_i$ have a monomial form in x_1, y_1, z_1.

Now we will verify that f_1 is pre-τ-quasi-well prepared.

We first verify that 1 of Definition 13.1 holds. We have that $U(\overline{R}_i^1) = \Psi^{-1}(U(R_i))$ for all i (as Ψ is the blow up of a 2-curve).

$$\begin{aligned} T(R^1) &= \cup_i \left[G_{X_1}(f_1, \tau) \cap \Phi^{-1}(T(R_i)) \cap f_1^{-1}(U(R_i^1))\right] \\ &= G_{X_1}(f_1, \tau) \cap \left[\cup_i(\Phi^{-1}(T(R_i)) \cap f_1^{-1}(U(R_i^1)))\right] \\ &= G_{X_1}(f_1, \tau) \cap \left[\cup_i \Phi^{-1}(T(R_i))\right] \\ &= G_{X_1}(f_1, \tau) \cap \left[\Phi^{-1}(T(R))\right] \\ &= G_{X_1}(f_1, \tau) \cap \left[\Phi^{-1}(f^{-1}(U(R))\right] \\ &= G_{X_1}(f_1, \tau) \cap f^{-1}(U(R)). \end{aligned}$$

From the above and the local calculations of the proof of Lemma 10.4, we see that 2, 3 and 4 of Definition 13.1 hold. To prove 3, observe that if $p \in \Phi^{-1}(T(R_i)) \cap f_1^{-1}(U(R_i^1))$ and 3 does not hold at p, then $\tau_{f_1}(p) < \tau_f(\Phi(p))$ implies $p \notin T(R_i^1)$.

It remains to verify that 5 of Definition 13.1 holds.

Suppose that $q_1 \in U(\overline{R}_i^1)$ and $p_1 \in f_1^{-1}(q_1)$. Then

$$u = u_{\overline{R}_i}(q), v = v_{\overline{R}_i}(q), w = w_{\overline{R}_i}(q)$$

are super parameters at $q = \Psi(q_1)$, and $p = \Phi(p_1) \in f^{-1}(q)$ has permissible parameters x, y, z for u, v, w such that one of the forms of Definition 10.1 hold for u, v, w and x, y, z. We will verify that 5 of Definition 13.1 holds when q_1 and q are 2-points. The other cases are similar.

After possibly interchanging u and v, we have

$$(14.2)\qquad \begin{array}{ll} u_{\overline{R}_i^1}(q_1) & = \overline{u} = u \\ v_{\overline{R}_i^1}(q_1) & = \overline{v} = \frac{v}{u} \\ w_{\overline{R}_i^1}(q_1) & = w. \end{array}$$

We can verify that there exist permissible parameters x_1, y_1, z_1 at p_1 such that $\overline{u}, \overline{v}, w$ have one of the forms 1 - 4 of Definition 10.1 in x_1, y_1, z_1. The most difficult case to verify is when p_1 is a 1-point and p is a 3-point. Then since Lemma 5.2 is used to construct Φ, $\hat{\mathcal{O}}_{X_1,p_1}$ has regular parameters $\overline{x}_1, \overline{y}_1, \overline{z}_1$ defined by

$$\begin{array}{ll} x & = \overline{x}_1^{\overline{a}}(\overline{y}_1+\overline{\alpha})^{\overline{b}}(\overline{z}_1+\overline{\beta})^{\overline{c}} \\ y & = \overline{x}_1^{\overline{d}}(\overline{y}_1+\overline{\alpha})^{\overline{e}}(\overline{z}_1+\overline{\beta})^{\overline{f}} \\ w & = \overline{x}_1^{\overline{g}}(\overline{y}_1+\overline{\alpha})^{\overline{h}}(\overline{z}_1+\overline{\beta})^{\overline{i}} \end{array}$$

where $\overline{\alpha}, \overline{\beta} \in \mathbf{k}$ are nonzero and

$$\mathrm{Det}\begin{pmatrix} \overline{a} & \overline{b} & \overline{c} \\ \overline{d} & \overline{e} & \overline{f} \\ \overline{g} & \overline{h} & \overline{i} \end{pmatrix} = \pm 1.$$

We substitute into

$$\begin{array}{ll} u & = x^a y^b z^c \\ v & = x^d y^e z^f \\ w & = x^g y^h z^i \gamma + x^j y^k z^l \end{array}$$

of 4 of Definition 10.1. Using the fact that

$$\mathrm{rank}\begin{pmatrix} a & b & c \\ d & e & f \\ j & k & l \end{pmatrix} = 3,$$

we can make a change of variables in $\overline{x}_1, \overline{y}_1, \overline{z}_1$ to get permissible parameters x_1, y_1, z_1 at p_1 satisfying

$$(14.3)\qquad \begin{array}{ll} u & = x_1^{\overline{a}} \\ v & = x_1^{\overline{d}}(\overline{\gamma}+y_1) \\ x^j y^k z^l & = x_1^{\overline{g}}(\overline{\epsilon}+z_1) \end{array}$$

with $\overline{d} > \overline{a}$ and $0 \neq \overline{\epsilon}, \overline{\gamma} \in \mathbf{k}$.

For each of the monomials M in the series $x^g y^h z^i \gamma$ we have a relation

$$(14.4)\qquad M^{\tilde{e}} = u^{\tilde{a}} v^{\tilde{b}}$$

with $\tilde{e}, \tilde{a}, \tilde{b} \in \mathbf{Z}$. On substitution of (14.2) and (14.3) into (14.4) we see that

$$M = x_1^{e'}\phi_1(y_1)$$

where ϕ_1 is a unit series. Thus $\overline{u}, \overline{v}, w$ have an expansion of the form 1 of Definition 10.1 in terms of x_1, y_1, z_1.

The other cases can be verified by a similar but simpler argument to show that $\overline{u}, \overline{v}, w$ are super parameters at q_1. Thus 5 of Definition 13.1 holds for f_1, so that f_1 is pre-τ-quasi-well prepared.

Suppose that f is τ-well prepared. We will verify that f_1 is τ-well prepared. 1 and 2 of Definition 13.5 are immediate. We must verify that 3 and 4 of Definition 13.5 hold for f_1.

Suppose that $q_1 \in U(\overline{R}_i^1) \cap U(\overline{R}_j^1)$. Let $q = \Phi(q_1) \in U(\overline{R}_i) \cap U(\overline{R}_j)$, and

$$\begin{aligned} u &= u_{\overline{R}_i}(q) = u_{\overline{R}_j}(q), \\ v &= v_{\overline{R}_i}(q) = v_{\overline{R}_j}(q), \\ w_i &= w_{\overline{R}_i}(q), w_j = w_{\overline{R}_j}(q). \end{aligned}$$

We will assume that q and q_1 are 2-points, which is the most difficult case.

We have a relation

$$w_j = w_i + u^{a_{ij}} v^{b_{ij}} \phi_{ij}(u, v), \tag{14.5}$$

from (13.7) for f. After possibly interchanging u and v we have permissible parameters

$$\begin{aligned} \overline{u} &= u_{\overline{R}_i^1}(q_1) = u_{\overline{R}_j^1}(q_1) \\ \overline{v} &= v_{\overline{R}_i^1}(q_1) = v_{\overline{R}_j^1}(q_1) \\ w_i &= w_{\overline{R}_i^1}(q_1), w_j = w_{\overline{R}_j^1}(q_1) \end{aligned}$$

at q_1, where $u = \overline{u}$, $v = \overline{uv}$. We have

$$w_j = w_i + \overline{u}^{a_{ij}+b_{ij}} \overline{v}^{b_{ij}} \phi_{ij}(\overline{u}, \overline{uv}) \tag{14.6}$$

so 3 of Definition 13.5 holds for f_1.

Since the set (13.9) of Definition 13.5 is totally ordered for $q \in U(R)$, it follows from (14.6) that the corresponding set (13.9) for $q_1 \in U(R^1)$ is totally ordered. Thus 4 of Definition 13.5 holds for f_1 and R^1 and f_1 is τ-well prepared.

We now verify 2 of Lemma 14.1. Let $E = \Psi^{-1}(C)$ be the exceptional divisor of Ψ. Suppose that $q \in U(\overline{R}_i^1) \cap E$. Let $\tilde{q} = \Psi(q)$. There exist permissible parameters

$$u = u_{\overline{R}_i(\tilde{q})}, v = v_{\overline{R}_i(\tilde{q})}, w_i = w_{\overline{R}_i(\tilde{q})}$$

at $\tilde{q}$ such that $u = v = 0$ are local equations of C, and after possibly interchanging u and v,

$$u = u_{\overline{R}_i^1(q)}, v = u_{\overline{R}_i^1(q)}(v_{\overline{R}_i^1(q)} + \alpha), w_i = w_{\overline{R}_i^1(q)}$$

for some $\alpha \in \mathbf{k}$.

Since $\overline{u}$ is a local equation of E at q, $\gamma_i = \overline{S_{\overline{R}_i^1}(q) \cdot E} = \Psi^{-1}(q)$. Let $\gamma = \gamma_i$.

We will verify that γ_i is prepared for R^1 of type 6. Since $\tilde{q} \in \Omega(\overline{R}_i)$, C intersects $\Omega(\overline{R}_i)$ transversally at q (and possibly a finite number of other points), and $\Omega(\overline{R}_i^1)$ is the strict transform of $\Omega(\overline{R}_i)$ by Ψ, we have that $\gamma_i \subset \Omega(\overline{R}_i^1)$. Thus 2 of Definition 13.8 holds. Suppose that for some j, component E_α of D_{Y_1} and $q_\beta \in U(\overline{R}_j^1)$,

$$\gamma' = \overline{E_\alpha \cdot S_{\overline{R}_j^1}(q_\beta))} \subset \Omega(\overline{R}_j^1),$$

$\gamma \neq \gamma'$ and there exists $\overline{q} \in \gamma \cap \gamma'$. Let $E_1 = \Psi(E_\alpha)$, a component of D_Y. Then $\overline{\gamma} = \Psi(\gamma')$ is a curve on D_Y through $\tilde{q}$. Since $\gamma' \subset \Omega(\overline{R}_j^1)$, we must have $\overline{\gamma} \subset \Omega(\overline{R}_j) \cap E_1$, so that $\tilde{q} \in U(R) \cap \overline{\gamma} = U(\overline{R}_j) \cap \overline{\gamma}$, and thus $\overline{\gamma} = \overline{E_1 \cdot S_{\overline{R}_j}(q)}$. We thus have that E_α is the strict transform of E_1, $\overline{q} \in U(\overline{R}_i^1) \cap U(\overline{R}_j^1)$ and $\gamma' = \overline{E_\alpha \cdot S_{\overline{R}_j^1}(\overline{q})}$. Thus 3 of Definition 13.8 holds.

Suppose that $\gamma' = \overline{E_\alpha \cdot S_{\overline{R}_j^1}(q_\delta)}$ and $\gamma = \gamma'$. Then we must have $\gamma' = \overline{E \cdot S_{\overline{R}_j^1}(q)}$ and 4 of Definition 13.8 holds.

Since for j such that $\tilde{q} \in U(\overline{R}_j)$, $U(\overline{R}_j^1)$ contains $\gamma_j = \Psi^{-1}(q) \cap E$, 5 of Definition 13.8 holds. Thus $\gamma = \gamma_i$ is prepared for R^1 of type 6.

Suppose that $\gamma \neq C$ is a prepared curve for R on Y. The verification that the strict transform γ' of γ on Y_1 is prepared for f_1 follows from a local calculation at points of $C \cap \gamma$.

3 of the conclusions of Lemma 14.1 follows from a computation of $\tau_{f_1}(p)$ and $\tau_f(\Phi(p))$ for $p \in D_{X_1}$ using local forms of the mappings Φ and Ψ.

Suppose that f is τ-very-well prepared. We have seen that f_1 is τ-well prepared, so that 1 of Definition 13.9 holds for f_1, and 2 of Definition 13.9 holds for f_1 by our verification of 2 of this lemma. Let $\overline{\Psi}_i : \Omega(\overline{R}_i^1) \to \Omega(\overline{R}_i)$ be the restriction of Ψ to $\Omega(\overline{R}_i^1)$. Then $\overline{\Psi}_i$ is the blow up of the union of nonsingular points $C \cdot \Omega(\overline{R}_i)$ on the nonsingular surface $\Omega(\overline{R}_i)$. Thus since

$$F_i = \sum_{\gamma \in V_i(Y)} \gamma$$

is a SNC divisor on $\Omega(\overline{R}_i)$ whose intersection graph is a forest,

$$\overline{\Psi}_i^{-1}(F_i) = \sum_{\gamma' \in V_i(Y_1)} \gamma'$$

is a SNC divisor on $\Omega(\overline{R}_i^1)$ whose intersection graph is a forest and 3 of Definition 13.9 holds for f_1. Thus f_1 is τ-very-well prepared. $\square$

The proofs of Remarks 14.2 and 14.3 follow easily from the methods of the proof of Lemma 14.1 and Lemma 10.2.

REMARK 14.2. *Suppose that $f : X \to Y$ is pre-τ-quasi-well prepared (or τ-well prepared or τ-very-well prepared) and $C \subset D_X$ is a 2-curve or a 3-point. Let $\Phi_1 : X_1 \to X$ be the blow up of C, $f_1 = f \circ \Phi_1 : X_1 \to Y$. Then*

1. *f_1 is pre-τ-quasi-well prepared (or τ-well prepared or τ-very well prepared).*
2. *Suppose that $p_1 \in X_1$, $p = \Phi_1(p_1)$, $q = f_1(p)$, u, v, w are permissible parameters at q such that w is good (weakly good) at p for f. If w is not good (not weakly good) at p_1 for f, then $\tau_{f_1}(p_1) < \tau_f(p)$.*
3. *$(f \circ \Phi_1)^{-1}(\Theta(f, Y)) \subset \Theta(f_1, Y_1)$.*

REMARK 14.3. *The proof of Lemma 14.1 shows that if $f : X \to Y$ is pre-τ-quasi-well prepared (or τ-well prepared or τ-very-well prepared), $C \subset Y$ is a 2-curve, $\Psi : Y_1 \to Y$ is the blow up of C and $\Phi : X_1 \to X$ is a sequence of blow ups of 2-curves and 3-points such that the rational map $f_1 : X_1 \to Y_1$ is a morphism, then*

$$\begin{array}{ccc} X_1 & \overset{f_1}{\to} & Y_1 \\ \Phi \downarrow & & \downarrow \Psi \\ X & \overset{f}{\to} & Y \end{array}$$

is pre-τ-quasi-well prepared (or τ-well prepared or τ-very-well prepared) for R and Ψ. In fact, with the above notation, if f satisfies 1 – 3 of Definition 13.1, then f_1 satisfies 1 – 3 of Definition 13.1. Further, 2 and 3 of the conclusions of Lemma 14.1 hold.

THEOREM 14.4. *Suppose that $f : X \to Y$ is prepared, pre-τ-quasi-well prepared and $\tau = \tau_f(X) \ge 1$. Then there exists a pre-τ-quasi-well prepared diagram*

$$\begin{array}{ccc} X_1 & \stackrel{f_1}{\to} & Y_1 \\ \Phi\downarrow & & \downarrow\Psi \\ X & \stackrel{f}{\to} & Y \end{array} \tag{14.7}$$

such that f_1 is pre-τ-well prepared, Φ, Ψ are products of blowups of 2-curves, and $G_{Y_1}(f_1,\tau)$ contains no 3-points and no 2-curves, so that f_1 is τ-prepared.

PROOF. By Lemma 7.1 and Lemma 14.1, there exists a diagram (14.7) such that f_1 is pre-τ-quasi-well prepared, and $G_{Y_1}(f,\tau)$ contains no 3-points. We may thus assume that $G_Y(f,\tau)$ contains no 3-points.

Suppose that C is a 2-curve of Y such that $C \subset G_Y(f,\tau)$. Let H be a general hyperplane section of Y. H intersects C transversally in general points of C, and $f^*(H)$ is nonsingular and makes SNCs with D_X by Bertini's theorem. Thus at a general point q of C there exist uniformizing parameters u, v, w such that for $p \in f^{-1}(q)$, u, v, w have a form 2 (b) or 2 (c) of Definition 4.6 at p. Let $\Psi_1 : Y_1 \to Y$ be the blow up of C, and

$$\begin{array}{ccc} X_1 & \stackrel{f_1}{\to} & Y_1 \\ \Phi_1\downarrow & & \downarrow\Psi_1 \\ X & \stackrel{f}{\to} & Y \end{array}$$

bet the pre-τ-quasi-well prepared diagram of Lemma 14.1. Φ_1 is a product of blow ups of 2-curves and from the local forms 2 (b) and 2 (c) of Definition 4.6, we see that Φ_1 is an isomorphism over a general point of C. Since C contains no 3-points of Y, we see that $G_{Y_1}(f_1,\tau)$ contains no 3-points. We can iterate this construction to produce a diagram (14.7) satisfying the conclusions of the theorem.

□

LEMMA 14.5. *Suppose that $f : X \to Y$ is pre-τ-quasi-well prepared (or τ-well prepared), $\overline{q} \in Y$ is a 1-point such that $\overline{q} \notin U(R)$, and $C \subset D_Y$ is a resolving curve for f at $\overline{q}$. Then there exists a pre-τ-quasi-well prepared (τ-well prepared) diagram*

$$\begin{array}{ccc} X_1 & \stackrel{f_1}{\to} & Y_1 \\ \Phi_1\downarrow & & \downarrow\Psi_1 \\ X & \stackrel{f}{\to} & Y \end{array}$$

such that

1. *Ψ_1 is the blow up of C,*
2. *Φ_1 is a sequence of blow ups of 2-curves and 3-points.*
3. *$\tau_{f_1}(p_1) \le \tau_f(\Phi_1(p_1))$ for $p_1 \in X_1$. Thus $\tau_{f_1}(p_1) < \tau$ if $p_1 \in (\Psi_1 \circ f_1)^{-1}(C - \{\overline{q}\})$.*
4. *Suppose that $C_1 \subset Y_1$ is a section over C, $q \in C$ is a 1-point, u, v, w are permissible parameters at q such that $u = v = 0$ are local equations of C, u, v are toroidal forms at p for all $p \in f^{-1}(q)$, and $q_1 \in C_1 \cap \Psi_1^{-1}(q)$ is a 1-point. Then there exist permissible parameters $\overline{u}_1, \overline{v}_1, w$ at q_1 such that $\overline{u}_1, \overline{v}_1$ are torodial forms at p_1 for all $p_1 \in f_1^{-1}(q_1)$, and $\overline{u}_1 = \overline{v}_1 = 0$ is a local equation of C_1.*

5. *Suppose that* $\overline{p} \in f^{-1}(\overline{q})$, $p' \in \Phi_1^{-1}(\overline{p})$, u, v, w *are permissible parameters at* $\overline{q}$ *such that* $u = v = 0$ *are local equations of* C, *and* w *is good at* $\overline{p}$ *for* f. *If* w *is not good at* p' *for* f_1, *then* $\tau_{f_1}(p') < \tau_f(\overline{p})$.
6. *If* f *is* τ*-prepared then* f_1 *is* τ*-prepared.*

PROOF. Suppose that $q \in C$ is a 1-point. Then there exist permissible parameters u, v, w at q such that $u = v = 0$ are local equations of C, and if $p \in f^{-1}(q)$, then u, v are toroidal forms at p. Thus there exist permissible parameters x, y, z at p for u, v, w such that one of the forms (4.5) or (4.6) hold.

If $q \in C$ is a 2-point, then there exist super parameters u, v, w at q such that $u = w = 0$ are local equations of C. Thus if $p \in f^{-1}(q)$, then there exist permissible parameters x, y, z at p for u, v, w such that one of the forms (10.1) - (10.4) hold.

By Lemma 5.3 and Remark 14.2 (a detailed proof of a similar statement is given in the proof of Lemma 14.7), after blowing up 2-curves and 3-points by a morphism $\Phi_0 : X_0 \to X$ we obtain that if $q \in C$ is a 2-point, u, v, w are the above permissible parameters at q, and $p \in (f \circ \Phi_0)^{-1}(q)$, then $(x^a y^b, x^e y^f, x^g y^h)$ is a principal ideal if (10.2) holds at p, $((x^a y^b)^k, (x^a y^b)^l, x^c y^d)$ is principal if (10.3) holds at p, $(x^a y^b z^c, x^g y^h z^i, z^j y^k z^l)$ is principal if (10.4) holds at p. We further have, by Lemma 10.2, that $f \circ \Phi_0$ is pre-τ-quasi-well prepared (τ-well prepared),

$$\tau_{f \circ \Phi_0}(p) \leq \tau_f(\Phi_0(p))$$

for $p \in D_{X_0}$ and C is a resolving curve for $f \circ \Phi_0$ at $\overline{q}$.

We now analyze the points $p \in X_0$ where $\mathcal{I}_C \mathcal{O}_{X_0}$ is not principal.

First suppose that $q \in C$ is a 1-point, $p \in X_0$, and $f \circ \Phi_0(p) = q$.

If p is a 1-point then we have an expression

$$u = x^a, v = y$$

and $u = v = 0$ are local equations of C. The non principal locus has local equations $x = y = 0$

If p is a 2-point then

$$u = (x^a y^b)^k, v = z$$

and $u = v = 0$ are local equations of C. The non principal locus has local equations $\{x = z = 0\} \cup \{y = z = 0\}$

Now suppose that $q \in C$ is a 2-point and $f \circ \Phi_0(p) = q$.

If p is a 1-point of the form (10.1), then

$$\begin{aligned} u &= x^a \\ v &= x^b(\alpha + y) \\ w &= x^c \gamma(x, y) + x^d(z + \beta). \end{aligned}$$

$u = w = 0$ are local equations of C. If p is in the non principal locus then we have (after possibly making a change of variables in z) $w = x^d z$ with $d < a$ and $x = z = 0$ are local equations of the non principal locus.

If p is a 2-point of the form (10.2), then

$$\begin{aligned} u &= x^a y^b \\ v &= x^c y^d \\ w &= x^e y^f \gamma(x, y) + x^g y^h(z + \beta). \end{aligned}$$

$u = w = 0$ are local equations of C. If p is in the non principal locus then (after possibly making a change of variables in z), we have $w = x^g y^h z$ with $(g, h) < (a, b)$.

Local equations of the non principal locus are $x = z = 0$ (if $g < a$, $h = b$), $y = z = 0$ (if $g = a$, $h < b$) and $\{x = z = 0\} \cup \{y = z = 0\}$ if $g < a$ and $h < b$.

If p is a 2-point of the form (10.3), then

$$\begin{aligned} u &= (x^a y^b)^k \\ v &= (x^a y^b)^t(\alpha + z) \\ w &= (x^a y^b)^l \gamma(x^a y^b, z) + x^c y^d \end{aligned}$$

$u = w = 0$ are local equations of C, and p is in the principal locus.

If p is a 3-point, of the form (10.4), then

$$\begin{aligned} u &= x^a y^b z^c \\ v &= x^d y^e z^f \\ w &= x^g y^h z^i \gamma + x^j y^k z^l \end{aligned}$$

$u = w = 0$ are local equations of C, and p is in the principal locus.

We see that the non principal locus of $\mathcal{I}_C \mathcal{O}_{X_0}$ is a union of nonsingular curves which are possible centers and are not 2-curves.

Let $U_0 \subset X_0$ be the largest open set on which the rational map $\overline{f}_0 = \Psi_1^{-1} \circ f \circ \Phi_0 : X_0 \to Y_1$ is a morphism. We will now show that $\overline{f}_0 : U_0 \to Y_1$ is prepared, and $\tau_{\overline{f}_0}(p) \le \tau_f(\Phi_0(p))$ for $p \in U_0$.

Suppose that $p \in U_0 \cap (f \circ \Phi_0)^{-1}(C)$. Then $(f \circ \Phi_0)(p) = q$ is a 2-point, and there exist super parameters u, v, w at q such that $u = w = 0$ are local equations of C, and one of the forms (10.1) - (10.4) hold for u, v, w at p. Let $q_1 = \overline{f}_0(p)$. We have permissible parameters u_1, v, w_1 in $\mathcal{O}_{Y_1, q_1}$ such that either

$$u = u_1, w = u_1(w_1 + \delta)$$

with $\delta \in \mathbf{k}$, or

$$u = u_1 w_1, w = w_1.$$

The most difficult case to analyze is when p is a 3-point, (so that u, v, w satisfy (10.4)), $\tau_{f \circ \Phi_0}(p) \ge 1$, and $u = u_1 w_1, w = w_1$. We will work out this case in detail. Since q is a 2-point, we will then have that

$$\tau_{\overline{f}_0}(p) \le \tau_{f \circ \Phi_0}(p) \le \tau_f(\Phi_0(p)) < \tau.$$

In this case q_1 is a 3-point. With the notation of (9.1),

$$u = x^a y^b z^c, v = x^d y^e z^f, w = \sum_{i \ge 0} \alpha_i x^{a_i} y^{b_i} z^{c_i} + x^g y^h z^i$$

with $0 \ne \alpha_i \in \mathbf{k}$ for all i. Set

$$\gamma = \sum \alpha_i x^{a_i - a_0} y^{b_i - b_0} z^{c_i - c_0} + x^{g - a_0} y^{h - a_0} z^{i - a_0}.$$

$\gamma \in \hat{\mathcal{O}}_{X,p}$ is a unit series, and

$$w = x^{a_0} y^{b_0} z^{c_0} \gamma,$$

with $(a_0, b_0, c_0) < (a, b, c)$.

Suppose that (a, b, c) and (a_0, b_0, c_0) are linearly independent.

After possibly interchanging x, y, z, we may assume that $a_0 b - a b_0 \ne 0$. There exist $\lambda_1, \lambda_2 \in \mathbf{Q}$ such that if we set

(14.8) $$x = \overline{x} \gamma^{\lambda_1}, y = \overline{y} \gamma^{\lambda_2},$$

then

$$\begin{aligned} w &= \overline{x}^{a_0}\overline{y}^{b_0}z^{c_0} \\ u &= \overline{x}^{a}\overline{y}^{b}z^{c} \\ v &= \overline{x}^{d}\overline{y}^{e}z^{f}\gamma^{\lambda} \end{aligned}$$

for some $0 \neq \lambda \in \mathbf{Q}$.

There exists a series $P(x,y,z)$ where the monomials in x,y,z in P with nonzero coefficients are monomials in $x^{a_i-a_0}y^{b_i-b_0}z^{c_i-c_0}$ for $i \geq 0$, such that

$$\gamma^{\lambda} = P + x^{g-a_0}y^{h-b_0}z^{i-c_0}\Omega. \tag{14.9}$$

Iterating, by substituting (14.8) into successive iterations of (14.9), we see that there exists a series $\overline{P}(\overline{x},\overline{y},\overline{z})$, where the monomials in $\overline{x},\overline{y},\overline{z}$ in $\overline{P}$ with nonzero coefficients are monomials in $\overline{x}^{a_i-a_0}\overline{y}^{b_i-b_0}\overline{z}^{c_i-c_0}$ for $i \geq 0$ such that

$$\gamma^{\lambda} = \overline{P} + \overline{x}^{g-a_0}\overline{y}^{h-b_0}z^{i-c_0}\overline{\Omega}.$$

Comparing the Jacobian determinants

$$\mathrm{Det}\left(\frac{\partial(u,v,w)}{\partial(x,y,z)}\right)$$

and

$$\mathrm{Det}\left(\frac{\partial(w,u,v)}{\partial(\overline{x},\overline{y},z)}\right),$$

we see that $\overline{\Omega}$ is a unit series.

There exist rational numbers β_1,β_2,β_3 such that we can make a formal change of variables, setting

$$\tilde{x} = \overline{x}\overline{\Omega}^{\beta_1}, \tilde{y} = \overline{y}\overline{\Omega}^{\beta_2}, \tilde{z} = z\overline{\Omega}^{\beta_3}$$

to get an expression

$$\begin{aligned} w &= \tilde{x}^{a_0}\tilde{y}^{b_0}\tilde{c}^{c_0} \\ u &= \tilde{x}^{a}\tilde{y}^{b}\tilde{z}^{c} \\ v &= \tilde{x}^{d}\tilde{y}^{e}\tilde{z}^{f}\left(\overline{P}(\tilde{x},\tilde{y},\tilde{z}) + \tilde{x}^{g-a_0}\tilde{y}^{h-b_0}\tilde{z}^{i-c_0}\right). \end{aligned}$$

Thus

$$\begin{aligned} w_1 &= \tilde{x}^{a_0}\tilde{y}^{b_0}\tilde{c}^{c_0} \\ u_1 &= \tilde{x}^{a-a_0}\tilde{y}^{b-b_0}\tilde{z}^{c-c_0} \\ v &= \tilde{x}^{d}\tilde{y}^{e}\tilde{z}^{f}\left(\overline{P}(\tilde{x},\tilde{y},\tilde{z}) + \tilde{x}^{g-a_0}\tilde{y}^{h-b_0}\tilde{z}^{i-c_0}\right). \end{aligned}$$

is an expression of the form of (9.1).

We see that $\overline{f}_0$ is prepared at p, and

$$\begin{aligned} \tau_{\overline{f}_0}(p) &\leq |((a_0,b_0,c_0)\mathbf{Z} + (a-a_0,b-b_0,c-c_0)\mathbf{Z} + (d,e,f)\mathbf{Z} \\ &\quad + \textstyle\sum_{i\geq 1}(a_i-a_0,b_i-b_0,c_i-c_0)\mathbf{Z}) \\ &\quad /((a_0,b_0,c_0) + (a-a_0,b-b_0,c-c_0)\mathbf{Z} + (d,e,f)\mathbf{Z})| \\ &= |((a,b,c)\mathbf{Z} + (d,e,f)\mathbf{Z} + \textstyle\sum_{i\geq 0}(a_i,b_i,c_i)\mathbf{Z}) \\ &\quad /((a,b,c)\mathbf{Z} + (d,e,f)\mathbf{Z} + (a_0,b_0,c_0)\mathbf{Z})| \\ &= \tau_{f\circ\Phi_0}(p) \leq \tau_f(\Phi_0(p). \end{aligned}$$

Suppose that (a,b,c) and (a_0,b_0,c_0) are linearly dependent.

Then (d,e,f) and (a_0,b_0,c_0) are linearly independent. After possibly interchanging x,y,z, we may assume that $db_0 - ea_0 \neq 0$. There exist $\lambda_1,\lambda_2 \in \mathbf{Q}$ such that if

$$x = \overline{x}\gamma^{\lambda_1}, y = \overline{y}\gamma^{\lambda_2},$$

then

$$\begin{aligned} w &= \overline{x}^{a_0}\overline{y}^{b_0}z^{c_0} \\ v &= \overline{x}^{d}\overline{y}^{e}z^{f} \\ u &= \overline{x}^{a}\overline{y}^{b}z^{c}\gamma^{\lambda} \end{aligned}$$

for some $0 \neq \lambda \in \mathbf{Q}$.

As in the case when (a, b, c) and (a_0, b_0, c_0) are linearly independent, we can make a change of variables to get an expression where the monomials in $\tilde{x}, \tilde{y}, \tilde{z}$ in $\tilde{P}$ with nonzero coefficients are monomials in $\tilde{x}^{a_i-a_0}\tilde{y}^{b_i-b_0}\tilde{z}^{c_i-c_0}$ for $i \geq 0$.

$$\begin{aligned} w &= \tilde{x}^{a_0}\tilde{y}^{b_0}\tilde{c}^{c_0} \\ v &= \tilde{x}^{d}\tilde{y}^{e}\tilde{z}^{f} \\ u &= \tilde{x}^{a}\tilde{y}^{b}\tilde{z}^{c}\left(\overline{P}(\tilde{x},\tilde{y},\tilde{z}) + \tilde{x}^{g-a_0}\tilde{y}^{h-b_0}\tilde{z}^{i-c_0}\right). \end{aligned}$$

Thus

$$\begin{aligned} w_1 &= \tilde{x}^{a_0}\tilde{y}^{b_0}\tilde{c}^{c_0} \\ v &= \tilde{x}^{d}\tilde{y}^{e}\tilde{z}^{f} \\ u_1 &= \tilde{x}^{a-a_0}\tilde{y}^{b-b_0}\tilde{z}^{c-c_0}\left(\overline{P}(\tilde{x},\tilde{y},\tilde{z}) + \tilde{x}^{g-a_0}\tilde{y}^{h-b_0}\tilde{z}^{i-c_0}\right). \end{aligned}$$

is an expression of the form of (9.1).

We see that $\overline{f}_0$ is prepared at p, and

$$\begin{aligned} \tau_{\overline{f}_0}(p) &\leq |((a_0,b_0,c_0)\mathbf{Z} + (d,e,f)\mathbf{Z} \\ &\quad +(a-a_0,b-b_0,c-c_0)\mathbf{Z} + \textstyle\sum_{i\geq 1}(a_i-a_0,b_i-b_0,c_i-c_0)\mathbf{Z}) \\ &\quad /((a_0,b_0,c_0)+(d,e,f)\mathbf{Z} + (a-a_0,b-b_0,c-c_0)\mathbf{Z})| \\ &= \tau_{f\circ\Phi_0}(p) \leq \tau_f(\Phi_0(p)). \end{aligned}$$

We conclude that $\overline{f}_0 : U_0 \rightarrow Y_1$ is prepared, and $\tau_{\overline{f}_0}(p) \leq \tau_f(\Phi_0(p))$ for $p \in U_0$. Thus $\overline{f}_0$ is pre-τ-quasi-well prepared (τ-well prepared), as $C \cap U(R) = \emptyset$.

Let $Z_0 = X_0$.

We now construct a sequence of morphisms $\Lambda_i : Z_i \rightarrow Z_{i-1}$ which are the blow up of a possible curve C_{i-1} contained in the locus where $\mathcal{I}_C\mathcal{O}_{Z_{i-1}}$ is not invertible.

Let $h_i : Z_i \rightarrow Y_1$ be the rational map $h_i = \overline{f}_0 \circ \Lambda_1 \circ \cdots \circ \Lambda_i$. We will verify by induction that:

A. For all p_1 in the locus where h_i is a morphism, h_i is τ-prepared and

$$\tau_{h_i}(p_1) \leq \tau_f(\Phi_0 \circ \Lambda_1 \circ \Lambda_2 \circ \cdots \circ \Lambda_i(p_1)).$$

B. Suppose that $p_1 \in Z_i$ is a point where $\mathcal{I}_C\mathcal{O}_{Z_i,p_1}$ is not principal. Let $p = \Lambda_1 \circ \cdots \circ \Lambda_i(p_1)$, $q = f \circ \Phi_0(p)$. Then there exist permissible parameters u, v, w at q, permissible parameters x, y, z in $\hat{\mathcal{O}}_{Z_0,p}$ for u, v and regular parameters x_1, y_1, z_1 in $\hat{\mathcal{O}}_{Z_i,p_1}$ such that we have one of the following forms:

1. q a 1-point, p a 1-point, p_1 a 1-point. We have an expression

$$u = x^a, v = z, w = \sum_{i<n, a_{ij}\neq 0} a_{ij}x^i z^j + x^n(y+\beta)$$

in $\hat{\mathcal{O}}_{Z_0,p}$, where $u = v = 0$ are local equations of C.

We have permissible parameters x_1, y_1, z_1 at p_1 such that

$$x = x_1, y = y_1, z = x_1^b z_1$$

with $b < a$, and

$$
\begin{aligned}
u &= x_1^a \\
v &= x_1^b z_1 \\
w &= \sum a_{ij} x_1^{i+bj} z_1^j + x_1^n(y_1+\beta).
\end{aligned}
\tag{14.10}
$$

2. q a 1-point, p a 2-point. We have an expression

$$u = (x^a y^b)^k, v = z, w = \sum_{i,l\geq 0} a_{il} z^l (x^a y^b)^i + x^e y^f$$

in $\hat{O}_{Z_0,p}$, where the sum is over i, l such that $(ia, ib) \not\geq (e, f)$, $af - eb \neq 0$ and $u = v = 0$ are local equations of C.
We have permissible parameters x_1, y_1, z_1 at p_1 such that

$$x = x_1, y = y_1, z = x_1^c y_1^d z_1$$

with $(c, d) < (ak, bk)$, and

$$
\begin{aligned}
u &= (x_1^a y_1^b)^k \\
v &= x_1^c y_1^d z_1 \\
w &= \textstyle\sum_{i,l} a_{il} z_1^l x_1^{ai+cl} y_1^{bi+dl} + x_1^e y_1^f.
\end{aligned}
\tag{14.11}
$$

3. q a 2-point, p a 1-point, p_1 a 1-point. We have an expression

$$
\begin{aligned}
u &= x_1^a \\
v &= x_1^b(\beta + y_1) \\
w &= x_1^d z_1
\end{aligned}
\tag{14.12}
$$

in $\hat{O}_{Z_i,p_i}$, with $d < a$, $u = w = 0$ a local equation of C

4. q a 2-point, p a 2-point, p_1 a 2-point. We have an expression

$$
\begin{aligned}
u &= x_1^a y_1^b \\
v &= x_1^c y_1^d \\
w &= x_1^g y_1^h z_1
\end{aligned}
\tag{14.13}
$$

with $(g, h) < (a, b)$, $u = w = 0$ are local equation of C.

We have verified that A and B are true for $h_0 = \overline{f}_0$. Suppose that A and B are true for h_i. We will verify that A and B are true for h_{i+1}.

We may suppose that $p_1 \in C_i$ (recall that C_i is the curve blown up by Λ_{i+1}). Then C_i has local equations $x_1 = z_1 = 0$ if p_1 satisfies (14.10). C_i has local equations $x_1 = z_1 = 0$ (or $y_1 = z_1 = 0$) in (14.11). C_i has local equations $x_1 = z_1 = 0$ in (14.12). C_i has local equations $x_1 = z_1 = 0$ (or $y_1 = z_1 = 0$) in (14.13).

After possibly interchanging x_1 and y_1, we may assume that $x_1 = z_1 = 0$ is a local equation of C_i.

Suppose that $p_2 \in \Lambda_{i+1}^{-1}(p_1)$. Then $\hat{O}_{Z_{i+1},p_2}$ has regular parameters x_2, y_2, z_2 such that

$$x_1 = x_2, y_1 = y_2, z_1 = x_2(z_2 + \alpha) \tag{14.14}$$

with $\alpha \in \mathbf{k}$, or

$$x_1 = x_2 z_2, y_1 = y_2, z_1 = z_2. \tag{14.15}$$

Suppose that (14.10) holds. Then

$$\tau_{f\circ\Phi_0}(p) = |[\sum_{a_{ij}\neq 0} i\mathbf{Z} + a\mathbf{Z}]/a\mathbf{Z}|.$$

Suppose that (14.10) and (14.14) hold. Then

$$(14.16)\qquad \begin{aligned} u &= x_2^a\\ v &= x_2^{b+1}(z_2+\alpha)\\ w &= \sum a_{ij}x_2^{i+(b+1)j}(z_2+\alpha)^j + x_2^n(y_2+\beta). \end{aligned}$$

Suppose that $a = b+1$ in (14.16). Then $Z_{i+1}\rightarrow Y_1$ is a morphism at p_2. Let $q_1 = h_{i+1}(p_2)$. There exist permissible parameters u_1, v_1, w at q_1 defined by

$$u = u_1, v = u_1(v_1+\alpha).$$

We have an expression

$$(14.17)\qquad \begin{aligned} u_1 &= x_2^a\\ v_1 &= z_2\\ w &= \sum a_{ij}x_2^{i+aj}(z_2+\alpha)^j + x_2^n(y_2+\beta), \end{aligned}$$

of type (4.5) so that h_{i+1} is prepared at p_2. We have by (9.11) that

$$\tau_{h_{i+1}}(p_2) \le |(a\mathbf{Z} + \sum_{a_{ij}\neq 0}[i+aj]\mathbf{Z})/a\mathbf{Z}| = \tau_{f\circ\Phi_0}(p) \le \tau_f(\Phi_0(p)).$$

Suppose that $b+1 < a$ and $\alpha \neq 0$ in (14.16). Then $Z_{i+1}\rightarrow Y_1$ is a morphism at p_2. Let $q_1 = h_{i+1}(p_2)$. There exist permissible parameters u_1, v_1, w at q_1 defined by

$$u = u_1v_1, v = v_1.$$

There exist regular parameters $\overline{x}_2, \overline{y}_2, \overline{z}_2$ in $\hat{\mathcal{O}}_{Z_{i+1},p_2}$ such that

$$(14.18)\qquad \begin{aligned} u_1 &= x_2^{a-b-1}(z_2+\alpha)^{-1} = \overline{x}_2^{a-b-1}\\ v_1 &= x_2^{b+1}(z_2+\alpha) = \overline{x}_2^{b+1}(\overline{z}_2+\overline{\alpha})\\ w &= \textstyle\sum_s x_2^s(\sum_{i+(b+1)j=s} a_{ij}(z_2+\alpha)^j) + x_2^n(y_2+\beta)\\ &= \textstyle\sum_s \overline{x}_2^s(\overline{z}_2+\alpha)^{\frac{s}{a}}\left[\sum_j a_{ij}(\overline{z}_2+\alpha)^{j(\frac{a-b-1}{a})}\right] + \overline{x}_2^n(\overline{y}_2+\overline{\beta}). \end{aligned}$$

of type (4.1), so that h_{i+1} is prepared at p_2.
We observe that

$$\sum_{i+(b+1)j=s} a_{ij}(z_2+\alpha)^j \neq 0$$

if and only if some $a_{ij}\neq 0$ with $i+(b+1)j = s$. To show this, observe that

$$x^s[\sum_{i+(b+1)j=s} a_{ij}\left(\frac{y}{x^{b+1}}\right)^j] = \sum_{i+(b+1)j=s} a_{ij}x^iy^j.$$

We calculate from (9.9),

$$\begin{aligned}\tau_{h_{i+1}}(p_2) &= |[(a-b-1)\mathbf{Z}+(b+1)\mathbf{Z}\\ &\quad+\textstyle\sum_{a_{ij}\neq 0, i+(b+1)j\not\geq n}(i+(b+1)j)\mathbf{Z}]/[(a-b-1)\mathbf{Z}+(b+1)\mathbf{Z}]|\\ &\leq |[\textstyle\sum_{a_{ij}\neq 0} i\mathbf{Z}+a\mathbf{Z}+(b+1)\mathbf{Z}]/[a\mathbf{Z}+(b+1)\mathbf{Z}]|\\ &\leq \tau_{f\circ\Phi_0}(p)\leq\tau_f(\Phi_0(p))\end{aligned}$$

Suppose that $b+1<a$, $0=\alpha$ in (14.16). Then we are back in the form (14.10) with a decrease in $a-b$.

Suppose that (14.10) and (14.15) hold. Then

$$\begin{aligned}u &= x_2^a z_2^a\\ v &= x_2^b z_2^{b+1}\\ w &= \textstyle\sum a_{ij}x_2^{i+bj}z_2^{i+(b+1)j}+x_2^n z_2^n(y_2+\beta).\end{aligned}$$

$Z_{i+1}\to Y_1$ is a morphism at p_2. Let $q_1=h_{i+1}(p_2)$. There exist permissible parameters u_1,v_1,w at q_1 defined by

$$u=u_1v_1, v=v_1.$$

We have an expression

$$\begin{aligned}u_1 &= x_2^{a-b}z_2^{a-b-1}\\ v_1 &= x_2^b z_2^{b+1}\\ w &= \textstyle\sum a_{ij}x_2^{i+bj}z_2^{i+(b+1)j}+x_2^n z_2^n(y_2+\beta)\end{aligned} \tag{14.19}$$

of type (4.2), so that h_{i+1} is prepared at p_2.

From (9.3), we have

$$\begin{aligned}\tau_{h_{i+1}}(p_2) &= |((a-b,a-b-1)\mathbf{Z}+(b,b+1)\mathbf{Z}\\ &\quad+\textstyle\sum_{a_{ij}\neq 0,(i+bj,i+(b+1)j)\not\geq(n,n)}(i+bj,i+(b+1)j)\mathbf{Z})\\ &\quad/((a-b,a-b-1)\mathbf{Z}+(b,b+1)\mathbf{Z})|\\ &\leq |((a,a)\mathbf{Z}+\textstyle\sum_{a_{ij}\neq 0}(i,i)\mathbf{Z})/(a,a)\mathbf{Z})|\\ &= \tau_{f\circ\Phi_0}(p)\leq\tau_f(\Phi_0(p)).\end{aligned}$$

Suppose that (14.11) holds. Then

$$\tau_{f\circ\Phi_0}(p)=|(k\mathbf{Z}+\sum_{a_{il}\neq 0} i\mathbf{Z})/k\mathbf{Z}|.$$

Suppose that (14.11) and (14.14) hold. Then we have

$$\begin{aligned}u &= (x_2^a y_2^b)^k\\ v &= x_2^{c+1}y_2^d(z_2+\alpha)\\ w &= \textstyle\sum_{i,l}a_{il}(z_2+\alpha)^l x_2^{ai+(c+1)l}y_2^{bi+dl}+x_2^e y_2^f.\end{aligned} \tag{14.20}$$

Suppose that $(ak,bk)=(c+1,d)$ in (14.20). Then $Z_{i+1}\to Y_1$ is a morphism at p_2. Let $q_1=h_{i+1}(p_2)$. There exist permissible parameters u_1,v_1,w at q_1 defined by

$$\begin{aligned}u &= u_1\\ v &= u_1(v_1+\alpha).\end{aligned}$$

We have an expression

$$(14.21)\quad \begin{aligned} u_1 &= (x_2^a y_2^b)^k \\ v_1 &= z_2 \\ w &= \textstyle\sum_{i,l} a_{il}(z_2+\alpha)^l x_2^{ai+(c+1)l} y_2^{bi+dl} + x_2^e y_2^f \\ &= \textstyle\sum_{i,l} a_{il}(z_2+\alpha)^l (x_2^a y_2^b)^{i+lk} + x_2^e y_2^f \end{aligned}$$

of type (4.6), so that h_{i+1} is prepared at p_2. From (9.6), we have

$$\tau_{h_{i+1}}(p_2) \le |(k\mathbf{Z} + \sum_{a_{il}\ne 0, (i+lk)(a,b)\not\ge(e,f)} (i+lk)\mathbf{Z}/k\mathbf{Z}| \le \tau_{f\circ\Phi_0}(p) \le \tau_f(\Phi_0(p)).$$

Suppose that $(ak, bk) > (c+1, d)$ and $0 \ne \alpha$ in (14.20). Then $Z_{i+1} \to Y_1$ is a morphism at p_2. Let $q_1 = h_{i+1}(p_2)$. There exist permissible parameters u_1, v_1, w at q_1 defined by

$$u = u_1 v_1, v = v_1.$$

Suppose that $ad - b(c+1) \ne 0$ in (14.20). Then there exist regular parameters $\overline{x}_2, \overline{y}_2, \overline{z}_2$ in $\hat{\mathcal{O}}_{Z_{i+1},p_2}$ defined by

$$\begin{aligned} \overline{x}_2 &= x_2(z_2+\alpha)^{-\frac{bk}{\overline{h}}} \\ \overline{y}_2 &= y_2(z_2+\alpha)^{\frac{ak}{\overline{h}}} \\ \overline{z}_2 &= (z_2+\alpha)^{\frac{ebk-fak}{\overline{h}}} - \overline{\alpha} \end{aligned}$$

with $\overline{h} = adk - (c+1)bk$ and $\overline{\alpha} = \alpha^{\frac{ebk-fak}{\overline{h}}}$, such that

$$(14.22)\quad \begin{aligned} u_1 &= x_2^{ak-c-1} y_2^{bk-d}(z_2+\alpha)^{-1} = \overline{x}_2^{ak-c-1}\overline{y}_2^{bk-d} \\ v_1 &= x_2^{c+1} y_2^d (z_2+\alpha) = \overline{x}_2^{c+1}\overline{y}_2^d \\ w &= \textstyle\sum_{i,l} a_{il}\overline{x}_2^{ai+(c+1)l}\overline{y}_2^{bi+dl} + \overline{x}_2^e\overline{y}_2^f(\overline{z}_2+\overline{\alpha}) \end{aligned}$$

with $0 \ne \overline{\alpha} \in \mathbf{k}$, which is an expression of type (4.2), so that h_{i+1} is prepared at p_2. From (9.3), we see that

$$(14.23)\quad \begin{aligned} \tau_{h_{i+1}}(p_2) &= |(ak-c-1, bk-d)\mathbf{Z} + (c+1,d)\mathbf{Z} \\ &\quad + \textstyle\sum_{a_{il}\ne 0, i(a,b)+l(c+1,d)\not\ge(e,f)} [i(a,b)+l(c+1,d)]\mathbf{Z}]/ \\ &\quad [(ak-c-1, bk-d)\mathbf{Z} + (c+1,d)\mathbf{Z}]| \\ &\le |(k(a,b)\mathbf{Z} + \textstyle\sum_{a_{il}\ne 0} i(a,b)\mathbf{Z})/k(a,b)\mathbf{Z}| \\ &= |k\mathbf{Z} + \textstyle\sum_{a_{il}\ne 0} i\mathbf{Z}/k\mathbf{Z}| = \tau_{f\circ\Phi_0}(p) \le \tau_f(\Phi_0(p)). \end{aligned}$$

Suppose that $ad - b(c+1) = 0$ in (14.20), (and $(ak, bk) > (c+1, d)$, $0 \ne \alpha$ in (14.20)). There exist positive integers $\overline{a}, \overline{b}, \overline{t}, \overline{k}$ such that $\gcd(\overline{a}, \overline{b}) = 1$, and

$$(ak-c-1, bk-d) = \overline{t}(\overline{a}, \overline{b}), \quad (c+1, d) = \overline{k}(\overline{a}, \overline{b}).$$

From $k(a,b) = (\overline{t}+\overline{k})(\overline{a}, \overline{b})$ and $\gcd(a,b) = 1$, $\gcd(\overline{a}, \overline{b}) = 1$, we see that $(a,b) = (\overline{a}, \overline{b})$ and $\overline{t} + \overline{k} = k$.

There exist regular parameters $\overline{x}_2, \overline{y}_2, \overline{z}_2$ in $\hat{\mathcal{O}}_{Z_{i+1},p_2}$, defined by

$$\begin{aligned} \overline{x}_2 &= x_2(z_2+\alpha)^{-\frac{f}{\overline{h}\overline{t}}} \\ \overline{y}_2 &= y_2(z_2+\alpha)^{\frac{e}{\overline{h}\overline{t}}} \\ \overline{z}_2 &= (z_2+\alpha)^{1+\frac{\overline{k}}{\overline{t}}} - \overline{\alpha} \end{aligned}$$

where $\overline{h} = fa - eb$, $\overline{\alpha} = e^{1+\frac{\overline{k}}{\overline{t}}}$, such that

$$\begin{aligned} u_1 &= x_2^{ak-c-1} y_2^{bk-d}(z_2+\alpha)^{-1} = (\overline{x}_2^a \overline{y}_2^b)^{\overline{t}} \\ v_1 &= x_2^{c+1} y_2^d (z_2+\alpha) = (\overline{x}_2^a \overline{y}_2^b)^{\overline{k}} (\overline{z}_2 + \overline{\alpha}) \\ w &= \textstyle\sum_{i,l} a_{il} (\overline{z}_2+\overline{\alpha})^{l+\frac{i}{k}} (\overline{x}_2^a \overline{y}_2^b)^{i+\overline{k}l} + \overline{x}_2^e \overline{y}_2^f, \end{aligned} \tag{14.24}$$

which is an expression of type (4.3), so that h_{i+1} is prepared at p_2.

From (9.4), we have

$$\begin{aligned} \tau_{h_{i+1}}(p_2) &= |(\overline{t}\mathbf{Z} + \overline{k}\mathbf{Z} + \textstyle\sum_{a_{il}\neq 0, (i+\overline{k}l)(a,b)\not\geq(e,f)} (l\overline{k}+i)\mathbf{Z})/(\overline{t}\mathbf{Z}+\overline{k}\mathbf{Z})| \\ &\leq |(k\mathbf{Z} + \overline{k}\mathbf{Z} + \textstyle\sum_{a_{il}\neq 0} i\mathbf{Z})/(k\mathbf{Z}+\overline{k}\mathbf{Z})| \\ &\leq |(k\mathbf{Z} + \textstyle\sum_{a_{il}\neq 0} i\mathbf{Z})/k\mathbf{Z}| \\ &= \tau_{f\circ\Phi_0}(p) \leq \tau_f(\Phi_0(p)). \end{aligned} \tag{14.25}$$

Suppose that $(ak, bk) > (c+1, d)$ and $0 = \alpha$ in (14.20). Then we are back in the form (14.11), with a decrease in $(ak-c)+(bk-d)$.

Suppose that (14.11) and (14.15) hold. Then

$$\begin{aligned} u &= x_2^{ak} y_2^{bk} z_2^{ak} \\ v &= x_2^c y_2^d z_2^{c+1} \\ w &= \textstyle\sum_{i,l} a_{il} x_2^{ai+cl} y_2^{bi+dl} z_2^{l+ai+cl} + x_2^e y_2^f z_2^e. \end{aligned} \tag{14.26}$$

Then $Z_{i+1} \rightarrow Y_1$ is a morphism at p_2. Let $q_1 = h_{i+1}(p_2)$. There exist permissible parameters u_1, v_1, w at q_1 defined by

$$u = u_1 v_1, v = v_1.$$

We have an expression

$$\begin{aligned} u_1 &= x_2^{ak-c} y_2^{bk-d} z_2^{ak-c-1} \\ v_1 &= x_2^c y_2^d z_2^{c+1} \\ w &= \textstyle\sum_{i,l} a_{il} x_2^{ai+cl} y_2^{bi+dl} z_2^{l+ai+cl} + x_2^e y_2^f z_2^e. \end{aligned} \tag{14.27}$$

of type (4.4), so that h_{i+1} is prepared at p_2. We calculate $\tau_{h_{i+1}}(p_2)$ from (9.1).

Set

$$\begin{aligned} G &= ((ak, bk, ak)\mathbf{Z} + (c, d, c+1)\mathbf{Z} \\ &\quad + \textstyle\sum_{a_{il}\neq 0} (ai+cl, bi+dl, l+ai+cl)\mathbf{Z})/ \\ &\quad ((ak, bk, ak)\mathbf{Z} + (c, d, c+1)\mathbf{Z}) \\ &\cong ((ak, bk, ak)\mathbf{Z} + \textstyle\sum_{a_{il}\neq 0} (ai, bi, ai)\mathbf{Z})/(ak, bk, ak)\mathbf{Z}). \end{aligned}$$

$$\begin{aligned} \tau_{h_{i+1}}(p_2) &= |(ak-c, bk-d, ak-c-1)\mathbf{Z} + (c, d, c+1)\mathbf{Z} \\ &\quad + \textstyle\sum_{a_{il}\neq 0, (ai+cl, bi+dl, l+ai+cl)\not\geq(e,f,e)} (ai+cl, bi+dl, l+ai+cl)\mathbf{Z})/ \\ &\quad ((ak-c, bk-d, ak-c-1)\mathbf{Z} + (c, d, c+1)\mathbf{Z})| \\ &\leq |G|. \end{aligned}$$

The homomorphism $\mathbf{Z} \rightarrow \mathbf{Z}^3$ defined by $x \mapsto (xa, xb, xa)$ induces an isomorphism

$$(k\mathbf{Z} + \sum_{a_{il}\neq 0} i\mathbf{Z})/k\mathbf{Z} \rightarrow H. \tag{14.28}$$

Thus $\tau_{h_{i+1}}(p_2) \leq \tau_{f\circ\Phi_0}(p) \leq \tau_f(\Phi_0(p))$.

Suppose that (14.12) and (14.14) hold. Then

$$
\begin{aligned}
u &= x_2^a \\
v &= x_2^b(\beta + y_2) \\
w &= x_2^{d+1}(z_2 + \alpha).
\end{aligned}
\tag{14.29}
$$

Suppose that $d + 1 = a$ in (14.29). Then $Z_{i+1} \rightarrow Y_1$ is a morphism at p_2. Let $q_1 = h_{i+1}(p_2)$. There exist permissible parameters u_1, v, w_1 at q_1 defined by

$$u = u_1, w = u_1(w_1 + \alpha).$$

We have an expression

$$u_1 = x_2^a, v = x_2^b(\beta + y_2), w_1 = z_2$$

which is toroidal, so that h_{i+1} is prepared at p_2, and

$$\tau_{h_{i+1}}(p_2) = -\infty \leq \tau_f(\Phi_0(p)).$$

Suppose that $d + 1 < a$ and $0 \neq \alpha$ in (14.29). Then $Z_{i+1} \rightarrow Y_1$ is a morphism at p_2. Let $q_1 = h_{i+1}(p_1)$. There exist permissible parameters u_1, v, w_1 at q_1 defined by

$$u = u_1 w_1, w = w_1,$$

so that q_1 is a 3-point.

There exist regular parameters $\overline{x}_2, \overline{y}_2, \overline{z}_2$ in $\hat{\mathcal{O}}_{Z_{i+1}, p_2}$ and $0 \neq \overline{\alpha}, \overline{\beta} \in \mathbf{k}$ such that

$$
\begin{aligned}
u_1 &= x_2^{a-d-1}(z_2 + \alpha)^{-1} = \overline{x}_2^{a-d-1} \\
v &= \overline{x}_2^b(\overline{\beta} + \overline{y}_2) \\
w_1 &= \overline{x}_2^{d+1}(\overline{z}_2 + \overline{\alpha})
\end{aligned}
$$

which is toroidal, so that h_{i+1} is prepared at p_2, and

$$\tau_{h_{i+1}}(p_2) = -\infty \leq \tau_f(\Phi_0(p)).$$

Suppose that $d + 1 < a$ and $0 = \alpha$ in (14.29). Then we are back in the form (14.12) with a decrease in $a - d$.

Suppose that (14.12) and (14.15) hold. Then

$$
\begin{aligned}
u &= x_2^a z_2^a \\
v &= x_2^b z_2^b(\alpha + y_2) \\
w &= x_2^d z_2^{d+1}.
\end{aligned}
$$

Thus $Z_{i+1} \rightarrow Y_1$ is a morphism at p_2. Let $q_1 = h_{i+1}(p_2)$. There exist permissible parameters u_1, v, w_1 at q_1 defined by

$$u = u_1 w_1, w = w_1.$$

We have a 2-point mapping to a 3-point, and an expression

$$
\begin{aligned}
u_1 &= x_2^{a-d} z_2^{a-d-1} \\
v &= x_2^b z_2^b(\alpha + y_2) \\
w_1 &= x_2^d z_2^{d+1}
\end{aligned}
$$

which is a toroidal form, so that h_{i+1} is prepared at p_2, and

$$\tau_{h_{i+1}}(p_2) = -\infty \leq \tau_f(\Phi_0(p)).$$

Suppose that (14.13) and (14.14) hold. Then

$$\begin{aligned} u &= x_2^a y_2^b \\ v &= x_2^c y_2^d \\ w &= x_2^{g+1} y_2^h (z_2+\alpha). \end{aligned} \tag{14.30}$$

Suppose that $(a,b)=(g+1,h)$ in (14.30). Then $Z_{i+1}\rightarrow Y_1$ is a morphism at p_2. Let $q_1=h_{i+1}(p_2)$. There exist permissible parameters u_1,v,w_1 at q_1 defined by

$$u=u_1, w=u_1(w_1+\alpha).$$

We have an expression

$$\begin{aligned} u_1 &= x_2^a y_2^b \\ v &= x_2^c y_2^d \\ w_1 &= z_2 \end{aligned}$$

which is toroidal, so that h_{i+1} is prepared at p_2, and

$$\tau_{h_{i+1}}(p_2)=-\infty\le\tau_f(\Phi_0(p)).$$

Suppose that $(g+1,h)<(a,b)$ and $0\ne\alpha$ in (14.30). Then $Z_{i+1}\rightarrow Y_1$ is a morphism at p_2. Let $q_1=h_{i+1}(p_2)$. There exist permissible parameters u_1,v,w_1 at q_1 defined by

$$u=u_1w_1, w=w_1.$$

We have an expression

$$\begin{aligned} u_1 &= x_2^{a-g-1} y_2^{b-h}(z_2+\alpha)^{-1} \\ v &= x_2^c y_2^d \\ w_1 &= x_2^{g+1} y_2^h (z_2+\alpha) \end{aligned}$$

which is toroidal after a change of variable in $\hat{\mathcal{O}}_{Z_{i+1},p_2}$, so that h_{i+1} is prepared at p_2, and

$$\tau_{h_{i+1}}(p_2)=-\infty\le\tau_f(\Phi_0(p)).$$

Suppose that $(g+1,h)<(a,b)$ and $\alpha=0$ in (14.30). Then we are back in the form (14.13), with a decrease in $(a-g)+(b-h)$.

Suppose that (14.13) and (14.15) hold. Then

$$\begin{aligned} u &= x_2^a y_2^b z_2^a \\ v &= x_2^c y_2^d z_2^c \\ w &= x_2^g y_2^h z_2^{g+1}. \end{aligned}$$

Thus $Z_{i+1}\rightarrow Y_1$ is a morphism at p_2. Let $q_1=h_{i+1}(p_2)$. There exist permissible parameters u_1,v,w_1 at q_1 defined by

$$u=u_1w_1, w=w_1.$$

We have a 3-point mapping to a 3-point, and an expression

$$\begin{aligned} u_1 &= x_2^{a-g} y_2^{b-h} z_2^{a-g-1} \\ v &= x_2^c y_2^d z_2^c \\ w_1 &= x_2^g y_2^h z_2^{g+1} \end{aligned}$$

which is toroidal, so that h_{i+1} is prepared at p_2, and

$$\tau_{h_{i+1}}(p_2)=-\infty\le\tau_f(\Phi_0(p)).$$

We have thus verified A and B for all h_i.

Suppose that $p \in Z_i$ is in the locus Σ_i where $\mathcal{I}_C\mathcal{O}_{Z_i}$ is not locally principal. Define

$$C(p) = \begin{cases} a-b & \text{if (14.10) holds at } p \\ ak-c+bk-d & \text{if (14.11) holds at } p \\ a-d & \text{if (14.12) holds at } p \\ a-g+b-h & \text{if (14.13) holds at } p. \end{cases}$$

Let

$$C(Z_i) = \max\{C(p) \mid p \in \Sigma_i\}.$$

We have shown in the above analysis that

$$0 \le C(Z_{i+1}) < C(Z_i)$$

if the rational map h_i is not a morphism. Thus there exists a finite n such that $f_1 = h_n$, $\Phi_1 = \Phi_0 \circ \Lambda_1 \circ \cdots \circ \Lambda_n$, $X_1 = Z_n$ satisfy 1 - 3 of the conclusions of Lemma 14.5.

We now prove 4. If $p \in f^{-1}(q)$, then we have permissible parameters x, y, z for u, v, w in $\hat{\mathcal{O}}_{X,p}$ such that there is one of the forms:

p a 1-point

$$u = x^a, v = y, w = g(x,y) + x^n(z+\beta) \tag{14.31}$$

or p a 2-point

$$u = (x^a y^b)^k, v = z, w = h(x^a y^b, z) + x^c y^d. \tag{14.32}$$

By assumption, $u = v = 0$ are local equations of C.

Since q_1 is a 1-point, at q_1 we have permissible parameters u_1, v_1, w_1 defined by

$$u = u_1, v = u_1(v_1 + \alpha), w = w_1$$

for some $\alpha \in \mathbf{k}$. Since C_1 is a section over C, there exists a series $\lambda(v_1, w)$ such that $u_1 = \lambda(v_1, w_1) = 0$ are local equations of C_1 (in $\hat{\mathcal{O}}_{Y_1,q_1}$), and

$$\mathbf{k}[[u,v,w]]/(u,v) \to \mathbf{k}[[u_1,v_1,w]]/(u_1,\lambda)$$

is an isomorphism, which implies that $\lambda = (v_1 - \phi(w))\gamma$ where ϕ is a series and γ is a unit series in $\hat{\mathcal{O}}_{Y_1,q_1}$. Set

$$\overline{u}_1 = u_1, \overline{v}_1 = v_1 - \phi(w).$$

$\overline{u}_1, \overline{v}_1, w$ are permissible parameters at q_1, and $\overline{u}_1 = \overline{v}_1 = 0$ are local equations of C_1 at q_1.

Suppose that $p_1 \in f_1^{-1}(q_1)$. First suppose that $p = \Phi_1(p_1)$ has the form (14.31). Then $\hat{\mathcal{O}}_{X_1,p_1}$ has regular parameters x_1, y_1, z_1 with $x = x_1, y = x_1^a(y_1 + \alpha)$, since $X_0 \to X$ is an isomorphism over p. p_1 is a 1-point, and substituting into (14.31), we have

$$u_1 = x_1^a, v_1 = y_1, w = g(x_1, x_1^a(y_1+\alpha)) + x_1^n(z+\beta)$$

and thus $\overline{u}_1, \overline{v}_1$ are toroidal forms at p_1.

Now suppose that $p = \Phi_1(p_1)$ has the form (14.32).

Let

$$p' = \Lambda_1 \circ \cdots \circ \Lambda_n(p) \in X_0.$$

There exist regular parameters x_1, y_1, z_1 in $\hat{\mathcal{O}}_{X_1,p'}$ such that

$$x = x_1^{\overline{a}} y_1^{\overline{b}}, y = x_1^{\overline{c}} y_1^{\overline{d}}, z = z_1 \tag{14.33}$$

or

$$x = x_1^{\overline{a}}(y_1+\alpha)^{\overline{b}}, y = x_1^{\overline{c}}(y_1+\overline{\alpha})^{\overline{d}}, z = z_1 \tag{14.34}$$

with $0 \neq \overline{\alpha} \in \mathbf{k}$. If (14.33) holds, then we have

$$\begin{aligned} u &= (x_1^{a'}y_1^{b'})^k \\ v &= z_1 \\ w &= \overline{h}(x_1^{a'}y_1^{b'}, z_1) + x_1^{c'}y_1^{d'}. \end{aligned} \tag{14.35}$$

If (14.34) holds, then we have $0 \neq \alpha' \in \mathbf{k}$ and permissible parameters $\overline{x}_1, \overline{y}_1, z_1$ in $\hat{\mathcal{O}}_{X_1,p'}$ such that

$$\begin{aligned} u &= \overline{x}_1^{a'k} \\ v &= z_1 \\ w &= \overline{h}(\overline{x}_1^{a'}, z_1) + \overline{x}_1^{b'}(\overline{y}_1+\alpha') \end{aligned} \tag{14.36}$$

with $\alpha' \in \mathbf{k}$.

Suppose that a form (14.35) holds at p'. Then $\hat{\mathcal{O}}_{X_1,p_1}$ has regular parameters x_2, y_2, z_2 with

$$x_1 = x_2, y_1 = y_2, z_1 = x_2^{a'k}y_2^{b'k}(z_2+\alpha).$$

p_1 is a 2-point, and substituting into (14.35), we have

$$u_1 = (x_2^{a'}y_2^{b'})^k, v_1 = z_2, w = \overline{h}(x_2^{a'}y_2^{b'}, x_2^{a'k}y_2^{b'k}(z_2+\alpha)) + x_2^{c'}y_2^{d'}.$$

Thus $\overline{u}_1, \overline{v}_1$ are toroidal forms at p_1.

Suppose that a form (14.36) holds at p'. Then $\hat{\mathcal{O}}_{X_1,p_1}$ has regular parameters x_2, y_2, z_2 with

$$\overline{x}_1 = x_2, \overline{y}_1 = y_2, z_1 = x_2^{a'k}(z_2+\alpha).$$

p_1 is a 1-point, and substituting into (14.36), we have

$$u_1 = x_2^{a'k}, v_1 = z_2, w = h(x_2^{a'}, x_2^{a'k}(z_2+\alpha)) + x_2^{c'}y_2^{d'}.$$

Thus $\overline{u}_1, \overline{v}_1$ are toroidal forms at p_1.

We now prove 5. We may assume that $\tau_{f_1}(p') = \tau_f(\overline{p})$. There exists a smallest i such that $X_1 \to Z_{i+1}$ is an isomorphism at p'. Let p_2 be the image of p' in Z_{i+1}. Let p be the image of p_2 in Z_0.

If $Z_0 \to X$ is not an isomorphism at p, then $\overline{p}$ is a 2-point, and we have permissible parameters x, y, z at $\overline{p}$ such that

$$\begin{aligned} u &= (x^a y^b)^k \\ v &= z \\ w &= \textstyle\sum_{i,l\geq 0} a_{il}z^l(x^a y^b)^i + x^e y^f, \end{aligned} \tag{14.37}$$

where $\gcd(a,b)=1$ and the sum is over i such that $(ia, ib) \not\geq (e,f)$. Since we are assuming w is good at $\overline{p}$ for f, we have $a_{il}=0$ if k divides i.

Since $Z_0 \to X$ is a sequence of blow ups of 2-curves above $\overline{p}$, we have regular parameters x_1, y_1, z in $\hat{\mathcal{O}}_{X_0,p}$ such that

$$x = x_1^{\overline{a}}y_1^{\overline{b}}, y = x_1^{\overline{c}}y_1^{\overline{d}} \tag{14.38}$$

with $\overline{a}\overline{d} - \overline{b}\overline{c} = \pm 1$, (and $\gcd(a\overline{a}+b\overline{c}, a\overline{b}+b\overline{d}) = 1$) or

$$x = \overline{x}_1^{\overline{a}}(\overline{y}_1+\alpha)^{\overline{b}}, y = \overline{x}_1^{\overline{c}}(\overline{y}_1+\alpha)^{\overline{d}}$$

where $\overline{a}, \overline{c} \in \mathbf{N}$, $\overline{b}, \overline{d} \in \mathbf{Q}$, $0 \neq \alpha \in \mathbf{k}$ are such that

$$x^a y^b = \overline{x}_1^{\overline{a}a+\overline{c}b}, x^e y^f = \overline{x}_1^{\overline{a}e+\overline{c}f}(\overline{y}_1+\alpha). \tag{14.39}$$

If either (14.38) or (14.39) holds, we see from substitution into (14.37) that w is good at p for $f \circ \Phi_0$.

Since $\overline{q}$ is a 1-point, $\overline{f}_0 = \Psi_1^{-1} \circ f \circ \Phi_0$ is not a morphism at p. Let p_1 be the image of p_2 in Z_i. At p_1 we have an expression (14.10) or (14.11).

First suppose that (14.10) holds at p_1. Since w is good at p for $f \circ \Phi_0$, we have $a_{ij} = 0$ if a divides i.

Suppose that (14.10) holds at p_1 and at p_2 and there is an expression (14.17).

Suppose that $i + aj \in a\mathbf{Z}$ for some i. Then $i \in a\mathbf{Z}$, and since w is good at p for $f \circ \Phi_0$, we have that $a_{ij} = 0$. Thus w is good at p_2 for h_{i+1} (and w is good at p' for f_1).

Suppose that (14.10) holds at p_1 and at p_2 there is an expression (14.18).

If w is not good at p_2 for h_{i+1}, there exists $a_{ij} \neq 0$ with $i + (b+1)j < n$ and $i + (b+1)j \in a\mathbf{Z} + (b+1)\mathbf{Z}$. Thus $i \in a\mathbf{Z} + (b+1)\mathbf{Z}$ which implies that i is in the kernel of the surjective projection homomorphism

$$[\sum_{a_{ij}\neq 0} i\mathbf{Z} + a\mathbf{Z}]/a\mathbf{Z}] \rightarrow [\sum_{a_{ij}\neq 0} i\mathbf{Z} + a\mathbf{Z} + (b+1)\mathbf{Z}]/[a\mathbf{Z} + (b+1)\mathbf{Z}].$$

Thus

$$\tau_{f_1}(p') = \tau_{h_{i+1}}(p_2) < \tau_{f\circ\Phi_0}(p) \leq \tau_f(\overline{p}),$$

a contradiction.

Suppose that (14.10) holds at p_1 and at p_2 there is an expression (14.19).

Suppose that w is not good at p_2 for h_{i+1}. Then there exists $a_{ij} \neq 0$ with

$$(i + bj, i + (b+1)j) \not\geq (n, n)$$

and

$$(i + bj, i + (b+1)j) \in (a - b, a - b - 1)\mathbf{Z} + (b, b+1)\mathbf{Z} = (a, a)\mathbf{Z} + (b, b+1)\mathbf{Z},$$

which implies that $(i, i) \in (a, a)\mathbf{Z} + (b, b+1)\mathbf{Z}$. Thus $i \in a\mathbf{Z}$, a contradiction, and we conclude that w is good at p_2 for h_{i+1}.

Now suppose that (14.11) holds at p_1. Since w is good at p for $f \circ \Phi_0$, we have $a_{il} = 0$ if k divides i.

Suppose that (14.11) holds at p_1 and at p_2 there is an expression (14.21). Suppose that $(i + lk)(a, b) \not\geq (e, f)$ and $i + lk \in k\mathbf{Z}$. then $i \in k\mathbf{Z}$, and since w is good at p for $f \circ \Phi_0$, we have that $a_{il} = 0$. Thus w is good at p_2 for h_{i+1}.

Suppose that (14.11) holds at p_1 and at p_2 there is an expression (14.22). Then by a similar calculation to the above analysis of (14.18), (14.23) is a strict inequality if w is not good at p_2 for h_{i+1}.

Suppose that (14.11) holds at p_1 and at p_2 there is an expression (14.24). Then (14.25) is a strict inequality if w is not good at p_2 for h_{i+1}.

Suppose that (14.11) holds aat p_1 and p_2 has an expression (14.27). The homomorphism (14.28) has a nontrivial kernel if w is not good at p_2 for h_{i+1}.

In all of these cases, we have a contradiction to our assumption that $\tau_{f_1}(p') = \tau_f(\overline{p})$.

□

LEMMA 14.6. *Suppose that $f : X \rightarrow Y$ is pre-τ-quasi-well prepared with relation R, $q \in Y - U(R)$ is a 2-point such that $q \in G_Y(f, \tau)$, and u, v, w are permissible parameters at q. Let E be the 2-curve of Y containing q. Then there exists a* *pre-τ-quasi-well diagram*

$$\begin{array}{ccc} X_1 & \xrightarrow{f_1} & Y_1 \\ \Phi \downarrow & & \downarrow \Psi \\ X & \xrightarrow{f} & Y \end{array}$$

such that Φ and Ψ are products of blow ups of 2-curves, f_1 is pre-τ-quasi-well prepared, Φ is an isomorphism over $f^{-1}(Y-E)$, and if $q_1 \in \Psi^{-1}(q)$ (with permissible parameters u_1, v_1, w), then if $p_1 \in f_1^{-1}(q_1)$ is such that $\tau_{f_1}(p_1) = \tau$, then there exists a series $\phi_{p_1}(u_1, v_1)$ such that $w - \phi_{p_1}(u_1, v_1)$ is good at p_1. If f is τ-prepared, then f_1 is τ-prepared.

PROOF. Let u, v, w be permissible parameters at q. By Lemma 10.5 and Lemma 14.1, there exist sequences of blow ups of 2-curves $\Psi_0 : Y_0 \rightarrow Y$ and $\Phi_0 : X_0 \rightarrow X$ making a pre-τ-quasi-well diagram

$$\begin{array}{ccc} X_0 & \xrightarrow{f_0} & Y_0 \\ \Phi_0 \downarrow & & \downarrow \Psi_0 \\ X & \xrightarrow{f} & Y \end{array}$$

such that if $q_0 \in \Psi_0^{-1}(q)$ is a 2-point, then there are permissible parameters (u_0, v_0, w_0) at q_0 such that

$$u = u_0^a v_0^b, v = u_0^c, v_0^d$$

with $ad - bc = \pm 1$, and there are no points of the form 2 (b) or 2 (c) of Definition 4.6 for f_0 above q_0.

We construct an infinite commutative diagram of morphisms

$$(14.40) \qquad \begin{array}{ccc} \vdots & & \vdots \\ \downarrow & & \downarrow \\ X_n & \xrightarrow{f_n} & Y_n \\ \Phi_n \downarrow & & \downarrow \Psi_n \\ \vdots & & \vdots \\ \Phi_2 \downarrow & & \downarrow \Psi_2 \\ X_1 & \xrightarrow{f_1} & Y_1 \\ \Phi_1 \downarrow & & \downarrow \Psi_1 \\ X_0 & \xrightarrow{f_0} & Y_0 \end{array}$$

as follows. Order the 2-curves of Y_0 which intersect $\Psi_0^{-1}(q)$, and let $\Psi_1 : Y_1 \rightarrow Y_0$ be the blow up of the 2-curve C of smallest order. Then construct (by Lemma 14.1) a pre-τ-quasi-well diagram

$$(14.41) \qquad \begin{array}{ccc} X_1 & \xrightarrow{f_1} & Y_1 \\ \Phi_1 \downarrow & & \downarrow \Psi_1 \\ X_0 & \xrightarrow{f_0} & Y_0 \end{array}$$

where Ψ_1 is a product of blow up of 2-curves and Φ_1 is an isomorphism above $f_0^{-1}(Y_0 - C)$. Order the 2-curves of Y_1 which intersect $(\Psi_0 \circ \Psi_1)^{-1}(q)$ so that the 2-curves contained in the exceptional divisor of Ψ_1 have larger order than the order of the (strict transforms of the) 2-curves of Y_0.

Let $\Psi_2 : Y_2 \to Y_1$ be the blow up of the 2-curve C_1 on Y_1 of smallest order, and construct a pre-τ-quasi-well diagram

$$\begin{array}{ccc} X_2 & \stackrel{f_2}{\to} & Y_2 \\ \Phi_2 \downarrow & & \downarrow \Psi_2 \\ X_1 & \stackrel{f_1}{\to} & Y_1 \end{array}$$

as in (14.41). We now iterate to construct (14.40). Let

$$\overline{\Psi}_n = \Psi_0 \circ \Psi_1 \circ \cdot \circ \Psi_n : Y_n \to Y,$$

$$\overline{\Phi}_n = \Phi_0 \circ \Phi_1 \circ \cdots \circ \Phi_n : X_n \to X.$$

For all $q_n \in \overline{\Psi}_n^{-1}(q)$ there exist permissible parameters u_n, v_n, w at q_n such that

$$u = u_n^a (v_n + \alpha)^b, v = u_n^c (v_n + \alpha)^d$$

with $ad - bc \neq 0$ and $\alpha \in \mathbf{k}$. q_n is a 2-point if $\alpha = 0$, and a 1-point if $\alpha \neq 0$

Suppose that $p_n \in (\overline{\Psi}_n \circ f_n)^{-1}(q)$. We will say that p_n is good for $q_n = f_n(p_n)$ if $\tau_{f_n}(p_n) < \tau$ or if $\tau_{f_n}(p_n) = \tau$ and there exists a series $\phi_{p_n}(u_n, v_n)$ such that $w - \phi_{p_n}(u_n, v_n)$ is good for f_n at p_n.

We first observe that if p_n is good for q_n and $p_{n+1} \in \Phi_{n+1}^{-1}(p_n)$, then p_{n+1} is good for q_{n+1}. (This follows from 3 of Lemma 14.1).

Let ν be a zero-dimensional valuation of $\mathbf{k}(X)$ whose center on Y is q, and let p_n be the center of ν on X_n, $q_n = f_n(p_n)$.

We will show that there exists n_0 such that $n \geq n_0$ implies p_n is good for q_n.

Once we have established this, it will follow from compactness of the Zariski-Riemann manifold of X [**Z**] that there exists n' such that all $p \in (\overline{\Psi}_{n'} \circ f_{n'})^{-1}(q)$ are good for $q' = f_{n'}(p)$, so that the conclusions of the theorem hold.

We may identify ν with an extension of ν to the quotient field of $\hat{\mathcal{O}}_{X_n,p_n}$ which dominates $\hat{\mathcal{O}}_{X_n,p_n}$.

If $\nu(u)$ and $\nu(v)$ are rationally dependent, then there exists n_0 such that q_{n_0} is a 1-point, which implies that p_{n_0} is good (by Remark 11.4), and thus p_n is good for all $n \geq n_0$.

So we may assume that $\nu(u)$ and $\nu(v)$ are rationally independent. We then have that

$$u = u_n^{a_n} v_n^{b_n}, v = u_n^{c_n} v_n^{d_n}$$

with $a_n d_n - b_n c_n = \pm 1$ for all n. We thus have (for $n >> 0$) that q_n is a 2-point, and p_n has one of the forms (14.42) or (14.43) below:

p_n is a 2-point

$$\begin{aligned} u_n &= x_n^a y_n^b \\ v_n &= x_n^c y_n^d \\ w &= \gamma_n + x_n^e y_n^f (z_n + \beta) \end{aligned} \tag{14.42}$$

with

$$\gamma_n = \sum \alpha_i M_i,$$

$\beta, \alpha_i \in \mathbf{k}$,

$$M_i = x_1^{a_i} y_1^{b_i} \text{ and } M_i^{e_i} = u_1^{k_i} v_1^{l_i}$$

with $k_i, l_i, e_i \in \mathbf{Z}$, $e_i > 0$ and $a_i, b_i \in \mathbf{N}$, or

p_n is a 3-point

$$
\begin{aligned}
u_n &= x_n^a y_n^b z_n^c \\
v_n &= x_n^d y_n^e z_n^f \\
w &= \gamma_n + N
\end{aligned}
\tag{14.43}
$$

with $N = x_n^{g_n} y_n^{h_n} z_n^{i_n}$, $\mathrm{rank}(u_n, v_n, N) = 3$,

$$\gamma_n = \sum \alpha_i M_i,$$

$\alpha_i \in \mathbf{k}$,

$$M_i = x_1^{a_i} y_1^{b_i} z_1^{c_i} \text{ and } M_i^{e_i} = u_1^{k_i} v_1^{l_i}$$

with $k_i, l_i, e_i \in \mathbf{Z}$, $e_i > 0$ and $a_i, b_i, c_i \in \mathbf{N}$.

Further, for $n >> 0$, either all points p_n have the form (14.42) or all p_n have the form (14.43).

It is shown in the proof of Lemma 10.6 that if (14.43) holds, then for $n >> 0$, we have $k_i, l_i \in \mathbf{N}$ for all i, which implies there exists a good form for p_n at q_n. Essentially the same argument shows that the same statement holds if (14.42) holds for $n >> 0$. □

LEMMA 14.7. *Suppose that $f : X \to Y$ is pre-τ-quasi-well prepared (or τ-well prepared or τ-very-well prepared) and $q \in U(R)$ is a 2-point (prepared of type 1 in Definition 13.7). Then q is a permissible center for R, and there exists a pre-τ-quasi-well prepared (or τ-well prepared or τ-very-well prepared) diagram (13.10) of R and the blow up $\Psi : Y_1 \to Y$ of q such that:*

1. *Φ is an isomorphism over $f^{-1}(Y - \Sigma(Y))$.*
2. *Suppose that f is τ-well prepared. Then*
   (a) *Let E be the exceptional divisor of Ψ. Suppose that $q_1 \in U(\overline{R}_i^1) \cap E$. Let $\gamma_i = \overline{S_{\overline{R}_i^1}(q_1) \cdot E}$. Then γ_i is a prepared curve for R^1 of type 6. Suppose that $q' \in U(\overline{R}_j^1) \cap E$. Let $\gamma_j = \overline{S_{\overline{R}_j^1}(q') \cdot E}$. Then either*
      (i) *$\gamma_i = \gamma_j$ or*
      (ii) *γ_i, γ_j intersect transversally at a 2 point on E (their tangent spaces have distinct directions at this point) and γ_i, γ_j are otherwise disjoint.*
   (b) *If γ is a prepared curve on Y then the strict transform of γ is a prepared curve on Y_1.*
3. *If f is τ-prepared then f_1 is τ-prepared.*

PROOF. There exists a pre-relation $\overline{R}_i$ associated to R such that $q \in U(\overline{R}_i) \subset U(R)$. Fix such an i. Let

$$u = u_{\overline{R}_i}(q), v = v_{\overline{R}_i}(q), w_i = w_{\overline{R}_i}(q).$$

u, v, w_i are super parameters at q, and $w_i = 0$ is a local equation of $S_{\overline{R}_i}(q)$. Let $m_q \subset \mathcal{O}_{Y,q}$ be the maximal ideal.

By Lemma 5.1, there exists a morphism $\Phi_0 : X_0 \to X$ which is a sequence of blow ups of 2-curves such that $(u, v)\mathcal{O}_{X,p}$ is invertible for all $p \in (f \circ \Phi_0)^{-1}(q)$, Φ_0 is an isomorphism over $f^{-1}(Y - \Sigma(Y))$, $f \circ \Phi_0$ is prepared and u, v, w_i are super parameters for $f \circ \Phi_0$ at q (by Lemma 10.2).

We will next show that there exists a sequence of blow ups of 2-curves and 3-points $\Phi_1 : X_1 \to X_0$ such that $(u, v)\mathcal{O}_{X_1,p}$ is invertible at all $p \in (f \circ \Phi_0 \circ \Phi_1)^{-1}(q)$

and if $m_q\mathcal{O}_{X_1,p}$ is not invertible, then we have permissible parameters x, y, z for u, v, w at p of one of the following forms:

p is a 1-point

$$u = x^a, v = x^b(\alpha + y), w_i = x^d z \tag{14.44}$$

with $\alpha \neq 0$ and $d < \min\{a, b\}$ or,

p is a 2-point of the type of (4.2) of Definition 4.1

$$u = x^a y^b, v = x^c y^d, w_i = x^g y^h z \tag{14.45}$$

with $ad - bc \neq 0$, and $(g, h) < \min\{(a, b), (c, d)\}$.

Suppose that $p \in (f \circ \Phi_0)^{-1}(q)$ is a 3-point, so that u, v, w_i have a form 4 of Definition 10.1 at p. There exist $\overline{x}, \overline{y}, \overline{z} \in \mathcal{O}_{X_0,p}$ and series $\lambda_1, \lambda_2, \lambda_3 \in \hat{\mathcal{O}}_{X_0,p}$ such that $x = \overline{x}\lambda_1$, $y = \overline{y}\lambda_2$, $z = \overline{z}\lambda_3$. Let $I^p \subset \mathcal{O}_{X_0,p}$ be the ideal

$$I^p = (u, v, \overline{x}^g\overline{y}^h\overline{z}^i, \overline{x}^j\overline{y}^k\overline{z}^l).$$

Observe that $\mathcal{O}_{X_0,p}/I^p$ is supported on $(f \circ \Phi_0)^{-1}(q)$. By Lemma 5.3, there exists a sequence of blow ups of 2-curves and 3-points $\Phi_1 : X_1 \to X_0$ such that $\Phi_0 \circ \Phi_1$ is an isomorphism above $f^{-1}(Y - \Sigma(Y))$ and $I^p\mathcal{O}_{X_1,p_1}$ is invertible for all 3-points $p \in (f \circ \Phi_0)^{-1}(q)$ and $p_1 \in \Phi_1^{-1}(p)$. Thus if $p \in (f \circ \Phi_0)^{-1}(q)$ is a 3-point and $p_1 \in (f \circ \Phi_0 \circ \Phi_1)^{-1}(p)$, then $m_q\mathcal{O}_{X_1,p_1}$ is invertible. $f \circ \Phi_0 \circ \Phi_1$ is prepared, and u, v, w_i are super parameters for $f \circ \Phi_0 \circ \Phi_1$ at q by Lemma 10.2.

We construct an infinite sequence of morphisms

$$\cdots \to X_n \overset{\Phi_n}{\to} \cdots \overset{\Phi_3}{\to} X_2 \overset{\Phi_2}{\to} X_1 \tag{14.46}$$

as follows. Order the 2-curves C of X_1 such that $q \in (f \circ \Phi_0 \circ \Phi_1)(C) \subset \Sigma(Y)$. Let $\Phi_2 : X_2 \to X_1$ be the blow up of the 2-curve C_1 on X_1 of smallest order. Order the 2-curves C' of X_2 such that $q \in (f \circ \Phi_0 \circ \Phi_1 \circ \Phi_2)(C') \subset \Sigma(Y)$ so that the 2-curves contained in the exceptional divisor of Φ_2 have order larger than the order of the (strict transform of the) 2-curves C of X_1 such that $q \in f \circ \Phi_0 \circ \Phi_1(C) \subset \Sigma(Y)$. Let $\Phi_3 : Y_3 \to Y_2$ be the blow ups of the 2-curve C_2 on Y_3 of smallest order. Let $\overline{\Phi}_n = \Phi_2 \circ \cdots \Phi_n : X_n \to X_1$ be constructed by iterating this procedure. The morphisms $f \circ \Phi_0 \circ \Phi_1 \circ \overline{\Phi}_n$ are prepared, and u, v, w_i are super parameters for $f \circ \Phi_0 \circ \Phi_1 \circ \overline{\Phi}_n$.

Suppose that ν is a 0-dimensional valuation of $\mathbf{k}(X)$. Let p_n be the center of ν on X_n. Say that ν is resolved on X_n if (at least) one of the following holds:

1. $m_q\mathcal{O}_{X_n,p_n}$ is invertible.
2. p_n is a 1-point of the form (14.44).
3. p_n is a 2-point of the form (14.45).

The condition of begin resolved on X_n is an open condition on the Zariski-Riemann manifold of X. Further, if ν is resolved on X_n and $n' > n$, then $X_{n'}$ is also resolved at ν.

Suppose that ν is a 0-dimensional valuation of $\mathbf{k}(X)$. Suppose that the center of ν on X_n is a 3-point, then the center of ν on X_1 is also a 3-point, so $m_q\mathcal{O}_{X_n,p_n}$ is invertible.

Suppose that the center of ν on X_n is a 1-point. Then u, v, w_i have a form 1 of Definition 10.1 at p_n, and thus have a form (14.44) at p_n if $m_q\mathcal{O}_{X_n,p_n}$ is not invertible.

Suppose that the center p of ν on X_1 is a 2-point such that u, v, w_i have a form 2 of Definition 10.1 at p. There exist $\overline{x}, \overline{y} \in \mathcal{O}_{X_1,p}$ and series $\lambda_1, \lambda_2 \in \hat{\mathcal{O}}_{X_1,p}$ such that $x = \overline{x}\lambda_1$, $y = \overline{y}\lambda_2$. Let $I^p \subset \mathcal{O}_{X_1,p}$ be the ideal $(u, v, \overline{x}^e\overline{y}^f, \overline{x}^g\overline{y}^h)$. There exists an n such that $I^p\mathcal{O}_{X_n,p_1}$ is invertible for all $p_1 \in \overline{\Phi}_n^{-1}(p)$. Let p_1 be the center of ν on X_n. u, v, w_i are super parameters for $f \circ \Phi_0 \circ \Phi_1 \circ \overline{\Phi}_n$ at q. Then p_1 is either a 1-point (so that ν is resolved on X_n) or a 2-point of the form 2 of Definition 10.1. If p_1 is a 2-point and $m_q\mathcal{O}_{X_n,p_1}$ is not invertible, then u, v, w_i have a form (14.45). Thus ν is resolved on X_n.

Suppose that the center p of ν on X_1 is a 2-point such that u, v, w_i have a form 3 of Definition 10.1 at p. Then there exist $\overline{x}, \overline{y} \in \mathcal{O}_{X,p}$ and series $\lambda_1, \lambda_2 \in \hat{\mathcal{O}}_{X,p}$ such that $x = \overline{x}\lambda_1$, $y = \overline{y}\lambda_2$. Let $I^p \subset \mathcal{O}_{X,p}$ be the ideal

$$I^p = (u, v, (\overline{x}^a\overline{y}^b)^l, \overline{x}^c\overline{y}^d).$$

There exists an n such that $I^p\mathcal{O}_{X_n,p_1}$ is invertible for all $p_1 \in \overline{\Phi}_n^{-1}(p)$. Let p_1 be the center of ν on X_n. Then $m_q\mathcal{O}_{X_n,p_1}$ is invertible. Thus ν is resolved on X_n.

By compactness of the Zariski-Riemann manifold [**Z**], there exists an n such that every valuation ν of $\mathbf{k}(X)$ is resolved on X_n. Rename $X_1 = X_n$ and

$$f_1 = f \circ \Phi_0 \circ \Phi_1 \circ \overline{\Phi}_n : X_1 \to Y.$$

If $m_q\mathcal{O}_{X_1,p_1}$ is not invertible at some $p_1 \in f_1^{-1}(q)$, then one of the forms (14.44) or (14.45) must hold at p_1.

The locus of points p on X_1 where $m_q\mathcal{O}_{X_1,p}$ is not invertible is a (possibly not irreducible) curve $\overline{E}$ which makes SNCs with the toroidal structure of X. $\overline{E}$ is supported at points of the form (14.44) (with $d < \min(a,b)$) and (14.45) (with $(g,h) < \min\{(a,b),(c,d)\}$). $x = z = 0$ is a local equation of $\overline{E}$ in (14.44). $x = z = 0$, $y = z = 0$ or $xy = z = 0$ are the possible local equations of $\overline{E}$ in (14.45).

For an irreducible component C of $\overline{E}$, define an invariant

$$A(C) = \min\{a, b\} - d > 0$$

computed at a 1-point $p \in C$ (which has an expression (14.44)). Let C be a component of $\overline{E}$ such that $A(C) = \max_{\overline{C} \subset E} A(\overline{C})$. C is nonsingular and makes SNCs with D_{X_1}. Let $\Phi_2 : X_2 \to X_1$ be the blow up of C.

Suppose that $p \in C$, so that u, v, w_i have the form (14.44) or (14.45) at p. We may assume that $x = z = 0$ are local equations of C at p. Suppose that $p_1 \in \Phi_2^{-1}(p)$. p_1 has (formal) regular parameters x_1, y_1, z_1 defined by

$$x = x_1, y = y_1, z = x_1(z_1 + \beta) \tag{14.47}$$

with $\beta \in \mathbf{k}$ or

$$x = x_1z_1, z = z_1. \tag{14.48}$$

Suppose that $p \in C$ is a 1-point, so that (14.44) holds for u, v, w_i at p. Under (14.47) we have $m_q\hat{\mathcal{O}}_{X_2,p_1}$ and thus $m_q\mathcal{O}_{X_2,p_1}$ is invertible except possibly if $\beta = 0$. If $m_q\mathcal{O}_{X_2,p_1}$ is not invertible we have

$$u = x_1^a, v = x_1^b(\alpha + y), w_i = x_1^{d+1}z_1$$

with

$$d + 1 < \min\{a, b\}.$$

Then the curve C_1 with local equations $x_1 = z_1$ is a component of the locus where $m_q\mathcal{O}_{X_2}$ is not invertible. We have

$$A(C_1) = \min\{a, b\} - (d + 1) < A(C).$$

Under (14.48) we have a 2-point

$$u = (x_1z_1)^a, v = (x_1z_1)^b(\alpha + y), w_i = x_1^d z_1^{d+1}$$

and a local equation of the toroidal structure D_{X_2} is $x_1z_1 = 0$. Since

$$d + 1 \le \min\{a, b\},$$

$m_q\mathcal{O}_{X_2,p_1}$ is invertible.

Suppose that $p \in C$ is a 2-point, so that (14.45) holds at p. Since we are assuming that $x = z = 0$ is a local equation of C at p, we have that $g < \min\{a, c\}$. Let $p_1 \in \Phi_2^{-1}(p)$. p_1 has regular parameters x_1, y_1, z_1 defined by (14.47) or (14.48).

Under the substitution (14.47), $m_q\mathcal{O}_{X_2,p_1}$ is invertible, unless we have $\beta = 0$ and

$$u = x_1^a y_1^b, v = x_1^c y_1^d, w_i = x_1^{g+1} y_1^h z_1$$

which is back in the form (14.45).

Under the substitution (14.48),

$$u = x_1^a y_1^b z_1^a, v = x_1^c y_1^d z_1^c, w_i = x_1^g y_1^h z_1^{g+1} \tag{14.49}$$

so that p_1 is a 3-point, and since

$$(g + 1, h) \le \min\{(a, b), (c, d)\},$$

$m_q\mathcal{O}_{X_2,p_1}$ is invertible.

Observe that u, v, w_i are super parameters for $f_1 \circ \Phi_2$.

By descending induction on $\max_{\overline{C}\subset E}\{A(\overline{C})\}$, we construct a sequence of blow ups $\Phi_4 : X_4 \to X_2$ such that for $\overline{f}_4 = f_1 \circ \Phi_2 \circ \Phi_4 : X_4 \to Y$, u, v, w_i are super parameters at q for $\overline{f}_4$, and $m_q\mathcal{O}_{X_4}$ is invertible. Thus $\overline{f}_4 : X_4 \to Y$ factors through the blow up $\Psi : Y_1 \to Y$ of q. Let $f_4 : X_4 \to Y_1$, $\overline{\Phi} : X_4 \to X$ be the resulting maps.

Let $q_1 \in \Psi^{-1}(q)$. We obtain permissible parameters $\overline{u}, \overline{v}, \overline{w}$ at q_1 of one of the following forms:

1. q_1 a 1-point

$$u = \overline{u}, v = \overline{u}(\overline{v} + \alpha), w_i = \overline{u}(\overline{w} + \beta)$$

with $\alpha, \beta \in \mathbf{k}$, $\alpha \neq 0$.

2. q_1 a 2-point

$$u = \overline{u}, v = \overline{uv}, w_i = \overline{u}(\overline{w}_i + \alpha) \tag{14.50}$$

with $\alpha \in \mathbf{k}$, or

$$u = \overline{uv}, v = \overline{v}, w_i = \overline{v}(\overline{w}_i + \alpha) \tag{14.51}$$

with $\alpha \in \mathbf{k}$, or

3. q_1 a 3-point

$$u = \overline{uw}_i, v = \overline{vw}_i, w_i = \overline{w}_i. \tag{14.52}$$

If q_1 has the form (14.50) or (14.51) and $p \in f_4^{-1}(q_1)$ then $\overline{u}, \overline{v}$ are toroidal forms at p.

Suppose that q_1 has the form (14.52). Let $p \in f_4^{-1}(q_1)$. u, v, w_i have one of the forms 1 - 4 of Definition 10.1 at p. Since w_i must divide u and v, we certainly have that $\overline{u}, \overline{v}, \overline{w}_i$ are monomials in the local equations of the toroidal structure at p, times unit series in $\hat{\mathcal{O}}_{X_4,p}$.

Suppose that u, v, w_i have a form 1 of Definition 10.1 at p. We have an expression

$$\begin{aligned} u &= x^a \\ v &= x^b(\alpha + y) \\ w_i &= x^c(\tilde{\gamma}(x,y) + x^{\tilde{d}}z) \end{aligned}$$

where $0 \neq \alpha \in \mathbf{k}$, $c < a$, $c < b$, $\tilde{\gamma}$ is a unit series and $\tilde{d} \geq 0$. Set $\overline{x} = x(\tilde{\gamma} + x^{\tilde{d}}z)^{\frac{1}{c}}$. We have expansions

$$\begin{aligned} w_i &= \overline{x}^c \\ u &= \overline{x}^a(\tilde{\gamma} + x^{\tilde{d}}z)^{-\frac{a}{c}} \\ v &= \overline{x}^b(\tilde{\gamma} + x^{\tilde{d}}z)^{-\frac{b}{c}}(\alpha + y) \end{aligned}$$

at p.

If $\tilde{d} > 0$ and $\frac{\partial \tilde{\gamma}}{\partial y}(0,0) = 0$, then there exist $\overline{y}, \overline{z} \in \hat{\mathcal{O}}_{X_4,p}$ such that $\overline{x}, \overline{y}, \overline{z}$ are regular parameters in $\hat{\mathcal{O}}_{X_4,p}$ and

$$\begin{aligned} w_i &= \overline{x}^c \\ v &= \overline{x}^b(\overline{\alpha} + \overline{y}) \\ u &= \overline{x}^a\hat{\gamma}(\overline{x}, \overline{y}, \overline{z}) \end{aligned}$$

where $0 \neq \overline{\alpha} \in \mathbf{k}$ and $\hat{\gamma}$ is a unit series. Then

$$\begin{aligned} \overline{w}_i &= \overline{x}^c \\ \overline{v} &= \overline{x}^{b-c}(\overline{\alpha} + \overline{y}) \\ \overline{u} &= \overline{x}^{a-c}\hat{\gamma}(\overline{x}, \overline{y}, \overline{z}) \end{aligned}$$

and $\overline{w}_i$, $\overline{v}$ are toroidal forms at p.

If $\tilde{d} = 0$ or $\frac{\partial \tilde{\gamma}}{\partial y}(0,0) \neq 0$, then there exist $\overline{y}, \overline{z} \in \hat{\mathcal{O}}_{X_4,p}$ such that $\overline{x}, \overline{y}, \overline{z}$ are regular parameters in $\hat{\mathcal{O}}_{X_4,p}$ and

$$\begin{aligned} w_i &= \overline{x}^c \\ u &= \overline{x}^a(\overline{\alpha} + \overline{y}) \\ v &= \overline{x}^b\hat{\gamma}(\overline{x}, \overline{y}, \overline{z}) \end{aligned}$$

where $0 \neq \overline{\alpha} \in \mathbf{k}$ and $\hat{\gamma}$ is a unit series. Then

$$\begin{aligned} \overline{w}_i &= \overline{x}^c \\ \overline{u} &= \overline{x}^{a-c}(\overline{\alpha} + \overline{y}) \\ \overline{v} &= \overline{x}^{b-c}\hat{\gamma} \end{aligned}$$

and $\overline{w}_i, \overline{u}$ are toroidal forms at p.

Suppose that u, v, w_i have a form 3 of Definition 10.1 at p. There are two cases. Either

$$\begin{aligned} u &= (x^a y^b)^k \\ v &= (x^a y^b)^t(\alpha + z) \\ w_i &= x^c y^d \end{aligned} \tag{14.53}$$

with $0 \neq \alpha \in \mathbf{k}$ and $ad - bc \neq 0$, or

$$\begin{aligned} u &= (x^a y^b)^k \\ v &= (x^a y^b)^t(\alpha + z) \\ w_i &= (x^a y^b)^l\tilde{\gamma} \end{aligned} \tag{14.54}$$

with $0 \neq \alpha \in \mathbf{k}$, $l \leq \min\{k, t\}$ and $\tilde{\gamma}$ is a unit series.

If (14.53) holds then $\overline{u}, \overline{w}_i$ are toroidal forms at p of the form of (4.2) of Definition 4.1.

Assume that (14.54) holds. We then have

$$\begin{aligned}\overline{u} &= (x^ay^b)^{k-l}\tilde{\gamma}^{-1}\\ \overline{v} &= (x^ay^b)^{t-l}\tilde{\gamma}^{-1}(\alpha+z)\\ \overline{w}_i &= (x^ay^b)^l\tilde{\gamma}.\end{aligned}$$

If $\frac{\partial\tilde{\gamma}}{\partial z}(0,0,0)\neq 0$, then there exist regular parameters $\overline{x},\overline{y},\overline{z}$ at p such that

$$\begin{aligned}\overline{u} &= (\overline{x}^a\overline{y}^b)^{k-l}\\ \overline{w}_i &= (\overline{x}^a\overline{y}^b)^l(\overline{\beta}+\overline{z})\\ \overline{v} &= (\overline{x}^a\overline{y}^b)^{t-l}\hat{\gamma}(\overline{x},\overline{y},\overline{z})\end{aligned}$$

where $0\neq\overline{\beta}\in\mathbf{k}$ and $\hat{\gamma}$ is a unit series. Thus $\overline{u},\overline{w}_i$ are toroidal forms at p

If $\frac{\partial\tilde{\gamma}}{\partial z}(0,0,0)=0$ then there exist regular parameters $\overline{x},\overline{y},\overline{z}$ at p such that

$$\begin{aligned}\overline{u} &= (\overline{x}^a\overline{y}^b)^{k-l}\\ \overline{v} &= (\overline{x}^a\overline{y}^b)^{t-l}(\overline{\beta}+\overline{z})\\ \overline{w}_i &= (\overline{x}^a\overline{y}^b)^l\hat{\gamma}(\overline{x},\overline{y},\overline{z})\end{aligned}$$

where $0\neq\overline{\beta}\in\mathbf{k}$ and $\hat{\gamma}$ is a unit series. Thus $\overline{u},\overline{v}$ are toroidal forms at p.

If u,v,w_i have a form 2 or 4 of Definition 10.1 at p, then a simpler analysis shows that two of $\overline{u},\overline{v},\overline{w}_i$ are toroidal forms at p, and $\overline{u},\overline{v},\overline{w}_i$ are monomials in local equations of the toroidal structure at p times unit series.

We conclude that the morphism f_4 is prepared.

We have that for $p\in\overline{f}_4^{-1}(q)$, $\tau_{f_4}(p)=-\infty$ unless $\overline{\Phi}$ factors as a sequence of blow ups of 2-curves and 3-points at p.

If $\overline{\Phi}$ factors as a sequence of blowups of 2-curves and 3-points at p, we calculate from the above local analysis that

$$\tau_{f_4}(p)\le\tau_f(\overline{\Phi}(p)). \tag{14.55}$$

Suppose that f is τ-prepared. We will verify that f_4 is τ-prepared. Let $q_1\in\Psi^{-1}(q)$ be the 3-point. By upper semi-continuity of τ_{f_4} (Lemma 9.3) it suffices to show that $\tau_{f_4}(p)<\tau$ for all $p\in f_4^{-1}(q_1)$.

Suppose that $p\in f_4^{-1}(q_1)$. Then $\overline{\Phi}(p)\in T(R_i)$ for some i. Let

$$\overline{u}=u_{\overline{R}_i^1(q_1)},\overline{v}=v_{\overline{R}_i^1(q_1)},\overline{w}_i=w_{\overline{R}_i^1(q_1)}.$$

$\overline{u},\overline{v},\overline{w}_i$ are defined by (14.52), and we have permissible parameters $\overline{x},\overline{y},\overline{z}$ in $\hat{\mathcal{O}}_{X_4,p}$ such that one of the expressions following (14.52) hold.

We will verify that $\tau_{f_4}(p)<\tau$ if u,v,w_i have the form 1 of Definition 10.1 in terms of $\overline{x},\overline{y},\overline{z}$ and $\tau>1$. We leave the other cases to the reader.

We may assume that $\tau_{\overline{f}_4}(p)=\tau$. We have an expression

$$\begin{aligned}u &= \overline{x}^a\\ v &= \overline{x}^b(\alpha+\overline{y})\\ w_i &= \textstyle\sum_{j=0}^m\alpha_{i_j}(\overline{y})\overline{x}^{i_j}+\overline{x}^n(\overline{z}+\beta)\\ &= \overline{x}^{i_0}(\gamma(x,y)+x^{n-i_0}(z+\beta))\end{aligned}$$

where $i_m<n$, $\alpha_{i_j}(\overline{y})\neq 0$ for all j and γ is a unit in $\hat{\mathcal{O}}_{X_4,p}$.

$$\tau_{\overline{f}_4}(p)=|(a\mathbf{Z}+b\mathbf{Z}+\sum_{j=0}^m i_j\mathbf{Z})/(a\mathbf{Z}+b\mathbf{Z})|=\tau.$$

We further assume that we are in the case $n-i_0>0$ and $\frac{\partial\gamma}{\partial y}(0,0)=0$.

Let

$$\tilde{x} = \overline{x}(\sum_{j=0}^{m} \alpha_{i_j}(\overline{y})\overline{x}^{i_j - i_0} + \overline{x}^{n-i_0}(\overline{z}+\beta))^{\frac{1}{i_0}},$$

$$\tilde{y} = (\sum_{j=0}^{m} \alpha_{i_j}(\overline{y})\overline{x}^{i_j - i_0} + \overline{x}^{n-i_0}(\overline{z}+\beta))^{-\frac{b}{i_0}}(\alpha+\overline{y}) - \tilde{\alpha}.$$

We have an expression

$$\begin{aligned} u &= \tilde{x}^a(P_1(\overline{y}_0, \overline{x}^{i_1-i_0}, \ldots, \overline{x}^{i_m-i_0}) + \overline{x}^{n-i_0}\Omega_1) \\ v &= \tilde{x}^b(\tilde{\alpha}+\tilde{y}) \\ w_i &= \tilde{x}^{i_0} \end{aligned}$$

where P_1 is a series, $\Omega_1 \in \hat{\mathcal{O}}_{X_4,p}$.

By iteration, substituting

$$\overline{x} - \tilde{x}(\sum_{j=0}^{m} \alpha_{i_j}(\overline{y})\overline{x}^{i_j - i_0} + \overline{x}^{n-i_0}(\overline{z}+\beta))^{-\frac{1}{i_0}}$$

and

$$\overline{y} = (\sum_{j=0}^{m} \alpha_{i_j}(\overline{y})\overline{x}^{i_j - i_0} + \overline{x}^{n-i_0}(\overline{z}+\beta))^{\frac{b}{i_0}}(\tilde{y}+\tilde{\alpha}) - \alpha$$

into $P_1(\overline{y}, \overline{x}^{i_1-i_0}, \ldots, \overline{x}^{i_m-i_0})$, we see that there exists a unit series $P(\tilde{y}, \tilde{x}^{i_1-i_0}, \ldots, \tilde{x}^{i_m-i_0})$ and a series $\Omega_2 \in \hat{\mathcal{O}}_{X_4,p}$ such that

$$u = \tilde{x}^a(P(\tilde{y}, \tilde{x}^{i_1-i_0}, \ldots, \tilde{x}^{i_m-i_0}) + \tilde{x}^{n-i_0}\Omega_2.$$

Now comparing the jacobian determinants

$$\mathrm{Det}\left(\frac{\partial(u,v,w_i)}{\partial(\overline{x},\overline{y},\overline{z})}\right)$$

and

$$\mathrm{Det}\left(\frac{\partial(u,v,w_i)}{\partial(\tilde{x},\tilde{y},\overline{z})}\right),$$

we see that if we set $\tilde{\beta} = \Omega_2(0,0,0)$ and $\tilde{z} = \Omega_2 - \tilde{\beta}$, then $\tilde{x}, \tilde{y}, \tilde{z}$ are regular parameters in $\hat{\mathcal{O}}_{X_4,p}$ and we have an expression

$$\begin{aligned} u &= \tilde{x}^a\left(P(\tilde{y}, \tilde{x}^{i_1-i_0}, \ldots, \tilde{x}^{i_m-i_0}) + \tilde{x}^{n-i_0}(\tilde{z}+\tilde{\beta})\right) \\ v &= \tilde{x}^b(\tilde{\alpha}+\tilde{y}) \\ w_i &= \tilde{x}^{i_0}. \end{aligned}$$

Thus we have an expression

$$\begin{aligned} \overline{w}_i &= \tilde{x}^{i_0} \\ \overline{v} &= \tilde{x}^{b-i_0}(\tilde{\alpha}+\tilde{y}) \\ \overline{u} &= \tilde{x}^{a-i_0}\left(P(\tilde{y}, \tilde{x}^{i_1-i_0}, \ldots, \tilde{x}^{i_m-i_0}) + \tilde{x}^{n-i_0}(\tilde{z}+\tilde{\beta})\right). \end{aligned}$$

We compute (since q_1 is a 3-point)

$$\begin{aligned} \tau_{f_4}(p) &\le |\,(i_0\mathbf{Z} + (b-i_0)\mathbf{Z} + \textstyle\sum(a-i_j)\mathbf{Z})\,/\,(i_0\mathbf{Z} + (b-i_0)\mathbf{Z} + (a-i_0)\mathbf{Z})\,| \\ &= |\,(a\mathbf{Z} + b\mathbf{Z} + \textstyle\sum i_j\mathbf{Z})\,/\,(a\mathbf{Z} + b\mathbf{Z} + i_0\mathbf{Z})\,| \\ &< |\,(a\mathbf{Z} + b\mathbf{Z} + \textstyle\sum i_j\mathbf{Z})\,/\,(a\mathbf{Z} + b\mathbf{Z})\,| \\ &= \tau_{\overline{f}_4}(p) = \tau \end{aligned}$$

since we have

$$|\,(a\mathbf{Z}+b\mathbf{Z}+i_0\mathbf{Z})\,/\,(a\mathbf{Z}+b\mathbf{Z})\,| = e_{\overline{R}_i}(q) > 1.$$

We conclude that f_4 is τ-prepared.

For j such that $q \in U(\overline{R}_j)$, let $w_j = w_{\overline{R}_j}(q)$. We will analyze our construction of $X_4 \to X$ to show that u, v, w_j are super parameters for $\overline{f}_4$.

Since f is pre-τ-quasi-well prepared, for j such that $q \in U(\overline{R}_j)$, there exists a series $\lambda_{ij}(u,v)$ such that $w_j = w_i + \lambda_{ij}(u,v)$.

The morphism $\Phi_0 \circ \Phi_1 : X_1 \to X$ which we constructed is a product of blow ups of 2-curves and 3-points. Since u, v, w_j are super parameters for f, u, v, w_j are thus also super parameters for $f \circ \Phi_0 \circ \Phi_1$.

$X_4 \to X_1$ is a sequence of blow ups of curves C such that $A(C) > 0$. Suppose that C is such a curve, and $p \in C$. It suffices to analyze the blow up $\Phi_2 : X_2 \to X_1$ of C in our construction. We saw that u, v, w_i have a form (14.44) at p with $d < \min\{a,b\}$ or a form (14.45) with $(g,h) \leq \min\{(a,b),(c,d)\}$ and $g < \min\{a,c\}$. In either case $x = z = 0$ are local equations of C. We will show that for all j such that $q \in U(\overline{R}_j)$, u, v, w_j are super parameters for $f \circ \Phi_0 \circ \Phi_1 \circ \Phi_2$.

First assume that $p \in C$ is a 1-point. Then, we have

$$u = x^a, v = x^b(\alpha + y), w_i = x^d z$$

with $0 \neq \alpha \in \mathbf{k}$ and $d < \min\{a,b\}$. Thus, since

$$w_j = w_i + \lambda_{ij}(u,v),$$

$w_j = x^d\overline{z}$ where $\overline{z} = z + x\Omega(x,y)$ for some series $\Omega \in \hat{\mathcal{O}}_{X_2,p}$. It follows that $x = \overline{z} = 0$ are local equations of C at p, and u, v, w_j have a form (10.1) of Definition 10.1 at p.

Now assume that $p \in C$ is a 2-point. Then we have

$$u = x^a y^b, v = x^c y^d, w_i = x^g y^h z$$

where after possibly interchanging u and v, we have $(g,h) < (a,b) \leq (c,d)$ (recall that $(u,v)\mathcal{O}_{X_1,p}$ is invertible). Since $w_j - w_i \in \mathbf{k}[[u,v]]$, we have an expression

$$w_j = x^g y^h \overline{z} \tag{14.56}$$

at p where $x = \overline{z} = 0$ are local equations of C at p. it follows that u, v, w_j have a form (10.2) of Definition 10.1 at p.

By induction, u, v, w_j are super parameters for $\overline{f}_4 = f \circ \Phi_0 \circ \Phi_1 \circ \Phi_2 \circ \Phi_4$.

q is an admissible center for all 2-point relations $\overline{R}_j$ associated to R (Definition 12.2). For all j, let $\overline{R}_j^1$ be the transform of $\overline{R}_j$ on Y_1.

Suppose that $q_1 \in U(\overline{R}_j^1) \cap \Psi^{-1}(q)$.

Since u, v, w_j are super parameters at all $p \in \overline{f}_4^{-1}(q)$, we have that u, v are toroidal forms at all points $p \in \overline{f}_4^{-1}(q)$. Let

$$\overline{u} = u_{\overline{R}_j^1(q_1)}, \overline{v} = v_{\overline{R}_j^1(q_1)}, \overline{w}_j = w_{\overline{R}_j^1(q_1)}$$

after possibly interchanging $\overline{u}$ and $\overline{v}$, we have an expression

$$u = \overline{u}, v = \overline{u}(\overline{v} + \alpha), w_j = \overline{u\overline{w}_j},$$

with $\alpha \in \mathbf{k}$. Thus $\overline{u}, \overline{v}$ are toroidal forms at all $p \in f_4^{-1}(q_1)$.

It follows (Definition 12.6) that the transform R^1 of R for f_4 is defined, and 2 of Definition 13.1 holds.

To verify 1 of Definition 13.1 for f_4, we must show that $p_1 \in G_{X_4}(f_4,\tau)\cap \overline{f}_4^{-1}(q)$ implies $p_1 \in T(R_j^1)$ for some j. We will verify that this is the case when p_1 is a 3-point. We leave the remaining cases to the reader. Let $q_1 = f_4(p_1)$.

Suppose that $p_1 \in \overline{f}_4^{-1}(q)$ is a 3-point and $p_1 \not\in T(R_j^1)$ for any j. We will verify that $\tau_{f_4}(p_1) < \tau$.

If $\overline{\Phi}(p_1) \not\in T(\overline{R}_j)$ for any j, then by (14.55),

$$\tau_{f_4}(p_1) \le \tau_f(\overline{\Phi}(p)) < \tau.$$

Now suppose that $p_1 \in \overline{f}_4^{-1}(q)\cap\overline{\Phi}^{-1}(T(\overline{R}_j))$. If $X_4 \rightarrow X_1$ is not an isomorphism at p_1, then our construction shows that f_4 is toroidal at p_1. Thus $X_4 \rightarrow X_1$ is an isomorphism at p_1, and Φ is a sequence of blow ups of 2-curves and 3-points at p_1. $p = \overline{\Phi}(p_1) \in T(\overline{R}_j)$ is then a 3-point with permissible parameters x, y, z such that

$$\begin{aligned} u &= x^a y^b z^c \\ v &= x^d y^e z^f \\ w_j &= M_0\gamma \end{aligned} \tag{14.57}$$

where $\mathrm{rank}(u,v) = 2$, $\gamma(x,y,z)$ is a unit series, M_0 is a monomial in x, y, z and $w_j = w_{\overline{R}_j}(q)$. Let $w_j^{e_j} = \overline{\lambda}_j u^{a_j} v^{b_j}$ define $\overline{R}_j(q)$ if $\tau > 1$, $w_j = 0$ define $\overline{R}_j(q)$ if $\tau = 1$. If $\tau > 1$, then

$$M_0^{e_j} = u^{a_j} v^{b_j} \text{and } \gamma^{e_j}(0,0,0) = \overline{\lambda}_j, \tag{14.58}$$

and $\mathrm{rank}(u,v,M_0) = 3$ if $\tau = 1$. As we have just observed, $\overline{\Phi}$ is a sequence of blow ups of 2-curves and 3-points at p_1. Since p_1 is a 3-point, we have permissible parameters x_1, y_1, z_1 at p_1 such that

$$\begin{aligned} x &= x_1^{a_{11}} y_1^{a_{12}} z_1^{a_{13}} \\ y &= x_1^{a_{21}} y_1^{a_{22}} z_1^{a_{23}} \\ z &= x_1^{a_{31}} y_1^{a_{32}} z_1^{a_{33}} \end{aligned}$$

and $\mathrm{Det}(a_{ij}) = \pm 1$. On substitution into (14.57) we see that an expression

$$\begin{aligned} u &= x_1^{\overline{a}} y_1^{\overline{b}} z_1^{\overline{c}} \\ v &= x_1^{\overline{d}} y_1^{\overline{e}} z_1^{\overline{f}} \\ w_j &= M_0\gamma \end{aligned} \tag{14.59}$$

where $\mathrm{rank}_{(x_1,y_1,z_1)}(u,v) = 2$, holds at p_1 for u, v, w_j, and the relation (14.58) holds at p_1 if $\tau > 1$, and $\mathrm{rank}_{(x_1,y_1,z_1)}(u,v,M_0) = 3$ if $\tau = 1$. We further have $\tau_{\overline{f}_4}(p_1) = \tau_f(\overline{\Phi}(p_1)) = \tau$.

Continuing to suppose that $p_1 \in \overline{f}_4^{-1}(q)\cap\overline{\Phi}^{-1}(T(\overline{R}_j))$ is a 3-point, further suppose that $p_1 \not\in f_4^{-1}(U(\overline{R}_j^1))$. Then $f_4(p_1) = q_1$ where (after possibly interchanging u and v) q_1 has permissible parameters

$$\begin{aligned} u &= \overline{u} \\ v &= \overline{u}(\overline{v}+\alpha) \\ w_j &= \overline{u}(\overline{w}_j+\beta) \end{aligned} \tag{14.60}$$

$\alpha, \beta \in \mathbf{k}$ and β is non zero, or

$$\begin{aligned} u &= \overline{u}\overline{w}_j \\ v &= \overline{v}\overline{w}_j \\ w_j &= \overline{w}_j. \end{aligned} \tag{14.61}$$

Suppose that (14.60) holds at q_1. Since $\mathrm{rank}(u,v) = 2$ in (14.59), we must have $\alpha = 0$ (and $0 \neq \beta$). But then $M_0 = u$, a contradiction to the assumption that $e_j > 1$ (and $\gcd(a_j, b_j, e_j) = 1$) in (14.58) if $\tau > 1$, or to the assumption that $\mathrm{rank}(u, v, M_0) = 3$ if $\tau = 1$.

Suppose that (14.61) holds at q_1. Then $q_1 = f_4(p)$ is a 3-point. From equations (14.61) and (14.59) we have (in the notation of Definition 9.1) that $\tau_{f_4}(p) = -\infty$ if $\tau = 1$ and if $\tau > 1$ then

$$H_{f_4,p} = H_{\overline{f}_4,p} = H_{f,\overline{\Phi}(p)},$$
$$A_{f_4,p} = A_{f,\overline{\Phi}(p)} + (a_0, b_0, c_0)\mathbf{Z}$$

since $q = \overline{f}_4(p) = f(\overline{\Phi}(p))$ is a 2-point. Thus, since $e_j > 1$ in (14.58), we have

$$\tau_{f_4}(p) = |H_{f_4,p}/A_{f_4,p}| < |H_{f,\overline{\Phi}(p)}/A_{f,\overline{\Phi}(p)}| = \tau.$$

We verify by a local calculation that 3 of Definition 13.1 holds, since we would have $\tau_{f_4}(p) < \tau$ if one of the equations of 3 of Definition 13.1 were to fail at $p \in T(R^1)$.

Now we verify 4 of Definition 13.1. Suppose that $q_1 \in U(R^1)$. We will verify 4 when q_1 is a 2-point. Let $q = \Psi(q_1) \in U(R)$. If $q_1 \in U(\overline{R}_i^1) \cap U(\overline{R}_j^1)$ then (after possibly interchanging u and v) we have

$$\begin{aligned} u &= \overline{u} \\ v &= \overline{u}\overline{v} \\ w_i &= \overline{u}\overline{w}_i \\ w_j &= \overline{u}\overline{w}_j \end{aligned}$$

where

$$u = u_{\overline{R}_i}(q) = u_{\overline{R}_j}(q), v = v_{\overline{R}_i}(q) = v_{\overline{R}_j}(q), w_i = w_{\overline{R}_i}(q), w_j = w_{\overline{R}_j}(q)$$

and

$$\overline{u} = u_{\overline{R}_i^1}(q_1) = u_{\overline{R}_j^1}(q_1), \overline{v} = v_{\overline{R}_i^1}(q_1) = u_{\overline{R}_j^1}(q_1), \overline{w}_i = w_{\overline{R}_i^1}(q_1), \overline{w}_j = w_{\overline{R}_j^1}(q_1).$$

Since f is pre-τ-quasi-well prepared, there exists a series $\lambda_{ij}(u,v)$ such that

$$w_j = w_i + \lambda_{ij}(u,v).$$

Since $\lambda_{ij}(0,0) = 0$, $\overline{u}$ divides $\lambda_{ij}(\overline{u}, \overline{u}\overline{v})$, and

$$\overline{w}_j = \overline{w}_i + \frac{\lambda_{ij}(\overline{u}, \overline{u}\overline{v})}{\overline{u}}.$$

Thus 4 of Definition 13.1 holds for f_4.

Earlier in the proof we verified that if $q \in U(\overline{R}_j)$, for some $\overline{R}_j$ associated to R, then

$$u = u_{\overline{R}_j}(q), v = v_{\overline{R}_j}(q), w_j = w_{\overline{R}_j}(q)$$

are super parameters for $\overline{f}_4$ at q. If $q_1 \in U(\overline{R}_j^1) \cap \Psi^{-1}(q)$, then (after possibly interchanging u and v) we have permissible parameters at q_1

$$\overline{u} = u_{\overline{R}_j^1}(q_1), \overline{v} = v_{\overline{R}_j^1}(q_1), \overline{w}_j = w_{\overline{R}_j^1}(q_1)$$

such that

$$u = \overline{u}, v = \overline{u}(\overline{v} + \alpha), w_j = \overline{u}\overline{w}_j$$

with $\alpha \in \mathbf{k}$.

Substituting into the forms of Definition 10.1 for $\overline{f}_4$, we see that $\overline{u}, \overline{v}, \overline{w}_j$ are super parameters for f_4 at q_1. Thus 5 of Definition 13.1 holds for f_4 and we have verified that f_4 is pre-τ-quasi-well prepared.

Now suppose that f is τ-well prepared. We will verify that f_4 is τ-well prepared. 1 and 2 of Definition 13.5 are immediate.

We will verify that 3 of Definition 13.5 holds for f_4. Suppose that $q_1 \in U(\overline{R}_i^1) \cap U(\overline{R}_j^1)$ and

$$\overline{u} = u_{\overline{R}_i^1}(q_1) = u_{\overline{R}_j^1}(q_1), \overline{v} = v_{\overline{R}_i^1}(q_1) = v_{\overline{R}_j^1}(q_1),$$
$$\overline{w}_i = w_{\overline{R}_i^1}(q_1), \overline{w}_j = w_{\overline{R}_j^1}(q_1).$$

Let $q = \Psi(q_1)$,

$$u = u_{\overline{R}_i}(q) = u_{\overline{R}_j}(q), v = v_{\overline{R}_i}(q) = v_{\overline{R}_j}(q),$$
$$w_i = w_{\overline{R}_i}(q), w_j = w_{\overline{R}_j}(q).$$

We will verify 3 in the case when q and q_1 are 2-points. The remaining cases are similar. Since f is τ-well prepared, there exist unit series $\phi_{ij}(u, v)$ such that

$$w_j = w_i + u^{a_{ij}} v^{b_{ij}} \phi_{ij}$$

(or $\phi_{ij} = 0$ and $a_{ij} = b_{ij} = -\infty$). After possibly interchanging u and v, we may assume that

$$u = \overline{u}, v = \overline{uv}, w_i = \overline{uw}_i, w_j = \overline{uw}_j.$$

Let

$$\overline{\phi}_{ij}(\overline{u}, \overline{v}) = \phi_{ij}(\overline{u}, \overline{uv}).$$

Then we have

$$\overline{w}_j = \overline{w}_i + \overline{u}^{\overline{a}_{ij}} \overline{v}^{\overline{b}_{ij}} \overline{\phi}_{ij} \tag{14.62}$$

where

$$\overline{a}_{ij} = a_{ij} + b_{ij} - 1, \overline{b}_{ij} = b_{ij}. \tag{14.63}$$

Thus 3 of Definition 13.5 holds for f_4.

Since the set (13.9) associated to q and R is totally ordered, the set (13.9) associated to q_1 and R^1 is also totally ordered. Thus 4 of Definition 13.5 holds for f_4, and we see that f_4 is τ-well prepared.

We now verify 2 of Lemma 14.7. Suppose that $q_1 \in U(\overline{R}_i^1) \cap E$ for some $\overline{R}_i^1$ associated to R^1. Continuing with the notation we used in the verification that f_4 is τ-well prepared, let $\gamma_i = \overline{S_{\overline{R}_i^1}(q_1) \cdot E}$. γ_i is covered by two affine charts, with uniformizing parameters $\overline{u}, \overline{v}, \overline{w}_i$ and $\tilde{u}, \tilde{v}, \tilde{w}_i$ defined by

$$u = \overline{u}, v = \overline{uv}, w_i = \overline{uw}_i \tag{14.64}$$

and

$$u = \tilde{u}\tilde{v}, v = \tilde{v}, w_i = \tilde{v}\tilde{w}_i. \tag{14.65}$$

In the chart (14.64), $\overline{u} = 0$ is a local equation for E, and $\overline{v} = 0$ is a local equation for the strict transform of the component E_2 of D_Y with local equation $v = 0$ at q. In (14.65), $\tilde{v} = 0$ is a local equation of E and $\tilde{u} = 0$ is a local equation of the strict transform of the component E_1 of D_Y with local equation $u = 0$ at q. In the chart defined by (14.64) $\overline{u} = \overline{w}_i = 0$ are local equations of γ_i and in the chart defined by (14.65) $\tilde{v} = \tilde{w}_i = 0$ are local equations of γ_i. Thus γ_i makes SNCs with D_{Y_1}, and γ_i is a line on $E \cong \mathbf{P}^2$.

If $q \in U(\overline{R}_j)$ for some $j \neq i$, then we see from (14.62) and (14.63) that $\gamma_j = \overline{S_{\overline{R}_j^1}(q') \cdot E}$ (where $q' \in \Psi^{-1}(q) \cap U(\overline{R}_j^1)$) has local equations $\overline{u} = \overline{w}_j = 0$ in the chart (14.64) where

$$\overline{w}_j = \overline{w}_i + \overline{u}^{a_{ij}+b_{ij}-1}\overline{v}^{b_{ij}}\overline{\phi}_{ij}.$$

In the chart (14.65), γ_j has local equations $\tilde{v} = \tilde{w}_j = 0$ where $\tilde{w}_j = \tilde{w}_i + \tilde{u}^{a_{ij}}\tilde{v}^{a_{ij}+b_{ij}-1}\tilde{\phi}_{ij}$. If $a_{ij} + b_{ij} > 1$, then $\gamma_i = \gamma_j$ so that 2 (a) (i) holds in the statement of Lemma 14.7. If $a_{ij} + b_{ij} = 1$ then 2 (a) (ii) of the statement of Lemma 14.7 holds.

γ_i is a prepared curve for R^1 of type 6 since $\gamma_i \subset U(\overline{R}_i^1)$ for all i.

We now verify 2 (b) of Lemma 14.7. Suppose that γ is prepared for R, $q \in \gamma$, and γ is prepared for R of type 6. Then $\gamma = \overline{E_2 \cdot S_{\overline{R}_i}(q)}$ for some $\overline{R}_i$ and component E_2 of D_Y containing q and $\gamma \subset \Omega(\overline{R}_i)$. Let $u = u_{\overline{R}_i}(q), v = v_{\overline{R}_i}(q), w_i = w_{\overline{R}_i}(q)$. We may assume that $v = 0$ is a local equation of E_2. Then $v = w_i = 0$ are local equations at q of γ. Let γ' be the strict transform of γ on Y_1. $q_1 = \gamma' \cdot E$ has permissible parameters

$$\overline{u} = u_{\overline{R}_i^1}(q_1), \overline{v} = v_{\overline{R}_i^1}(q_1), \overline{w}_i = w_{\overline{R}_i^1}(q_1),$$

where

$$u = \overline{u}, v = \overline{uv}, w_i = \overline{uw}_i,$$

and $\overline{v} = \overline{w}_i = 0$ are local equations of γ'. Let $\tilde{E}_2$ be the strict transform of E_2. Since $q_1 \in U(\overline{R}_i^1)$, we have $\gamma' = \overline{\tilde{E}_2 \cdot S_{\overline{R}_i^1}(q_1)}$ and $\gamma' \subset \Omega(\overline{R}_i^1)$. Thus the conditions of Definition 13.8 hold for γ', and γ' is prepared for R^1 of type 6. If γ is a 2-curve on Y through q, then the strict transform γ' of γ on Y_1 is a 2-curve so γ' is prepared for R^1.

Finally, suppose that f is τ-very-well prepared. We have shown that 1 and 2 of Definition 13.9 hold for f_4. Since whenever $\overline{R}_i$ is a pre-relation associated to R containing q, $\Omega(\overline{R}_i^1) \rightarrow \Omega(\overline{R}_i)$ is the blow up of a point on a nonsingular surface, and $V_i(Y)$ satisfies 3 of Definition 13.9, 3 of Definition 13.9 holds for $V_i(Y_1)$. Thus f_4 is τ-very-well prepared. □

LEMMA 14.8. *Suppose that $f : X \rightarrow Y$ is pre-τ-quasi-well prepared (or τ-well prepared or τ-very-well prepared), $q \in U(R)$ is a 1-point (which is prepared of type 4 of Definition 13.7). Then there exists a pre-τ-quasi-well prepared (or τ-well prepared or τ-very-well prepared) diagram*

$$\begin{array}{ccc} X_1 & \overset{f_1}{\rightarrow} & Y_1 \\ \Phi \downarrow & & \downarrow \Psi \\ X & \overset{f}{\rightarrow} & Y \end{array} \tag{14.66}$$

where Ψ is the blow up of q such that

1. *Φ is a sequence of blow ups of 2-curves over $f^{-1}(Y - \{q\})$.*
2. *Suppose that f is τ-well prepared.*
   (a) *Let E be the exceptional divisor of Ψ. Suppose that $q_1 \in U(\overline{R}_i^1) \cap E$. Let $\gamma_i = \overline{S_{\overline{R}_i^1}(q_1) \cdot E}$. Then γ_i is a prepared curve for R^1 of type 6. Suppose that $q' \in U(\overline{R}_j^1) \cap E$. Let $\gamma_j = \overline{S_{\overline{R}_j^1}(q') \cdot E}$. Then either*
      (i) *$\gamma_i = \gamma_j$ or*

(ii) *γ_i, γ_j intersect transversally at a 2 point on E (their tangent spaces have distinct directions at this point) and γ_i, γ_j are otherwise disjoint.*

(b) *If γ is a prepared curve on Y then the strict transform of γ is a prepared curve on Y_1.*

3. *If f is τ-prepared then f_1 is τ prepared.*

The proof of Lemma 14.8 is a variant of the proof of Lemma 14.7, keeping in mind the simpler forms 5 and 6 of Definition 10.1 of super parameters above a 1-point, and the simpler from (13.8) of 3 of Definition 13.5 at q.

LEMMA 14.9. *Suppose that $f : X \rightarrow Y$ is pre-τ-quasi-well prepared (or τ-well prepared or τ-very-well prepared) with relation R. Suppose that $q \in Y$ is a 1-point or a 2-point such that $q \notin U(R)$ and q is prepared of type 2 of Definition 13.7 for R. Then q is a permissible center for R and there exists a pre-τ-quasi-well prepared (or τ-well prepared or τ-very-well prepared) diagram (13.10) of R and the blow up $\Psi : Y_1 \rightarrow Y$ of q such that:*

1. *Φ is an isomorphism over $f^{-1}(Y - \Sigma(Y))$ if q is a 2-point.*
2. *Suppose that f is τ-well prepared. If γ is a prepared curve on Y then the strict transform of γ is a prepared curve on Y_1.*

The proof of Lemma 14.9 is a simplification of the proofs of Lemma 14.7 and Lemma 14.8.

LEMMA 14.10. *Suppose that $f : X \rightarrow Y$ is pre-τ-quasi-well prepared, $\overline{q} \in Y$ is a 1-point such that $\overline{q} \notin U(R)$, and $C \subset D_Y$ is an integral curve such that $\overline{q} \in C$. Suppose that C satisfies 2 and 4 of Definition 13.6 of a resolving curve. Then there exists a pre-τ-quasi-well prepared diagram*

$$\begin{array}{ccc} X_1 & \stackrel{f_1}{\rightarrow} & Y_1 \\ \Phi_1 \downarrow & & \downarrow \Psi_1 \\ X & \stackrel{f}{\rightarrow} & Y \end{array}$$

such that

1. *Ψ_1 is a product of blow ups of 2-curves and 2-points, Φ_1 is a product of possible blow ups, Φ_1 is an isomorphism over $f^{-1}(Y - \Sigma(Y))$.*
2. *Let $\overline{C}$ be the strict transform of C on Y_1. Then $\overline{C}$ is a resolving curve for f_1 and R^1 at $\overline{q}$.*
3. *If f is τ-prepared, then f_1 is τ-prepared.*

PROOF. There exists a sequence of blow ups of 2-curves $\Psi_1 : Y_1 \rightarrow Y$ such that the strict transform C_1 of C on Y_1 contains no 3-points. Let

$$\begin{array}{ccc} X_1 & \stackrel{f_1}{\rightarrow} & Y_1 \\ \Phi_1 \downarrow & & \downarrow \Psi_1 \\ X & \stackrel{f}{\rightarrow} & Y \end{array} \tag{14.67}$$

be the pre-τ-quasi-well prepared diagram obtained by iterating the construction of Lemma 14.1.

Now by Lemma 14.1 and Lemma 10.7, there exists a pre-τ-quasi-well prepared diagram

$$\begin{array}{ccc} X_2 & \stackrel{f_2}{\rightarrow} & Y_2\\ \Phi_2\downarrow & & \downarrow\Psi_2\\ X_1 & \stackrel{f_1}{\rightarrow} & Y_1\end{array} \tag{14.68}$$

obtained by iterating the construction of Lemma 14.1 such that Ψ_2 and Ψ_2 are products of blow ups of 2-curves, and for all 2-points q on the strict transform C_2 of C on Y_2, there exist super parameters u, v, w at q.

Let $\Psi_3 : Y_3 \rightarrow Y_2$ be the blow up of the 2-points on C_2. By our construction, these points are disjoint from $G_{Y_2}(f_2, \tau)$. By Lemma 14.9, there exists a pre-τ-quasi-well prepared diagram

$$\begin{array}{ccc} X_3 & \stackrel{f_3}{\rightarrow} & Y_3\\ \Phi_3\downarrow & & \downarrow\Psi_3\\ X_2 & \stackrel{f_2}{\rightarrow} & Y_2.\end{array} \tag{14.69}$$

By iterating the above construction, (and by embedded resolution of plane curve singularities, c.f. Section 3.4, Exercise 3.13 [**C6**]) we eventually construct a pre-τ-quasi-well prepared diagram

$$\begin{array}{ccc} X' & \stackrel{f'}{\rightarrow} & Y'\\ \Phi'\downarrow & & \downarrow\Psi'\\ X & \stackrel{f}{\rightarrow} & Y\end{array}$$

such that the strict transform C' of C on Y' is nonsingular and makes SNCs with $D_{Y'}$. If $q \in C'$ is a 2-point, let u, v, w be permissible parameters at q such that $u = w = 0$ are local equations of C'.

By Lemma 14.1 and Lemma 10.7, there exists a pre-τ-quasi-well prepared diagram

$$\begin{array}{ccc} X'' & \stackrel{f''}{\rightarrow} & Y''\\ \Phi''\downarrow & & \downarrow\Psi''\\ X' & \stackrel{f'}{\rightarrow} & Y'\end{array}$$

such that 3 of Definition 13.6 holds for the strict transform of C on Y''. Thus the conclusions of the lemma hold. □

LEMMA 14.11. *Suppose that $f : X \rightarrow Y$ is τ-very-well prepared and $C \subset Y$ is a prepared curve of type 6 (of Definition 13.8). Further suppose that $q_\delta \in C \cap U(\overline{R}_j)$ for some $\overline{R}_j$ associated to R implies $C = \overline{E \cdot S_{\overline{R}_j}(q_\delta)}$ for some component E of D_X. Then C is a *-permissible center for R, and there exists a τ-very-well prepared diagram*

$$\begin{array}{ccc} X_1 & \stackrel{f_1}{\rightarrow} & Y_1\\ \Phi_1\downarrow & & \downarrow\Psi_1\\ X & \stackrel{f}{\rightarrow} & Y\end{array}$$

of R of the form of (13.11).

PROOF. Let $C = \overline{E_\alpha \cdot S_{\overline{R}_i}(q_\beta)}$. For $\overline{q} \in C$, we have permissible parameters

$$u, v, \tilde{w}_i \tag{14.70}$$

such that $u = \tilde{w}_i = 0$ are local equations at $\overline{q}$ for C, with the notation of 5 of Definition 13.8, if $\overline{q} \notin U(\overline{R}_i)$, and $u = u_{\overline{R}_i}(\overline{q})$, $v = v_{\overline{R}_i}(\overline{q})$, $\tilde{w}_i = w_{\overline{R}_i}(\overline{q})$ if $\overline{q} \in U(\overline{R}_i)$. As in the proof of Lemma 14.7, after blowing up 2-curves and 3-points above X, by a morphism $\Phi_0 \circ \Phi_1 : X_1 \to X$, with associated morphism $f_1 = f \circ \Phi_0 \circ \Phi_1 : X_1 \to Y$, we have that the following holds.

1. If $\overline{q} \in C$ is a 2-point, and $\overline{p} \in f_1^{-1}(\overline{q})$ and $\mathcal{I}_C \mathcal{O}_{X_1,\overline{p}}$ is not invertible, then one of the forms (14.44) or (14.45) hold at $\overline{p} \in f_1^{-1}(\overline{q})$ (with $d < a$ in (14.44), $(g,h) < (a,b)$ in (14.45)).
2. If $\overline{q} \in C$ is a 1 point then a form (14.71) below holds at $\overline{p} \in f_1^{-1}(\overline{q})$ if $\mathcal{I}_C \mathcal{O}_{X_1,\overline{p}}$ is not invertible.

$$
(14.71) \qquad \begin{aligned} u &= x^a \\ v &= y \\ \tilde{w}_i &= x^d z \end{aligned}
$$

where $d < a$ and $\overline{p}$ is a 1-point.

As in the proof of Lemma 14.7, the locus of points in X_1 where $\mathcal{I}_C \mathcal{O}_{X_1}$ is not invertible is a (possibly reducible) curve $\overline{E}$ which makes SNCs with the toroidal structure of X_1. As in the proof of Lemma 14.7, we can construct a sequence of blow ups of sections over components of $\overline{E}$, $X_4 \to X_1$, such that the resulting map $\overline{f}_4 : X_4 \to Y$ factors through the blow up $\Psi_1 : Y_1 \to Y$ of C. Let $\overline{\Phi} : X_4 \to X$ be the composite map. By our construction, if $u, v, \tilde{w}_i$ are our permissible parameters at $\overline{q} \in C$ of (14.70), then $u, v, \tilde{w}_i$ are super parameters for $\overline{f}_4$ at $\overline{q}$.

Let $f_4 : X_4 \to Y_1$ be the resulting morphism. As in the proof of Lemma 14.7, we see that f_4 is prepared.

Suppose that $\overline{q} \in C \cap U(\overline{R}_j)$ for some j. Then $\overline{q} \in U(\overline{R}_i)$, since C is prepared of type 6. By the hypothesis of this lemma, the germ of C at $\overline{q}$ is contained in $S_{\overline{R}_j}(\overline{q})$. Since $C \subset E_\alpha$, the germ of C at $\overline{q}$ is $E_\alpha \cdot S_{\overline{R}_j}(\overline{q})$.

Suppose that $\overline{q}$ is a 2-point. The case where $\overline{q}$ is a 1-point is simpler. Then in the forms of (13.7) of Definition 13.5 for R (after possibly interchanging u and v), we have $a_{jk} > 0$ for all $j, k \in I_{\overline{q}}$.

$\Psi_1^{-1}(\overline{q})$ is covered by 2 affine charts. The first chart has uniformizing parameters $\overline{u}, \overline{v}, \overline{w}_i$ defined by

$$
(14.72) \qquad u = \overline{u}, v = \overline{v}, \tilde{w}_i = \overline{u}\overline{w}_i.
$$

The second chart has uniformizing parameters u', v', w_i' defined by

$$
(14.73) \qquad u = u'w_i', v = v', \tilde{w}_i = w_i'.
$$

For $j \in I_{\overline{q}}$ we have a relation

$$
\tilde{w}_j = \tilde{w}_i + u^{a_{ij}} v^{b_{ij}} \phi_{ij}(u, v)
$$

with $a_{ij} > 0$ (where $\tilde{w}_j = w_{\overline{R}_j}(\overline{q})$). As in the proof of Lemma 14.7, it follows that f_4 is τ-prepared, the transform R^1 of R for f_4 is defined, and f_4 is pre-τ-quasi-well prepared.

We now verify that f_4 is τ-well prepared. 1 of Definition 13.5 is immediate.

Suppose that $C \cap U(\overline{R}_j) \neq \emptyset$ for some j. Then $C = \overline{E_\alpha \cdot \overline{R}_j(q_\beta)} \subset \Omega(\overline{R}_j)$. Since $\Omega(\overline{R}_j)$ is nonsingular and makes SNCs with D_Y, $\Omega(\overline{R}_j^1)$ is nonsingular, makes SNCs with D_{Y_1}, contains $U(\overline{R}_j^1)$, and $\Omega(\overline{R}_j^1) \cap U(R^1) = U(\overline{R}_j^1)$. If $C \cap U(\overline{R}_j) = \emptyset$, then after possibly replacing $\Omega(\overline{R}_j)$ with a neighborhood of F_j in $\Omega(\overline{R}_j)$ (with the notation of Definition 13.9, and following our convention on $\Omega(\overline{R}_j)$ stated after Definition 13.9), we have that $\Omega(\overline{R}_j)\cap C=\emptyset$. Thus 2 of Definition 13.5 holds for f_4.

We now verify 3 and 4 of Definition 13.5 for f_4. Suppose that $\overline{q}\in C\cap U(\overline{R}_j)$ is a 2-point. The case where $\overline{q}$ is a 1-point is simpler. Then $q_j=U(\overline{R}_j^1)\cap f_4^{-1}(\overline{q})$ is a 2-point in the chart (14.72), and

$$w_{\overline{R}_j^1}(q_j)=\overline{w}_j=\frac{\tilde{w}_j}{u}=\overline{w}_i+\overline{u}^{a_{ij}-1}\overline{v}^{b_{ij}}\phi_{ij}(\overline{u},\overline{v}). \tag{14.74}$$

Thus $q_j\in U(\overline{R}_i^1)$ if and only if $a_{ij}-1+b_{ij}>0$. Let $q_i=U(\overline{R}_i^1)\cap f_4^{-1}(\overline{q})$,

$$I_{q_i}=\{j\mid q_i\in U(\overline{R}_j^1)\}.$$

$j\in I_{q_i}$ if and only if $j\in I_{\overline{q}}$ and $a_{ij}+b_{ij}>1$.

From equation (14.74) we see that the set (13.9) of Definition 13.5 corresponding to q_i and R^1 is totally ordered, since the set (13.9) of Definition 13.5 corresponding to $\overline{q}$ and R is totally ordered. In particular, we see that 3 and 4 of Definition 13.5 hold for f_4. We have completed the verification that f_4 is τ-well prepared.

Let $E=\Psi_1^{-1}(C)$ and if $C\cap U(\overline{R}_j)\neq\emptyset$, let $\gamma_j=\Omega(\overline{R}_j^1)\cdot E$. γ_j is also nonsingular and makes SNCs with D_{Y_1}. We have that $\gamma_j=\overline{E\cdot S_{\overline{R}_j^1}(q_1)}$ for all $q_1\in E\cap U(\overline{R}_j^1)$, and γ_j is a section of E over C. In particular, 1 and 2 of Definition 13.8 hold for γ_j.

Suppose that for some $\overline{R}_k$ associated to R, there exists $\gamma'\subset\Omega(\overline{R}_k^1)$, with $\gamma'=\overline{E_\beta\cdot S_{\overline{R}_k^1}(q_\delta)}$, and $\gamma'\cap\gamma_j\neq\emptyset$.

First suppose that $\gamma'\subset E$. Then $U(\overline{R}_k^1)\cap E\neq\emptyset$ and $\gamma'=\gamma_k=\overline{E\cdot S_{\overline{R}_k^1}(q_2)}$ for all $q_2\in U(\overline{R}_k^1)\cap E$. Let $F=E_\alpha$ be the component of D_Y containing C. Since C is prepared of type 6 for R and $\Psi_1(\gamma_j)=\Psi_1(\gamma_k)=C$, we have that $C\cap U(\overline{R}_j)=C\cap U(\overline{R}_k)$ and $C=\overline{F\cdot S_{\overline{R}_j}(\overline{q})}=\overline{F\cdot S_{\overline{R}_k}(\overline{q})}$ for $\overline{q}\in C\cap U(\overline{R}_j)$.

Suppose that $q_2\in\gamma_j\cap\gamma_k$. Let $\overline{q}=\Psi_1(q_2)\in C$. Let $u,v,\tilde{w}_i$ and $u,v,\tilde{w}_k$ be the permissible parameters at $\overline{q}$ of (14.70).

Suppose that $\overline{q}$ is a 1-point. We have

$$\tilde{w}_j=\tilde{w}_k+u^{c_{kj}}\phi_{kj}(u,v),$$

where ϕ_{kj} is a unit series (or $\phi_{kj}=0$) by 5 (d) of Definition 13.8, since C is prepared for R of type 6. We have that q_2 has permissible parameters $\overline{u},\overline{v},\overline{w}_j$ where

$$u=\overline{u},v=\overline{v},\tilde{w}_j=\overline{w}_j\overline{u}. \tag{14.75}$$

Define $\overline{w}_k$ by $\tilde{w}_k=\overline{w}_k\overline{u}$. We have that

$$\overline{w}_j=\overline{w}_k+\overline{u}^{c_{kj}-1}\phi_{kj}(\overline{u},\overline{v}).$$

$\overline{u}=\overline{w}_j=0$ are local equations of γ_j at q_2 and $\overline{u}=\overline{w}_k=0$ are local equations of γ_k at q_2. Thus q_2 is a 1-point with $c_{jk}-1>0$ and $\gamma_j=\gamma_k$.

Now suppose that $q_2\in\gamma_j\cap\gamma_k$ and $\overline{q}=\Psi_1(q_2)\in C$ is a 2-point. First suppose that $\overline{q}\notin C\cap U(\overline{R}_j)=C\cap U(\overline{R}_k)$. We have a relation

$$\tilde{w}_j=\tilde{w}_k+u^{a_{kj}}v^{b_{kj}}\phi_{kj}(u,v)$$

where ϕ_{jk} is a unit series (or $\phi_{kj} = 0$) by 5 (d) of Definition 13.8 for C. We have $a_{kj} \ge 1$. q_2 has permissible parameters $\overline{u}, \overline{v}, \overline{w}_j$ where

$$u = \overline{u}, v = \overline{v}, \tilde{w}_j = \overline{w}_j \overline{u}. \tag{14.76}$$

Define $\overline{w}_k$ by $\tilde{w}_k = \overline{w}_k \overline{u}$. We then have

$$\overline{w}_j = \overline{w}_k + \overline{u}^{a_{kj}-1} \overline{v}^{b_{kj}} \phi_{kj}(\overline{u}, \overline{v}). \tag{14.77}$$

Thus q_2 is a 2-point (and $q_2 \notin U(R^1)$).

Finally, suppose that $q_2 \in \gamma_j \cap \gamma_k$ and $\overline{q} = \Psi_1(q_2) \in C \cap U(\overline{R}_j) = C \cap U(\overline{R}_k)$. Then $q_2 \in U(\overline{R}_j^1) \cap U(\overline{R}_k^1)$, and we have equations (14.76) and (14.77).

Now suppose that $\gamma' = \overline{E_\beta \cdot S_{\overline{R}_k^1}(q_\delta)} \not\subset E$, and $\gamma' \cap \gamma_j \ne \emptyset$. Then there exists a component G of D_Y such that $\overline{\gamma} = \Psi_1(\gamma') \subset G$, and E_β is the strict transform of G. Suppose that $q_2 \in \gamma' \cap \gamma_j$. $\overline{q} = \Psi_1(q_2) \in \overline{\gamma} \cap C$ implies $\overline{q} \in U(\overline{R}_j) \cap U(\overline{R}_k)$ and $\overline{\gamma} = \overline{G \cdot S_{\overline{R}_k}(\overline{q})}$ by 3 of Definition 13.8 for R. Thus $q_2 \in U(\overline{R}_j^1) \cap U(\overline{R}_k^1)$.

Suppose that $q_2 \in \gamma_j$, $\overline{q} = \Psi_1(q_2) \in C$ and $p \in f_4^{-1}(q_2)$. Let $u, v, \tilde{w}_j$ be the permissible parameters at $\overline{q}$ of (14.70). Further suppose that q_2 is a 1-point. Let $\overline{u}, \overline{v}, \overline{w}_j$ be the permissible parameters at q_2 of (14.75). $u, v, \tilde{w}_j$ satisfy 5 (b) of Definition 13.8 at p. Substituting (14.75) into these forms, we see that $\overline{u}, \overline{v}, \overline{w}_j$ satisfy 5 (b) of Definition 13.8 at p. Now suppose that $q_2 \in \gamma_j$ is a 2-point, but $q_2 \notin U(\overline{R}_j^1)$. Let $\overline{u}, \overline{v}, \overline{w}_j$ be the permissible parameters at q_2 of (14.76). $u, v, \tilde{w}_j$ satisfy 5 (b) of Definition 13.8 at p. Substituting (14.76) into these forms, we see that $\overline{u}, \overline{v}, \overline{w}_j$ satisfy 5 (b) of Definition 13.8 at p. Thus 5 of Definition 13.8 holds for γ_j.

We have seen that the curves γ_j only fail to be prepared of type 6 for f_4 at a finite set of 2-points $T_1 \subset E$, where condition 3 of Definition 13.8 fails. If $q \in T_1$, then $q \notin U(R^1)$ and there exist γ_j and γ_k such that $\gamma_j \ne \gamma_k$ and $q \in \gamma_j \cap \gamma_k$. For $q \in T_1$, let

$$J_q = \{j \mid q \in \gamma_j = \overline{E \cdot S_{\overline{R}_j^1}(q_\epsilon)} \text{ for some } q_\epsilon \in U(\overline{R}_j^1)\}.$$

Observe that:

1. For $q \in T_1$, $j \in J_q$ and $p \in f_4^{-1}(q)$, the permissible parameters $\overline{u}, \overline{v}, \overline{w}_j$ at q defined by (14.72) have a form 5 (b) of Definition 13.8.
2. For $i, j \in J_q$ there exists a relation of the form 5 (d) of Definition 13.8 for $\overline{w}_i$ and $\overline{w}_j$, and the set $\{(a_{ij}, b_{ij})\}$ is totally ordered.

In particular, the points in T_1 are prepared for R^1 of type 2 of Definition 13.7. Let $\Psi_2 : Y_2 \to Y_1$ be the blow up of T_1, and let

$$\begin{array}{ccc} X_5 & \overset{f_5}{\to} & Y_2 \\ \downarrow & & \downarrow \Psi_2 \\ X_4 & \overset{f_4}{\to} & Y_1 \end{array}$$

be the τ-well prepared diagram of Lemma 14.9. Let

$$T_2 = \left\{ \begin{array}{l} q \in \Psi_2^{-1}(T_1) \text{ such that } q \in \gamma_i^2 \cap \gamma_k^2 \\ \text{where } \gamma_i^2, \gamma_j^2 \text{ are the strict transforms of some} \\ \gamma_i, \gamma_k \text{ such that } \gamma_i \ne \gamma_k \text{ and } \gamma_i \cap \gamma_k \ne \emptyset \end{array} \right\}.$$

The points of T_2 must again be prepared for the transform R^2 of R on X_5 of type 2 of Definition 13.7, and the points of T_2 satisfy the corresponding statements 1 and 2 above that the points of T_1 and J_q satisfy.

We can iterate this process a finite number of times to produce a τ-well prepared diagram

$$\begin{array}{ccc} \overline{X}_m & \overset{\overline{f}_m}{\to} & \overline{Y}_m \\ \downarrow & & \downarrow \\ \overline{X}_1 = X_4 & \overset{f_4}{\to} & \overline{Y}_1 = Y_1 \\ \downarrow & & \downarrow \\ X & \overset{f}{\to} & Y \end{array}$$

of the form of (13.12) such that the strict transform of the γ_i are disjoint on $\overline{Y}_m$ above T_1. It follows that the strict transforms of the γ_i are prepared of type 6 for the transform R^m of R on $\overline{X}_m$.

To show that $\overline{f}_m$ is τ-very-well prepared, it only remains to verify that the strict transform γ' of a curve $\gamma = \overline{E_\beta \cdot S_{\overline{R}_i}(q)}$ on Y (with $\gamma \neq C$ and $\gamma \cap C \neq \emptyset$) is prepared on $\overline{Y}_m$. Since γ is prepared of type 6, we have that $\gamma \cap C \subset U(R_i)$. By our previous analysis, we then know that $\gamma' \cap E \subset U(R^1)$, so that $\overline{Y}_m \to Y_1$ is an isomorphism in a neighborhood of γ'. It thus suffices to check that γ' is prepared of type 6 on Y_1. This follows by a local analysis.

□

REMARK 14.12. 1. *Suppose that $f : X \to Y$ is pre-τ-quasi-well prepared and $C \subset D_Y$ is a nonsingular (integral) curve which makes SNCs with D_Y and contains a 1-point such that*

(a) *$q \in C \cap U(\overline{R}_i)$ for some pre-relation $\overline{R}_i$ associated to R implies the (formal) germ of C at q is contained in $S_{\overline{R}_i}(q)$, and*

(b) *$q \in C - U(R)$ implies that one of the following holds*

   (i) *there exist super parameters u, v, w at q such that $u = w = 0$ are local equations of C at q.*

   (ii) *q is a 1-point with $\tau_f(p) < \tau$ for all $p \in f^{-1}(q)$ and there exist permissible parameters u, v, w at q such that u, v are toroidal forms at p for all $p \in f^{-1}(q)$ and $u = v = 0$ are local equations of C at q.*

Then there exists a pre-τ-quasi-well prepared diagram

$$\begin{array}{ccc} \overline{X}_1 & \overset{\overline{f}_1}{\to} & \overline{Y}_1 \\ \overline{\Phi}_1 \downarrow & & \downarrow \overline{\Psi}_1 \\ X & \overset{f}{\to} & Y \end{array}$$

where $\overline{\Psi}_1$ is the blow up of C. If $C \not\subset G_Y(f, \tau)$ and f is τ-prepared then $\overline{f}_1$ is τ-prepared.

2. *Further suppose that $f : X \to Y$ is τ-well prepared, and if $\gamma = \overline{E_\beta \cdot R_k(q_\alpha)}$ is prepared for R of type 6, then either $C = \gamma$ or $q \in C \cap \gamma$ implies $q \in U(\overline{R}_k)$ and the germ of C at q is contained in $S_{R_k}(q)$. Then there exists a τ-well prepared diagram*

$$\begin{array}{ccc} X_1 & \overset{f_1}{\to} & Y_1 \\ \Phi \downarrow & & \downarrow \Psi \\ X & \overset{f}{\to} & Y \end{array}$$

where Ψ *is the blow up* $\overline{\Psi}_1$ *of* C*, possibly followed by blow ups of 2-points which are prepared for the transform of* R *(of type 2 of Definition 13.7) if* C *is prepared of type 6 for* R*, such that*

(a) *If* $\gamma \subset Y$ *is prepared for* f*, (and* $\gamma \neq C$*) then the strict transform of* γ *is prepared for* f_1*.*

(b) *If* $C \subset Y$ *is prepared for* f *(of type 6) and* $q \in U(\overline{R}_i) \cap C$ *for some* i*, then* $\overline{E \cdot S_{\overline{R}_i^1}(q')}$ *is prepared for* f_1 *(of type 6) for all* $q' \in \Psi^{-1}(q) \cap U(\overline{R}_i^1)$*, where* E *is the component of* D_{Y_1} *dominating* C*.*

The proof of Remark 14.12 is a straight forward generalization of Lemma 14.11, Lemma 14.10 and Lemma 14.5.

CHAPTER 15

Construction of a τ-very well prepared morphism

Suppose that $f : X \rightarrow Y$ is a dominant, proper morphism of nonsingular 3-folds, with toroidal structures D_Y and $D_X = f^{-1}(D_Y)$, such that D_X contains the locus where f is not smooth.

THEOREM 15.1. *Suppose that $\tau \geq 1$, $f : X \rightarrow Y$ is τ-prepared, $q \in Y$ is a 1-point or a 2-point, and u, v, w are permissible parameters at q such that u, v are toroidal forms at p for all $p \in f^{-1}(q)$ and $u, v \in \mathcal{O}_{Y,q}$. Further suppose that the Zariski closure of $u = v = 0$ in Y intersects $G_Y(f, \tau)$ in a finite set of points. Then there exists an affine neighborhood V of q in Y and a commutative diagram*

$$
\begin{array}{ccc}
X_1 & \stackrel{f_1}{\rightarrow} & Y_1 \\
\Phi_1 \downarrow & & \downarrow \Psi_1 \\
f^{-1}(V) & \stackrel{f}{\rightarrow} & V
\end{array}
\tag{15.1}
$$

such that $u = v = 0$ are local equations of a nonsingular curve γ_0 in V, Ψ_1 is a product of possible blow ups of 2-curves and resolving curves which are sections over γ_0, Φ_1 is a product of possible blow ups, f_1 is τ-prepared, $\tau_{f_1}(p_1) \leq \tau_f(\Phi_1(p_1))$ for $p_1 \in X_1$, and there exist permissible parameters $u_{q'}, v_{q'}, w$ at all $q' \in \Psi_1^{-1}(q)$ such that u, v are related to $u_{q'}, v_{q'}$ birationally and $u_{q'}, v_{q'}, w$ are super parameters at q'. Further, $\tau_{f_1}(p) < \tau$ if $p \in (\Psi_1 \circ f_1)^{-1}(\gamma_0) - (\Psi_1 \circ f_1)^{-1}(q)$.

PROOF. Let V be an affine neighborhood of q with uniformizing parameters u, v, w such that u, v are toroidal forms at all p' above $\{u = v = 0\} \cap V$, the intersection of the nonfinite locus of f with the curve $u = w = 0$ on V is $\{q\}$, and $\{u = v = 0\} \cap G_Y(f, \tau) \cap V \subset \{q\}$. Let $W = f^{-1}(V)$ and $\overline{f} = f \mid W$. We have a smooth morphism $\pi_0 : V \rightarrow S_0 = \text{spec}(\mathbf{k}[u, v])$. Give S_0 the toroidal structure

$$
D_{S_0} = \begin{cases} uv = 0 & \text{if } q \text{ is a 2-point} \\ u = 0 & \text{if } q \text{ is a 1-point.} \end{cases}
$$

Let $\overline{q} = \pi_0(q)$. The morphism $\pi_0 \circ \overline{f} : W \rightarrow S_0$ is toroidal with respect to D_{S_0} and $(\pi_0 \circ \overline{f})^{-1}(D_{S_0})$. We construct a commutative diagram

$$
\begin{array}{ccccc}
\vdots & & \vdots & & \vdots\\
\downarrow & & \downarrow & & \downarrow\\
W_n & \overset{\overline{f}_n}{\rightarrow} & V_n & \overset{\pi_n}{\rightarrow} & S_n\\
\overline{\Phi}_n\downarrow & & \overline{\Psi}_n\downarrow & & \Lambda_n\downarrow\\
\vdots & & \vdots & & \vdots\\
\overline{\Phi}_2\downarrow & & \overline{\Psi}_2\downarrow & & \Lambda_2\downarrow\\
W_1 & \overset{\overline{f}_1}{\rightarrow} & V_1 & \overset{\pi_1}{\rightarrow} & S_1\\
\overline{\Phi}_1\downarrow & & \overline{\Psi}_1\downarrow & & \Lambda_1\downarrow\\
W & \overset{\overline{f}}{\rightarrow} & V & \overset{\pi_0}{\rightarrow} & S_0.
\end{array}
\tag{15.2}
$$

Here $\Lambda_1 : S_1 \rightarrow S_0$ is the blow up of $\overline{q}$, $\overline{\Psi}_1$ is the blow up of $\gamma_0 = \pi_0^{-1}(\overline{q})$, which is the curve with equations $u = v = 0$ in V. If q is a 1-point then γ_0 is a resolving curve for $\overline{f}$ at q. $\overline{\Phi}_1$ is the morphism of Lemma 14.1 (if q is a 2-point) or Lemma 14.5 (if q is a 1-point). $\pi_1 \circ \overline{f}_1 : W_1 \rightarrow S_1$ is toroidal.

Suppose that $\overline{q}_1 \in \Lambda_1^{-1}(q)$. Then $\mathcal{O}_{S_1,\overline{q}_1}$ has regular parameters u_1, v_1 defined by

$$u = u_1, v = u_1(v_1 + \alpha) \tag{15.3}$$

for some $\alpha \in \mathbf{k}$, or

$$u = u_1 v_1, v = v_1. \tag{15.4}$$

$\Lambda_2 : S_2 \rightarrow S_1$ is the blow up of the finitely many points $\overline{q}_1 = \pi_1(q_1)$ above $\overline{q}$ such that q_1 in $\overline{\Phi}_1^{-1}(q)$ and u_1, v_1, w are not super parameters at q_1. $\overline{\Psi}_2 : V_2 \rightarrow V_1$ is the blow up of the (disjoint) curves $\gamma_1 = \pi_1^{-1}(\overline{q}_1)$, and $\overline{\Phi}_2 : W_2 \rightarrow W_1$ is the morphism of Lemma 14.1 or Lemma 14.5. $\pi_2 \circ \overline{f}_2 : W_2 \rightarrow S_2$ is toroidal.

We continue in this way to construct (15.2) as long as $\overline{f}_n : W_n \rightarrow V_n$ does not satisfy the conclusions of the theorem.

Suppose that the algorithm never ends.

Let ν be a 0-dimensional valuation of $\mathbf{k}(X)$ whose center on Y is q, and let p_i be the center of ν on W_i, q_i the center of ν on V_i. Let $\overline{q}_i = \pi_i(q_i)$. We may suppose that $\overline{\Psi}_i$ is not an isomorphism at the center of ν for all i. There exist permissible parameters u_i, v_i, w_i at q_i for all i such that u_i, v_i are regular parameters in $\mathcal{O}_{S_i,\overline{q}_i}$, obtained by iteration of (15.3) and (15.4), as determined by ν. We will show that there exists j_0 such that u_j, v_j, w are super parameters at p_j for all $j \geq j_0$. If u_i, v_i, w are super parameters at p_i for some i, we have that u_j, v_j, w are super parameters at p_j for all $j \geq i$. Suppose that u_i, v_i, w are not super parameters at p_i for all i.

We may identify ν with an extension of ν to the quotient field of $\hat{\mathcal{O}}_{W_i,p_i}$ which dominates $\hat{\mathcal{O}}_{W_i,p_i}$.

There exist permissible parameters x_i, y_i, z_i in $\hat{\mathcal{O}}_{W_i,p_i}$ such that we have one of the following forms:

p_i a 1-point, q_i (and $\overline{q}_i$) a 1-point

$$u_i = x_i^{a_i}, v_i = y_i \tag{15.5}$$

or p_i a 2-point, q_i (and $\overline{q}_i$) a 1-point

$$u_i = (x_i^{a_i} y_i^{b_i})^k, v_i = z_i \tag{15.6}$$

p_i a 1-point, q_i (and $\overline{q}_i$) a 2-point

$$u_i = x_i^{a_i}, v_i = x_i^{b_i}(y_i + \alpha_i) \tag{15.7}$$

with $0 \neq \alpha_i$ or p_i a 2-point, q_i (and $\overline{q}_i$) a 2-point

$$u_i = x_i^{a_i} y_i^{b_i}, v_i = x_i^{c_i} y_i^{d_i} \tag{15.8}$$

p_i a 2-point, q_i (and $\overline{q}_i$) a 2-point

$$u_i = (x_i^{a_i} y_i^{b_i})^{k_i}, v_i = (x_i^{a_i} y_i^{b_i})^{t_i}(z_i + \alpha_i) \tag{15.9}$$

with $\alpha_i \neq 0$ or p_i a 3-point, q_i (and $\overline{q}_i$) a 2-point

$$u_i = x_i^{a_i} y_i^{b_i} z_i^{ci}, v_i = x_i^{d_i} y_i^{e_i} z_i^{f_i}. \tag{15.10}$$

Suppose that $f \in \hat{\mathcal{O}}_{S_0,\overline{q}}$. Then (by embedded resolution of plane curve singularities, c.f. Section 3.4, Exercise 3.13 [**C6**]) there exists an n_0 such that $uvf = 0$ is a SNC divisor in $\hat{\mathcal{O}}_{S_n,\overline{q}_n}$ for all $n \geq n_0$.

We further have that if $g \in \hat{\mathcal{O}}_{W,p}$ is a fractional series in u and v (a series in fractional rational powers of u and v), then $g \in \hat{\mathcal{O}}_{W_n,p_n}$ is a fractional series in u_n and v_n.

First suppose that q_i is a 2-point for all i. Then $\nu(u)$ and $\nu(v)$ are rationally independent, and we have that (15.8) or (15.10) hold for all i. By the algorithm of Lemma 14.1, we see that $\overline{\Phi}_{i+1}$ is a sequence of blow ups of 2-curves above p_i for all i.

We either have that p_i is a 3-point for all i (and a form (15.10) holds for all i) or some p_i is a 2-point, and thus (15.8) holds for all i sufficiently large.

Suppose that p_i is a 2-point for some i. We may then assume that p_i is a 2-point for all i, and (15.8) holds for all i. Then x_i and y_i are monomials in x_{i+1} and y_{i+1} for all i, and $z_{i+1} = z_i$ for all i.

We have an expression

$$\begin{aligned} u &= x^a y^b \\ v &= x^c y^d \\ w &= f_1(x,y) + x^l y^m z \end{aligned}$$

in $\hat{\mathcal{O}}_{W,p}$. There exists $n \in \mathbf{N}$ and $\overline{a}, \overline{b}, \overline{c}, \overline{d} \in \mathbf{Z}$ such that

$$x^n = u^{\overline{a}} v^{\overline{b}}, y^n = u^{\overline{c}} v^{\overline{d}}.$$

$\nu(u^{\overline{a}} v^{\overline{b}}) = \nu(x^n) > 0$, $\nu(u^{\overline{c}} v^{\overline{d}}) = \nu(y^n) > 0$ imply that for $i >> 0$,

$$u^{\overline{a}} v^{\overline{b}} = u_i^{\overline{a}_i} v_i^{\overline{b}_i}, u^{\overline{c}} v^{\overline{d}} = u_i^{\overline{c}_i} v_i^{\overline{d}_i}$$

with $\overline{a}_i, \overline{b}_i, \overline{c}_i, \overline{d}_i \in \mathbf{N}$ (by Theorem 2.7 [**C2**]).

We have

$$\begin{aligned} u_i &= x_i^{a_i} y_i^{b_i} \\ v_i &= x_i^{c_i} y_i^{d_i} \\ w &= f_1(x,y) + (x_i^{e_i} y_i^{f_i})^l (x_i^{g_i} y_i^{h_i})^m z. \end{aligned}$$

$$f_1(x,y) = f_1((u^{\overline{a}} v^{\overline{b}})^{\frac{1}{n}}, (u^{\overline{c}} v^{\overline{d}})^{\frac{1}{n}}) = g_1(u_i^{\frac{1}{n}}, v_i^{\frac{1}{n}}) \in \mathbf{k}[[u_i^{\frac{1}{n}}, v_i^{\frac{1}{n}}]].$$

Let $\omega\in\mathbf{k}$ be a primitive n-th of unity. Set

$$f=\prod_{i,j=1}^{n} g_1(\omega^i u_i^{\frac{1}{n}},\omega^j v_i^{\frac{1}{n}})\in\mathbf{k}[[u_i,v_i]].$$

Recall that for $i\ge n_0$, $fuv=0$ is a SNC divisor in $\hat{\mathcal{O}}_{S_i,\overline{q}_i}$. Since $\overline{q}_i$ is a 2-point for all i,

$$fuv=u_i^{\overline{a}}v_i^{\overline{b}}\gamma$$

where $\overline{a},\overline{b}>0$, γ is a unit series in $\hat{\mathcal{O}}_{S_i,\overline{q}_i}$. As $g_1\mid f$ in $\hat{\mathcal{O}}_{W_i,p_i}$, we have that u_i,v_i,w are super parameters at p_i.

Suppose that p_i is a 3-point for all i. Then x_i,y_i,z_i are monomials in x_{i+1},y_{i+1},z_{i+1} for all i. By Lemma 10.6, we obtain an expression of the form (10.19) and (10.20) for u_i,v_i,w in terms of x_i,y_i,z_i for $i\gg 0$. By a variant on the argument for the case when p_i is a 2-point for all i, we obtain that u_i,v_i,w_i are super parameters at p_i for $i\gg 0$.

The final case is when q_i is a 1-point for some i.

Suppose that q_i is a 1-point for some i. Without loss of generality, we may assume that $q_i=q$ and $p_i=p$.

If p is a 1-point (and q is a 1-point), we have permissible parameters x,y,z in $\hat{\mathcal{O}}_{W,p}$ such that

$$u=x^a,v=y,w=f_1(x,y)+x^rz$$

of the form (15.5).

Set

$$f=\prod_{i=1}^{a} f_1(\omega^i u^{\frac{1}{a}},v)\in\mathbf{k}[[u,v]]$$

where $\omega\in\mathbf{k}$ is a primitive a-th root of unity.

If p is a 2-point (and q is a 1-point), we have permissible parameters $x,y,z\in\hat{\mathcal{O}}_{W,p}$ such that

$$u=(x^ay^b)^k,v=z,w=f_1(x^ay^b,z)+x^cy^d$$

of the form (15.6). Set

$$f=\prod_{i=1}^{k} f_1(\omega^i u^{\frac{1}{k}},v)\in\mathbf{k}[[u,v]]$$

where $\omega\in\mathbf{k}$ is a primitive k-th root of unity.

Recall that $fuv=0$ is a SNC divisor in $\hat{\mathcal{O}}_{S_i,\overline{q}_i}$ for $i\ge n_0$.

We will now show that if $i\ge n_0$ and q_i is a 2-point, then u_i,v_i,w are super parameters at q_i. In these cases $\overline{q}_i$ is a 2-point, $u_iv_i=0$ is a local equation of D_{V_i} and $u=0$ is a local equation of D_{V_i}. Thus $fuv=u_i^mv_i^n\gamma$, where $m,n>0$ and $\gamma\in\hat{\mathcal{O}}_{S_i,\overline{q}_i}$ is a unit series. Since $f_1\mid f$ in $\hat{\mathcal{O}}_{X_i,p_i}$, and f_1 is a fractional series in u_i and v_i, we have the desired conclusion.

We have reduced to the case where q_i is a 1-point for all i.

Since $fuv=0$ is a SNC divisor for $i\ge n_0$, the only cases where u_i,v_i,w are not super parameters at p_i are if u_i,v_i satisfy (15.5) at p_i, and

$$u_i=x_i^a,v_i=y_i,w=x_i^s(y_i-\phi(x_i^a))^r\gamma(x_i,y_i)+x_i^c(z_i+\alpha)\tag{15.11}$$

where γ is a unit series, $\operatorname{ord}(\phi)>0$, $\alpha\in\mathbf{k}$, or p_i satisfies (15.6), and

$$u_i=(x_i^ay_i^b)^k,v_i=z_i,w=(x_i^ay_i^b)^s(z_1-\phi((x_i^ay_i^b)^k))^r\gamma(x_i^ay_i^b,z_i)+x_i^cy_i^d\tag{15.12}$$

where $\gcd(a,b)=1$, ord $\phi>0$ and γ is a unit series.

We consider the case when (15.12) holds at p_i. The case (15.11) is similar.

By the constructions of Lemma 14.1 and Lemma 14.5, we have a factorization of $W_{i+1}\rightarrow W_i$, by blow ups of possible centers.

$$W_{i+1}=Z_m\overset{\Omega_m}{\rightarrow}Z_{m-1}\overset{\Omega_{m-1}}{\rightarrow}\cdots\overset{\Omega_2}{\rightarrow}Z_1\overset{\Omega_1}{\rightarrow}Z_0=W_i$$

Let a_j be the center of ν on Z_j. We may assume that each morphism $Z_l\rightarrow Z_{l-1}$ is not an isomorphism at the center of ν.

If $j<m$, we have permissible parameters x_{ij},y_{ij},z_{ij} in $\hat{\mathcal{O}}_{Z_j,a_j}$ such that a_j is a 2-point and

$$\begin{aligned} u_i &= (x_{ij}^a y_{ij}^b)^k,\\ v_i &= z_{ij}x_{ij}^{e_i}y_{ij}^{f_i},\\ w &= (x_{ij}^a y_{ij}^b)^s(z_{ij}x_{ij}^{e_i}y_{ij}^{f_i}-\phi((x_{ij}^a y_{ij}^b)^k)^r\gamma(x_{ij}^a y_{ij}^b, z_{ij}x_{ij}^{e_i}y_{ij}^{f_i})+x_{ij}^c y_{ij}^d \end{aligned}\tag{15.13}$$

with $(e_i,f_i)<(ak,bk)$.

After possibly interchanging x_{ij} and y_{ij}, we may assume that $e_i<ak$ and there are regular parameters $x_{i,j+1},y_{i,j+1},z_{i,j+1}$ in $\hat{\mathcal{O}}_{Z_{j+1},a_{j+1}}$ and $\alpha\in\mathbf{k}$, defined by

$$x_{ij}=x_{i,j+1},\ z_{ij}=x_{i,j+1}(z_{i,j+1}+\alpha).\tag{15.14}$$

or

$$x_{ij}=x_{i,j+1}z_{i,j+1},\ z_{ij}=z_{i,j+1}.\tag{15.15}$$

Suppose that (15.14) holds. Then

$$\begin{aligned} u_i &= (x_{i,j+1}^a y_{i,j+1}^b)^k\\ v_i &= (z_{i,j+1}+\alpha)x_{i,j+1}^{e_i+1}y_{i,j+1}^{f_i}\\ w &= (x_{i,j+1}^a y_{i,j+1}^b)^s[x_{i,j+1}^{e_i+1}y_{i,j+1}^{f_i}(z_{i,j+1}+\alpha)-\phi((x_{i,j+1}^a y_{i,j+1}^b)^k)]^r\\ &\quad \gamma(x_{i,j+1}^a y_{i,j+1}^b, x_{i,j+1}^{e_i+1}y_{i,j+1}^{f_i}(z_{i,j+1}+\alpha))+x_{i,j+1}^c y_{i,j+1}^d. \end{aligned}\tag{15.16}$$

Suppose that $0\neq\alpha$, $(e_i+1,f_i)<(ak,bk)$ and $af_i-b(e_i+1)\neq 0$ in (15.16). Then the rational map $Z_{j+1}\rightarrow V_{i+1}$ is a morphism near $a_{j+1}=p_{i+1}$.

We make a change of variables, to get permissible parameters $\overline{x},\overline{y},\overline{z}\in\hat{\mathcal{O}}_{W_{i+1},p_{i+1}}$ and $0\neq\beta\in\mathbf{k}$ such that

$$\begin{aligned} u_{i+1} &= \tfrac{u_i}{v_i}=x_{i,j+1}^{ak-(e_i+1)}y_{i,j+1}^{bk-f_i}(z_{i,j+1}+\alpha)^{-1}=\overline{x}^{ak-(e_i+1)}\overline{y}^{bk-f_i}\\ v_{i+1} &= v_i=(z_{i,j+1}+\alpha)x_{i,j+1}^{e_i+1}y_{i,j+1}^{f_i}=\overline{x}^{e_i+1}\overline{y}^{f_i}\\ w &= (\overline{x}^a\overline{y}^b)^s[\overline{x}^{e_i+1}\overline{y}^{f_i}-\phi((\overline{x}^a\overline{y}^b)^k)]^r\gamma(\overline{x}^a\overline{y}^b,\overline{x}^{e_i+1}\overline{y}^{f_i})+\overline{x}^c\overline{y}^d(\overline{z}+\beta)\\ &= \overline{x}^{as+r(e_i+1)}\overline{y}^{bs+rf_i}(1-\tfrac{\phi((\overline{x}^a\overline{y}^b)^k)}{\overline{x}^{e_i+1}\overline{y}^{f_i}})^r\gamma(\overline{x}^a\overline{y}^b,\overline{x}^{e_i+1}\overline{y}^{f_i})+\overline{x}^c\overline{y}^d(\overline{z}+\beta) \end{aligned}$$

and u_{i+1},v_{i+1},w are super parameters at p_{i+1}.

Suppose that (15.14) holds, $0\neq\alpha$, $(e_i+1,f_i)<(ak,bk)$ and $(e_i+1,f_i)=t_i(a,b)$ for some integer t_i. Then we make a change of variable to get permissible parameters

$\overline{x}, \overline{y}, \overline{z} \in \hat{\mathcal{O}}_{W_{i+1}, p_{i+1}}$ such that

$$\begin{aligned} u_{i+1} &= \tfrac{u_i}{v_i} = (\overline{x}^a\overline{y}^b)^{k-t_i}\\ v_{i+1} &= v_i = (\overline{x}^a\overline{y}^b)^{t_i}(\overline{z}+\alpha)\\ w &= (\overline{x}^a\overline{y}^b)^s(\overline{z}+\alpha)^{\frac{s}{k}}[(\overline{x}^a\overline{y}^b)^{t_i}(\alpha+\overline{z}) - \phi((\overline{x}^a\overline{y}^b)^k(\overline{z}+\alpha))]^r\\ &\quad \gamma(\overline{x}^a\overline{y}^b(\overline{z}+\alpha)^{\frac{1}{k}}, (\overline{x}^a\overline{y}^b)^{t_i}(\overline{z}+\alpha)) + \overline{x}^c\overline{y}^d\\ &= (\overline{x}^a\overline{y}^b)^{s+t_ir}(\overline{z}+\alpha)^{\frac{s}{k}}(\alpha+\overline{z} - \tfrac{\phi((\overline{x}^a\overline{y}^b)^k(\overline{z}+\alpha))}{(\overline{x}^a\overline{y}^b)^{t_i}})^r\\ &\quad \gamma(\overline{x}^a\overline{y}^b(\overline{z}+\alpha)^{\frac{1}{k}}, (\overline{x}^a\overline{y}^b)^{t_i}(\overline{z}+\alpha)) + \overline{x}^c\overline{y}^d \end{aligned}$$

as $t_i < k$, and u_{i+1}, v_{i+1}, w are thus super parameters at p_{i+1}.

Suppose that (15.14) holds, and $(e_i+1, f_i) = (ak, bk)$. Then the rational map $Z_{j+1} \rightarrow V_{i+1}$ is a morphism near $a_{j+1} = p_{i+1}$, and the 1-point q_{i+1} has permissible parameters defined by

$$u_i = u_{i+1}, v_i = u_{i+1}(v_{i+1}+\alpha).$$

Substituting into (15.16), we see that

$$\begin{aligned} u_{i+1} &= (x_{i,j+1}^a y_{i,j+1}^b)^k\\ v_{i+1} &= z_{i,j+1}\\ w &= (x_{i,j+1}^a y_{i,j+1}^b)^{s+kr}[z_{i,j+1} + \alpha - \tfrac{\phi((x_{i,j+1}^a y_{i,j+1}^b)^k)}{(x_{i,j+1}^a y_{i,j+1}^b)^k}]^r\gamma + x_{i,j+1}^c y_{i,j+1}^d. \end{aligned}$$

Thus u_{i+1}, v_{i+1}, w are super parameters at p_{i+1}, or we have a form (15.12), with

$$(c-as, d-bs) > (c-(a(s+kr), d-b(s+kr)). \tag{15.17}$$

In fact, $(c-(a(s+kr), d-b(s+kr))$ must decrease from $(c-as, d-bs)$ by at least $(1,1)$.

If we have (15.14) with $(e_i+1, f_i) < (ak, bk)$ with $\alpha = 0$, then (15.16) is back in the form (15.13) (with a decrease in $(ak-e_i)+(bk-f_i)$.

Suppose that (15.15) holds. The rational map $Z_{j+1} \rightarrow V_{i+1}$ is a morphism near $a_{j+1} = p_{i+1}$, and the 2-point q_{i+1} has permissible parameters defined by $u_i = u_{i+1}v_{i+1}, v_i = v_{i+1}$. Substituting into (15.13), we have

$$\begin{aligned} u_{i+1} &= x_{i,j+1}^{ak-e_i} y_{i,j+1}^{bk-f_i} z_{i,j+1}^{ak-e_i-1}\\ v_{i+1} &= x_{i,j+1}^{e_i} y_{i,j+1}^{f_i} z_{i,j+1}^{e_i+1}\\ w &= (x_{i,j+1}^a y_{i,j+1}^b z_{i,j+1}^a)^s (x_{i,j+1}^{e_i} y_{i,j+1}^{f_i} z_{i,j+1}^{e_i+1} - \phi((x_{i,j+1}^a y_{i,j+1}^b z_{i,j+1}^a)^k))^r\gamma\\ &\quad + x_{i,j+1}^c y_{i,j+1}^d z_{i,j+1}^c\\ &= (x_{i,j+1}^a y_{i,j+1}^b z_{i,j+1}^a)^s (x_{i,j+1}^{e_i} y_{i,j+1}^{f_i} z_{i,j+1}^{e_i+1})^r (1 - \tfrac{\phi((x_{i,j+1}^a y_{i,j+1}^b z_{i,j+1}^a)^k)}{x_{i,j+1}^{e_i} y_{i,j+1}^{f_i} z_{i,j+1}^{e_i+1}})^r\gamma\\ &\quad + x_{i,j+1}^c y_{i,j+1}^d z_{i,j+1}^c \end{aligned}$$

Thus u_{i+1}, v_{i+1}, w are super parameters at the 3-point p_{i+1}.

We thus have that u_i, v_i, w are super parameters at p_i for $i >> 0$ unless each p_i has a form (15.12), and by (15.17) we eventually get

$$(c-as, d-bs) < (0,0).$$

We thus have that $w = x_i^c y_i^d \gamma$ where γ is a unit series. Thus u_i, v_i, w are super parameters at p_i.

We have shown that for any 0-dimensional valuation ν of $\mathbf{k}(X)$ whose center is q on Y, there exists j_1 such that u_j, v_j, w are super parameters at p_j for $j \geq j_1$. By compactness of the Zariski Riemann manifold $[\mathbf{Z}]$, it follows that the sequence

(15.2) must terminate after a finite number of steps, in $W_n \to V_n$ satisfying the conclusions of the theorem.

□

THEOREM 15.2. *Suppose that $\tau \geq 1$, $f : X \to Y$ is pre-τ-quasi-well prepared (or τ-well prepared) with relation R, f is τ-prepared and $C \subset Y$ is a reduced (but possibly not irreducible) curve consisting of components of the nonfinite locus of f which contain a 1-point of Y. Then there exist sequences of possible blow ups $\Phi_1 : X_1 \to X$ and $\Psi_1 : Y_1 \to Y$ such that there is a commutative diagram*

$$\begin{array}{ccc} X_1 & \xrightarrow{f_1} & Y_1 \\ \Phi_1 \downarrow & & \downarrow \Psi_1 \\ X & \xrightarrow{f} & Y \end{array} \tag{15.18}$$

satisfying

1. $$\tau_{f_1}(p_1) \leq \tau_f(\Phi_1(p_1))$$ *for $p_1 \in D_{X_1}$.*
2. *f_1 is pre-τ-quasi-well prepared with respect to a relation R^1 which extends the transform of R.*
3. $\Phi_1^{-1}(T(R)) \cap G_{X_1}(f_1, \tau) \subset T(R^1)$.
4. *If f is τ-quasi-well prepared (τ-well prepared) then (15.18) is a τ-quasi-well prepared (τ-well prepared) diagram.*
5. *The strict transform $\overline{C}$ of C on Y_1 is nonsingular and makes SNCs with D_{Y_1}.*
6. *If C_j is an irreducible component of $\overline{C}$ and $q \in U(\overline{R}_i^1)$ for some $\overline{R}_i^1$ associated to R^1 is such that $q \in C_j$ then the germ of C_j at q is contained in $S_{\overline{R}_i^1}(q)$.*
7. *If f is τ-well prepared, C_j is an irreducible component of $\overline{C}$, E is a component of D_{Y_1}, $q_\alpha \in U(R_k^1) \cap E$, and $\gamma = \overline{E \cdot R_k^1(q_\alpha)}$ is prepared for R^1 of type 6 (of Definition 13.8), then either $C_j = \gamma$ or $q \in C_j \cap \gamma$ implies $q \in U(\overline{R}_k^1)$ (and thus the germ of C_j at q is contained in $S_{R_k}(q)$ by 6).*
8. *The components C_j of $\overline{C}$ are permissible centers (or *-permissible centers if f is τ-well prepared and C_j is prepared of type 6) for R^1.*
9. *ψ_1 is a sequence of blow ups of prepared 1-points, prepared 2-points, 2-curves and resolving curves.*

There is a pre-τ-quasi-well prepared (or τ-well prepared) diagram

$$\begin{array}{ccc} X_2 & \xrightarrow{f_2} & Y_2 \\ \Phi_2 \downarrow & & \downarrow \Psi_2 \\ X_1 & \xrightarrow{f_1} & Y_1 \end{array}$$

where Ψ_2 is the blow up of $\overline{C}$, possibly followed by blow ups of 2-points which are prepared of type 2 of Definition 13.7 for the transform of R if f is τ-well prepared and C contains a component which is prepared of type 6 for R.

PROOF. Let $\{q_1, \ldots, q_m\}$ be the 1-points of C which are not contained in $U(R)$, and for which there do not exist super parameters $u, v, w \in \mathcal{O}_{Y,q}$ such that $u = w = 0$ are local equations of C at q. This set is finite by Lemma 11.6.

Step 1. Let γ_1 be a general curve through q_1 on D_Y.

If $q \in \gamma_1$ is a 1-point, then there exist permissible parameters $u, v, w \in \mathcal{O}_{Y,q}$, such that u, v have a toroidal form at p for all $p \in f^{-1}(q)$, $u = v = 0$ are local equations of γ_1 at q. This follows from (the proof of) Lemma 4.9 since γ_1 intersects the nonfinite locus of f transversally at general points of one dimensional components of the nonfinite locus.

Since γ_1 is a general curve through q_1, $(\gamma_1 - \{q_1\}) \cap G_Y(f,\tau) \subset \Theta(f,Y)$ by Remark 11.12. Suppose that $q \in (\gamma_1 - \{q_1\}) \cap (G_Y(f,\tau) - U(R))$.

Since q is perfect for f and by Lemma 11.5, there exist locally closed subsets $\overline{V}_1, \ldots, \overline{V}_n$ of X which are a partition of $G_X(f,\tau) \cap f^{-1}(q)$ and series

$$\phi_1(u,v), \ldots, \phi_n(u,v)$$

such that $u, v, w_i = w - \phi_i$ are super parameters for f at q and w_i is weakly good at p for $p \in \overline{V}_i$. Here $u, v, w \in \mathcal{O}_{Y,q}$ are permissible parameters at the 1-point q such that u, v have a toroidal form at p, $u = v = 0$ are local equations of γ_1 at q.

If $p \in f^{-1}(q)$ is a 1-point and $\tau_f(p) > 1$, then $w_i = 0$ is supported on D_X at p, since u, v, w_i are super parameters at q.

By blowing up 2-curves above X, by a map $\Phi_1 : X_1 \to X$, with induced map $f_1 = f \circ \Phi_1 : X_1 \to Y$, we obtain by Lemma 10.3 and Lemma 14.1 that $w_i = 0$ is supported on D_{X_1} at p_1 for all $p_1 \in f_1^{-1}(q)$ such that $\tau_{f_1}(p_1) > 1$. u, v, w_p are super parameters for f_1 at q.

By Lemma 10.3 and Remark 14.2, there exist locally closed subsets $V_1, \ldots, V_n$ of X_1 which are a partition of $G_{X_1}(f_1,\tau) \cap f_1^{-1}(q)$ such that

1. $u, v, w_i = w - \phi_i(u,v)$ are super parameters for f_1 at q
2. w_i is weakly good at p for $p \in V_i$
3. $w_i = 0$ is supported on D_{X_1} at p for $p \in V_i$ if $\tau > 1$.

Suppose that $\tau > 1$. Then there exists an expression $w_i^{e_i} - u^{a_i}\Lambda_i = 0$ for $p \in V_i \cap D_{X_1}$, where Λ_i is a unit on V_i, $e_i, a_i \in \mathbf{N}$, $e_i > 1$ and $\gcd(e_i, a_i) = 1$. Since u, v, w_i are super parameters at the 1-point q, we see from (10.5) and (10.6) that there exists $\lambda_q^i \in \mathbf{k}$ such that $\Lambda_i(p) = \lambda_q^i$ for $p \in V_i$.

Suppose that $\tau = 1$. Then $w_i = 0$ is a monomial form at p for $p \in V_i$ by Remark 11.3.

We may now define relations $R_{q,i}$ for f_1 by $T(R_{q,i}) = V_i$, $U(R_{q,i}) = \{q\}$, and define $R_{q,i}(p)$ for $p \in T(R_{q,i})$ by $R_{q,i} = w_i^{e_i} - \lambda_q^i u^{a_i}$ if $\tau > 1$ and by $R_{q,i} = w_i$ if $\tau = 1$.

We extend the transform R^1 of R on X_1 by adding in the new relations $R_{q,i}$ for all $q \in (\gamma_1 - \{q_1\}) \cap (G_Y(f,\tau) - U(R))$. $f_1 : X_1 \to Y$ is then τ-prepared and pre-τ-quasi-well prepared for R^1. We have that

$$(\gamma_1 - \{q_1\}) \cap G_Y(f_1,\tau) \subset U(R^1).$$

By Lemma 14.7 and Lemma 14.8, and since for $q \in \gamma_1 \cap U(R_i^1)$, the germ of γ_1 at q is not contained in $S_{R_i^1}(q)$ for any relation R_i^1 associated to R^1 (γ_1 is general), there exists a pre-τ-quasi-well prepared (τ-well prepared) diagram

$$\begin{array}{ccc} X_2 & \overset{f_2}{\to} & Y_2 \\ \Phi_2 \downarrow & & \downarrow \Psi_2 \\ X_1 & \overset{f_1}{\to} & Y \end{array}$$

such that f_2 is τ-prepared, Ψ_2 is a product of blow ups of prepared 1-points of type 4 (of Definition 13.7) and prepared 2-points of type 1 (of Definition 13.7) such that the strict transform γ_1^2 of γ_1 on Y_2 satisfies 1,2 and 4 of Definition 13.6 of a resolving curve for f_2 at q_1 (we may identify q_1 with a point of Y_2 since Ψ_2 is an isomorphism above q_1). Further, if $q \in \gamma_1^2$ is a 2-point, we have that $\tau_{f_2}(p) < \tau$ for $p \in f_2^{-1}(q)$. By Lemma 10.7 and Remark 14.3 there exists a pre-τ-quasi-well prepared (τ-well prepared) diagram

$$\begin{array}{ccc} X_3 & \stackrel{f_3}{\rightarrow} & Y_3 \\ \Phi_3 \downarrow & & \downarrow \Psi_3 \\ X_2 & \stackrel{f_2}{\rightarrow} & Y_2 \end{array}$$

where Φ_3 is a product of blow ups of 2-curves and 3-points and Ψ_3 is a product of blow ups of 2-curves such that if $q \in \gamma_1^3$ is a 2-point, where γ_1^3 is the strict transform of γ_1 on Y_3, then there exist super parameters u, v, w for f_3 at q such that $u = w = 0$ are local equations of γ_1^3 at q. Thus γ_1^3 is a resolving curve for f_3 at $\overline{q}$.

Let

$$\begin{array}{ccc} X_4 & \stackrel{f_4}{\rightarrow} & Y_4 \\ \Phi_4 \downarrow & & \downarrow \Psi_4 \\ X_3 & \stackrel{f_1}{\rightarrow} & Y_3 \end{array}$$

be the pre-τ-quasi-well prepared (τ-well prepared) diagram of Lemma 14.5 where Ψ_4 is the blow up of γ_1^3. The strict transform C^4 of C on Y_4 intersects $\Psi_4^{-1}(q_1)$ in a 2-point. $Y_4 \rightarrow Y$ is an isomorphism over a neighborhood of $\{q_2, \ldots, q_m\}$. f_4 is τ-prepared.

Step 2. Iterate the construction of Step 1 for the points $q_2, \ldots, q_m$. We obtain a commutative diagram

$$\begin{array}{ccc} \tilde{X} & \stackrel{\tilde{f}}{\rightarrow} & \tilde{Y} \\ \tilde{\Phi} \downarrow & & \downarrow \tilde{\Psi} \\ X & \stackrel{f}{\rightarrow} & Y \end{array}$$

such that $\tilde{f}$ is τ-prepared, pre-τ-quasi-well prepared (τ-well prepared) with respect to a relation $\tilde{R}$ which extends the transform of R. If $\overline{C}$ is the strict transform of C on $\tilde{Y}$, and if $q \in \overline{C} - U(\tilde{R})$ is a 1-point, then there exist super parameters u, v, w at q such that $u = w = 0$ are local equations of $\overline{C}$ at q.

Step 3. By Lemma 10.7 and Remark 14.3, there exists a pre-τ-quasi-well prepared (τ-well prepared) diagram

$$\begin{array}{ccc} \tilde{X}_1 & \stackrel{\tilde{f}_1}{\rightarrow} & \tilde{Y}_1 \\ \tilde{\Phi}_1 \downarrow & & \downarrow \tilde{\Psi}_1 \\ \tilde{X} & \stackrel{\tilde{f}}{\rightarrow} & \tilde{Y} \end{array}$$

where $\tilde{\Phi}_1$ is a product of blow ups of 2-curves and 3-points, and $\tilde{\Psi}_1$ is a product of blow ups of 2-curves such that if $\overline{C}_1$ is the strict transform of $\overline{C}$ on $\tilde{Y}_1$, and $q \in \overline{C}_1$ is a 2-point not in $U(\tilde{R}_1)$, then there exist super parameters u, v, w at q, so that q

is prepared of type 2 of Definition 13.7. Let

$$\begin{array}{ccc}\tilde X_2 & \overset{\tilde f_2}{\to} & \tilde Y_2\\ \tilde\Phi_2\downarrow & & \downarrow\tilde\Psi_2\\ \tilde X_1 & \overset{\tilde f_1}{\to} & \tilde Y_1\end{array}$$

be the pre-τ-quasi-well prepared (τ-well prepared) diagram obtained by blowing up the 2-points on $\overline{C}_1 - U(\tilde R^1)$ (by Lemma 14.9).

Step 4. By embedded resolution of plane curve singularities (c.f. Section 3.4 and Exercise 3.13 [**C6**]), we can iterate Step 3 to construct a pre-τ-quasi-well prepared (τ-well prepared) diagram

$$\begin{array}{ccc}\tilde X_3 & \overset{\tilde f_3}{\to} & \tilde Y_3\\ \tilde\Phi_3\downarrow & & \downarrow\tilde\Psi_3\\ \tilde X_2 & \overset{\tilde f_2}{\to} & \tilde Y_2\end{array}$$

such that if $\overline{C}_3$ is the strict transform of C on $\tilde Y_3$ and $q\in\overline{C}_3 - U(\tilde R^3)$ is a 2-point, then there exist super parameters u,v,w at q such that $u=w=0$ are local equations of $\overline{C}_3$.

Step 5. By embedded resolution of plane curve singularities (c.f. Section 3.4 and Exercise 3.13 [**C6**]), Lemma 14.1 (for 3-points in $\overline{C}_3$), Step 3, Lemma 14.7 and Lemma 14.8, there exists a pre-τ-quasi-well prepared (τ-well prepared) diagram

$$\begin{array}{ccc}\tilde X_4 & \overset{\tilde f_4}{\to} & \tilde Y_4\\ \tilde\Phi_4\downarrow & & \downarrow\tilde\Psi_4\\ \tilde X_3 & \overset{\tilde f_3}{\to} & \tilde Y_3\end{array}$$

such that the strict transform $\overline{C}_4$ of C on $\tilde Y_4$ satisfies the hypotheses of 1 of Remark 14.12.

Step 6. If f is τ-well-prepared, then we must perform a final sequence of blowups. Assume that f (and thus $\tilde f_4$) is τ-well prepared (for the transform $\tilde R^4$ of R). Let

$$\Sigma=\left\{\begin{array}{ll} q\in\overline{C}_4 & \text{such that } q\in\gamma \text{ and } q\not\in U(\tilde R^4)\\ & \text{where } \gamma \text{ is a curve which is prepared of type 6 for } \tilde R^4.\end{array}\right\}.$$

Σ is a finite set of points which are prepared of type 2 of Definition 13.7 (by 5 b) of Definition 13.8).

By Lemma 14.9, and embedded resolution of plane curve singularities, we can construct a τ-well prepared diagram

$$\begin{array}{ccc}\tilde X_5 & \overset{\tilde f_5}{\to} & \tilde Y_5\\ \tilde\Phi_5\downarrow & & \downarrow\tilde\Psi_5\\ \tilde X_4 & \overset{\tilde f_4}{\to} & \tilde Y_4\end{array}$$

such that the hypotheses of 2 of Remark 14.12 hold, and thus the conclusions of the theorem hold.

□

THEOREM 15.3. *Suppose that $f : X \to Y$ is τ-prepared and pre-τ-quasi-well-prepared (τ-well prepared) with relation R and R' is a restriction of R with $U(R') = G_Y(f,\tau) - \Theta(f,Y)$. Then there exists a pre-τ-quasi-well prepared (τ-well prepared) diagram of R*

$$\begin{array}{ccc} X_1 & \overset{f_1}{\to} & Y_1 \\ \Phi_1 \downarrow & & \downarrow \Psi_1 \\ X & \overset{f}{\to} & Y \end{array}$$

such that Φ_1, Ψ_1 are products of blow ups of possible centers and f_1 is τ-quasi-well prepared (τ-well prepared) with respect to the transform $(R^1)'$ of R'. Further, there exists a finite set of points $\Omega = \{q_1, \ldots, q_r\} \subset Y$ such that f_1 is toroidal on $f_1^{-1}(Y_1 - \Psi_1^{-1}(\Omega))$, and $\Psi_1 \circ f_1(T((R^1)')) \subset U(R')$.

PROOF. Let $C \subset Y$ be the union of the one dimensional components in the nonfinite locus of f which contain a 1-point. By Lemma 11.7, there exists a Zariski open subset $\overline{Y}$ of Y such that $\overline{Y} \cap G_Y(f,\tau) = \Theta(f,Y)$, $\overline{Y}$ contains a generic point of each component of C, and there exists a commutative diagram

$$(15.19)\qquad \begin{array}{ccc} \overline{X}_n & \overset{\overline{f}_n}{\to} & \overline{Y}_n \\ \overline{\Phi}_n \downarrow & & \downarrow \overline{\Psi}_n \\ \overline{X}_{n-1} & \overset{\overline{f}_{n-1}}{\to} & \overline{Y}_{n-1} \\ \overline{\Phi}_{n-1} \downarrow & & \downarrow \overline{\Psi}_{n-1} \\ \vdots & & \vdots \\ \downarrow & & \downarrow \\ \overline{X}_1 & \overset{\overline{f}_1}{\to} & \overline{Y}_1 \\ \overline{\Phi}_1 \downarrow & & \downarrow \overline{\Psi}_1 \\ f^{-1}(\overline{Y}) = \overline{X} & \overset{\overline{f}}{\to} & \overline{Y} \end{array}$$

where $\overline{f}_n$ is toroidal, each $\overline{\Psi}_i$ is the blow up of a nonsingular curve γ_i (in the fundamental locus of $\overline{f}_{i-1}$) dominating a component of C, and $\overline{\Phi}_i$ is a sequence of blow ups of nonsingular curves dominating γ_i. Further, we can choose $\overline{Y}$ so that $\overline{f}_n$ is τ-quasi-well prepared (τ-well prepared) for the transform of the restriction of R to $\overline{Y}$.

We have that $q \in (C - \Theta(f,Y)) \cap G_Y(f,\tau)$ implies $q \in U(R')$.

We apply Theorem 15.2 to C. We construct a pre-τ-quasi-well prepared (τ-well prepared) diagram of R. Let R^1 be the transform of R on X_1, and let $(R^1)'$ be the transform of R' on X_1 (which is a restriction of R^1). Let the diagram be

$$(15.20)\qquad \begin{array}{ccc} X_1 & \overset{f_1}{\to} & Y_1 \\ \Phi_1 \downarrow & & \downarrow \Psi_1 \\ X & \overset{f}{\to} & Y, \end{array}$$

which, after possibly replacing $\overline{Y}$ with a proper open subset of $\overline{Y}$, restricts to

$$(15.21)\qquad \begin{array}{ccc} \overline{X}_1 & \overset{\overline{f}_1}{\to} & \overline{Y}_1 \\ \overline{\Phi}_1 \downarrow & & \downarrow \overline{\Psi}_1 \\ \overline{X} & \overset{\overline{f}}{\to} & \overline{Y} \end{array}$$

over $\overline{Y}$. By Theorem 15.2, Remark 14.3 and Lemma 14.1, we may modify the diagrams (15.20) and (15.21) by blowing up 2-curves and 3-points above Y_1 and X_1 to achieve the condition that f_1 is τ-prepared. We may insert blow ups of 2-curves in the diagram (15.19) without changing the conclusion that $\overline{f}_n$ is toroidal (no 3-points are in (15.19)). We have that $G_{Y_1}(f_1,\tau) \subset U((R^1)') \cup \Phi_1^{-1}(f^{-1}(\Theta(f,Y)))$. We iterate this construction for the Zariski closure of γ_i in Y_i, using Theorem 15.2 if γ_i contains a 1-point, and Lemma 14.1 if γ_i is a 2-curve, for $2 \le i \le n$, to achieve the conclusions of the theorem. □

THEOREM 15.4. *Suppose that $f : X \to Y$ is τ-prepared. Then there exists a commutative diagram*

$$\begin{array}{ccc} X_1 & \overset{f_1}{\to} & Y_1 \\ \Phi\downarrow & & \downarrow\Psi \\ X & \overset{f}{\to} & Y \end{array}$$

such that Φ and Ψ are products of possible blow ups, and there exists a relation R^1 for f_1 such that f_1 is τ-quasi-well prepared with relation R^1.

PROOF. We will give an inductive construction. The set $G_Y(f,\tau) - \Theta(f,Y)$ is finite by Remark 11.12. Suppose that there exists a relation R on X such that $\Omega = U(R) \subset G_Y(f,\tau) - \Theta(f,Y)$. (Initially, $\Omega = \emptyset$). Let $\Lambda(Y) = G_Y(f,\tau) - \Theta(f,Y) - \Omega$.

Remark 11.4, Lemma 14.6 , Lemma 10.5 and Lemma 14.1 imply there exists a pre-τ-quasi-well prepared diagram (for R)

$$\begin{array}{ccc} X_1 & \overset{f_1}{\to} & Y_1 \\ \Phi_1\downarrow & & \downarrow\Psi_1 \\ X & \overset{f}{\to} & Y \end{array}$$

obtained by blowing up 2-curves such that if $q_1 \in \Psi_1^{-1}(\Lambda(Y))$, then there exist algebraic permissible parameters u_1, v_1, w_1 at q_1 such that if $p_1 \in f_1^{-1}(q_1) \cap G_{X_1}(f_1,\tau)$, then there exist good parameters $u_1, v_1, w_1 - \phi_{p_1}(u_1,v_1)$ at p_1 for f_1. Further, if q_1 is a 2-point, then u_1, v_1 are toroidal forms at all $p \in f_1^{-1}(q_1)$, and if C is the 2-curve containing q_1, then all points $q' \in C$ have permissible parameters $u_{q'}, v_{q'}, w_{q'}$ at q' such that $u_{q'}, v_{q'}$ are toroidal forms at all $p \in f_1^{-1}(q')$. In addition, we have that f_1 is τ-prepared.

We have $\Psi_1^{-1}(\Theta(f,Y)) \cap G_{Y_1}(f_1,\tau) \subset \Theta(f_1,Y_1)$. Let R^1 be the transform of R on X_1. We restrict R^1 by removing from $U(R^1)$ the points of $\Theta(f_1,Y_1)$, so that $U(R^1) \cap \Theta(f_1,Y_1) = \emptyset$. Let $\Omega_1 = U(R^1)$, a finite set of points (by Remark 11.12). We also have that a general point of each curve contained in $G_{Y_1}(f_1,\tau)$ is in $\Theta(f_1,Y_1)$, since f_1 is τ-prepared. Let

$$\Lambda(Y_1) = G_{Y_1}(f_1,\tau) - \Theta(f_1,Y_1) - \Omega_1 \subset \Psi_1^{-1}(\Lambda(Y)).$$

$\Lambda(Y_1)$ is a finite set of points. By Lemma 11.5, and our construction of f_1, for each $q \in \Lambda(Y_1)$, we can associate algebraic permissible parameters

$$u_q, v_q, w_q \tag{15.22}$$

at q, locally closed subsets $A_1, \ldots, A_{n(q)}$ of $f_1^{-1}(q) \cap G_{X_1}(f_1,\tau)$ such that

$$\{A_1, \ldots, A_{n(q)}\}$$

is a partition of $f_1^{-1}(q)\cap G_{X_1}(f_1,\tau)$, and permissible parameters

$$u_q, v_q, w_{qi} = w_q - \phi_i(u_q, v_q)$$

at q for $1 \le i \le n(q)$ such that for $p_1 \in A_i$, w_{qi} is good at p_1 for f_1. We can assume that $u_q = v_q = 0$ are local equations of a general curve through q on D_Y if q is a 1-point (by Bertini's theorem and Remark 11.4).

Now fix $q \in \Lambda(Y_1)$. By Theorem 15.1, applied to u_q, v_q, w_{qi} for $1 \le i \le n(q)$, there exists an affine neighborhood V_q of q and a commutative diagram

$$\begin{array}{ccc} \overline{X} & \xrightarrow{\overline{f}} & \overline{Y} \\ \overline{\Phi}\downarrow & & \downarrow\overline{\Psi} \\ f_1^{-1}(V_q) & \rightarrow & V_q \end{array} \tag{15.23}$$

satisfying the conclusions of Theorem 15.1 for all w_{qi}. If $\overline{q} \in \overline{\Psi}^{-1}(q)$, then there exist $\overline{u},\overline{v} \in \mathcal{O}_{\overline{Y},q}$ such that $\overline{u},\overline{v},w_{qi}$ are permissible parameters at $\overline{q}$, $\overline{u},\overline{v},w_{qi}$ are super parameters at $\overline{q}$ for all i, and by 3 of Lemma 14.1 and 5 of Lemma 14.5, for

$$p \in \overline{f}^{-1}(\overline{q}) \cap G_{\overline{X}}(\overline{f},\tau) \cap \overline{\Phi}^{-1}(A_i),$$

w_{qi} is good at p.

Observe that we can modify the construction of (15.1) in Theorem 15.1, by a diagram (15.2), by performing an arbitrary sequence of blow ups of 2-curves above each V_i and W_i before constructing V_{i+1} and W_{i+1}.

Suppose that our fixed $q \in \Lambda(Y_1)$ is 1-point. Recall that we have chosen u_q, v_q, w_{qi} so that $u_q = v_q = 0$ are local equations of a general curve C on D_{Y_1} through q, so that C makes SNCs with D_{Y_1}, C intersects 2-curves of Y_1 at general points, and $C - \{q\}$ intersects the nonfinite locus γ of f_1 transversally at general points of irreducible 1-dimensional components of γ. Thus we have $G_{Y_1}(f_1,\tau)\cap (C-\{q\}) \subset \Theta(f_1,Y_1)$ and C intersects $\Theta(f_1,Y_1)$ transversally at general points of curves in $G_{Y_1}(f_1,\tau)$. For $\overline{q} \in G_{Y_1}(f_1,\tau)\cap(C-\{q\})$, there exist algebraic permissible parameters

$$u_{\overline{q}}, v_{\overline{q}}, w_{\overline{q}}$$

at the 1-point $\overline{q}$ such that $u_{\overline{q}} = v_{\overline{q}} = 0$ are local equations of C, and since $\overline{q}$ is perfect for f, there exist finitely many series $\phi_i(u_{\overline{q}}, v_{\overline{q}}) \in \mathbf{k}[[u_{\overline{q}}, v_{\overline{q}}]]$ such that $u_{\overline{q}}, v_{\overline{q}}, w_{\overline{q}i} = w_{\overline{q}} - \phi_i$ are super parameters at $\overline{q}$ for all i, and for $p \in f_1^{-1}(\overline{q})$, some $w_{\overline{q}i}$ is weakly good at p.

Now by Lemma 11.5, for $\overline{q} \in G_{Y_1}(f_1,\tau)\cap(C-\{q\})$, there exist locally closed subsets $V_i \subset X_1$ for $1 \le i \le n(\overline{q})$ such that $f_1^{-1}(\overline{q})\cap G_{X_1}(f_1,\tau)$ is the disjoint union of $V_1,\dots,V_{n(\overline{q})}$, and for $p \in V_i$, $w_{\overline{q}i}$ is weakly good for f_1 at p.

By Lemma 10.3 and Remark 14.2, there exists a sequence of blow ups of 2-curves $\hat{\Phi} : \hat{X} \to X_1$ with induced pre-τ-quasi-well prepared morphism $\hat{f} = f \circ \hat{\Phi} : \hat{X} \to Y_1$ such that for all $\overline{q} \in G_{Y_1}(f_1,\tau)\cap(C-\{q\})$, if $p \in \hat{f}^{-1}(\overline{q})$, then for all i, $u_{\overline{q}}, v_{\overline{q}}, w_{\overline{q}i}$ are super parameters at p and if $\tau_{\hat{f}}(p) > 1$, then $w_{\overline{q}i} = 0$ is a local equation of a divisor supported on $D_{\hat{X}}$ at p. Further, if $p \in \hat{\Phi}^{-1}(V_i)\cap G_{\hat{X}}(\hat{f},\tau)$, then $w_{\overline{q}i}$ is weakly good for $\hat{f}$ at p. $\hat{f}$ is τ-prepared.

We define new primitive relations $R^1_{\overline{q},i}$ for the finitely many

$$\overline{q} \in G_{Y_1}(\hat{f},\tau)\cap(C-\{q\}) \subset G_{Y_1}(f_1,\tau)\cap(C-\{q\})$$

and $1 \le i \le n(\overline{q})$.

Let $U(R^1_{\overline{q},i}) = \{\overline{q}\}$, $T(R^1_{\overline{q},i}) = \hat{\Phi}^{-1}(V_i) \cap G_{\hat{X}}(\hat{f}, \tau)$.

Suppose that $\tau > 1$.

If $p \in \hat{\Phi}^{-1}(V_i)$ is a 1-point, then we have an expression

$$u_{\overline{q}} = x^a, v_{\overline{q}} = y, w_{\overline{q}i} = x^c\gamma$$

where x, y, z are regular parameters in $\hat{\mathcal{O}}_{\hat{X},p}$, $\gamma \in \hat{\mathcal{O}}_{\hat{X},p}$ is a unit series and $a \not| c$. Let $d = \gcd(a, c) < a$, $e = \frac{a}{d} > 1$ and $\overline{c} = \frac{c}{d}$.

We define $R^1_{\overline{q},i}(p) = w^e_{\overline{q}i} - \gamma(0,0,0)^e u^{\overline{c}}_{\overline{q}}$.

If $p \in \hat{\Phi}^{-1}(V_i)$ is a 2-point, then we have an expression

$$u_{\overline{q}} = (x^a y^b)^k, v_{\overline{q}} = z, w_{\overline{q}i} = (x^a y^b)^l \gamma$$

where x, y, z are regular parameters in $\hat{\mathcal{O}}_{\hat{X},p}$, $\gamma \in \hat{\mathcal{O}}_{\hat{X},p}$ is a unit series and $k \not| l$. Let $d = \gcd(k, l) < k$, $e = \frac{k}{d} > 1$ and $\overline{c} = \frac{l}{d}$. We define $R^1_{\overline{q},i}(p) = w^e_{\overline{q}i} - \gamma(0,0,0)^e u^{\overline{c}}_{\overline{q}}$.

Suppose that $\tau = 1$. We then define $R^1_{\overline{q},i}(p)$ by the relation $w_{\overline{q}i} = 0$ for $p \in \hat{\Phi}^{-1}(V_i)$ (by Remark 11.3).

We extend the transform $\hat{R}^1$ of R^1 on $\hat{X}$ to include the primitive relations $R^1_{\overline{q},i}$ for $\overline{q} \in G_{Y_1}(\hat{f}, \tau) \cap (C - \{q\})$ which we have just defined. Observe that we now have

$$(C - \{q\}) \cap G_{Y_1}(\hat{f}, \tau) \subset U(\hat{R}^1).$$

By embedded resolution of plane curve singularities (c.f. Section 3.4 and Exercise 3.13 [**C6**]), and Lemmas 14.7 and 14.8, there exists a pre-τ-quasi-well prepared diagram

$$\begin{array}{ccc} \tilde{X} & \xrightarrow{\tilde{f}} & \tilde{Y} \\ \tilde{\Phi} \downarrow & & \downarrow \tilde{\Psi} \\ \hat{X} & \xrightarrow{\hat{f}} & Y_1 \end{array}$$

such that $\tilde{\Psi}$ is a sequence of blow ups of prepared 1-points and 2-points (of types 4 and 1 of Definition 13.7) such that if $\tilde{C}$ is the strict transform of C on $\tilde{Y}$, then $\tilde{C} \cap G_{\tilde{Y}}(f, \tau) = \{q\}$ (since the germ of C at $\overline{q} \in G_{Y_1}(f_1, \tau) \cap (C - \{q\})$ is not contained in the surface germ $w_{\overline{q}i} = 0$ for any i). Further, $\tilde{f}$ is τ-prepared. Thus $\tilde{C}$ satisfies 1 and 2 of Definition 13.6 of a resolving curve. 4 of the definition holds since f_1 is τ-prepared and C intersects the nonfinite locus of $\tilde{f}$ transversally at general points.

By Lemma 10.7 and Lemma 14.1, there exists a pre-τ-quasi-well prepared diagram

$$\begin{array}{ccc} \tilde{X}_2 & \xrightarrow{\tilde{f}_2} & \tilde{Y}_2 \\ \tilde{\Phi}_2 \downarrow & & \downarrow \tilde{\Psi}_2 \\ \tilde{X} & \xrightarrow{\tilde{f}} & \tilde{Y} \end{array}$$

where $\tilde{\Psi}_2$ and $\tilde{\Phi}_2$ are products of blow ups of 2-curves such that the strict transform $\tilde{C}_2$ of C satisfies 3 of Definition 13.6. Observe that $\tilde{\Psi} \circ \Psi_2$ is an isomorphism over q, so that we may identify q with a point of $\tilde{Y}_2$. $\tilde{X}_2 \to X_1$ is a sequence of blow ups of 2-curves over $f_1^{-1}(q)$. Thus $\tilde{C}_2$ is a resolving curve for $\tilde{f}_2$ at q. We further have that $\tilde{f}_2$ is τ-prepared.

Let $\Phi' : \tilde{X}_2 \to X_1$ be our morphism $\Phi' = \hat{\Phi} \circ \tilde{\Phi} \circ \tilde{\Phi}_2$. Let $\tilde{A}_i = (\Phi')^{-1}(A_i) \cap G_{\tilde{X}_2}(\tilde{f}_2, \tau)$ for $1 \le i \le n(q)$. We have that u_q, v_q, w_{qi} are permissible parameters at q such that w_{qi} is good for $\tilde{f}_2$ at all $p \in \tilde{A}_i$ (by Remark 14.2).

After possibly replacing the neighborhood V_q of q in (15.23) with a smaller neighborhood of q, we may identify V_q with a neighborhood of q in $\tilde{Y}_2$.

Let

$$\begin{array}{ccc} X_2 & \stackrel{f_2}{\to} & Y_2 \\ \Phi_2 \downarrow & & \downarrow \Psi_2 \\ \tilde{X}_2 & \stackrel{\tilde{f}_2}{\to} & \tilde{Y}_2 \end{array}$$

be the pre-τ-quasi-well prepared diagram of the conclusions of Lemma 14.5, where Ψ_2 is the blow up of $\tilde{C}_2$. f_2 is τ-prepared.

As $(f_1 \circ \Phi')^{-1}(V_q) \to f_1^{-1}(V_q)$ is a sequence of blow ups of 2-curves, as commented after the construction of (15.23), we can assume that if we restrict the diagram

$$\begin{array}{ccc} X_2 & \to & Y_2 \\ \Phi'' \downarrow & & \downarrow \Psi'' \\ X_1 & \to & Y_1 \end{array}$$

that we have constructed to V_q, we obtain the diagram

$$\begin{array}{ccc} W_1 & \to & V_1 \\ \overline{\Phi}_1 \downarrow & & \downarrow \overline{\Psi}_1 \\ f_1^{-1}(V_q) = W & \to & V = V_q \end{array}$$

of (15.2) constructed in the proof of Theorem 15.1.

Now suppose that our fixed $q \in \Lambda(Y_1)$ is a 2-point. Then, by Lemma 14.1 we construct a τ-quasi-well prepared diagram for R^1 and the blow up Ψ'' of C

$$\begin{array}{ccc} X_2 & \stackrel{f_2}{\to} & Y_2 \\ \Phi'' \downarrow & & \downarrow \Psi'' \\ X_1 & \to & Y_1 \end{array}$$

where Φ'' is a sequence of blow ups of 2-curves. f_2 is τ-prepared.

We can assume that if we restrict the diagram

$$\begin{array}{ccc} X_2 & \to & Y_2 \\ \downarrow & & \downarrow \\ X_1 & \to & Y_1 \end{array}$$

to V_q, we obtain the diagram

$$\begin{array}{ccc} W_1 & \to & V_1 \\ \overline{\Phi}_1 \downarrow & & \downarrow \overline{\Psi}_1 \\ f_1^{-1}(V_q) = W & \to & V = V_q \end{array}$$

of (15.2) constructed in the proof of Theorem 15.1.

Let R^2 be our relation on X_2. We have (by Remark 14.2)

$$(\Psi'')^{-1}(\Theta(f_1, Y_1) - C) \subset \Theta(f_2, Y_2).$$

We restrict R^2 if necessary, so that $U(R^2) \cap \Theta(f_2, Y_2) = \emptyset$. Let $\Omega_2 = U(R^2)$

With the notation of the proof of Theorem 15.1, let C_2 be the Zariski closure of γ_1, the curve blown up in $V_2 \to V_1$, in Y_2. C_2 is a section over C. Either C_2 is a 2-curve or C_2 contains a 1-point.

By our assumptions on C, at all points $q' \in C$, there exist permissible parameters $u_{q'}, v_{q'}, w_{q'}$ at q' such that $u_{q'} = v_{q'} = 0$ are local equations of C and $u_{q'}, v_{q'}$ are toroidal forms at q'. Thus if $\overline{q}_1 \in (\Psi'')^{-1}(q') \cap C_2$, then there exist permissible parameters u_1, v_1, w such that $u_1 = v_1 = 0$ are local equations of C_2 at $\overline{q}_1$, and u_1, v_1 are toroidal forms at all $p \in f_2^{-1}(\overline{q}_1)$. In particular, if $\overline{q}_1$ is a 1-point, then C_2 satisfies 2 and 4 of Definition 13.6 of a resolving curve for f_2 at $\overline{q}_1$.

Suppose that C_2 contains a 1-point. Then $q_1 = (\Psi'')^{-1}(q) \cap C_2$ is a 1-point.

We can apply Lemma 14.10 to construct a pre-τ-quasi-well prepared diagram

$$\begin{array}{ccc} X_3 & \stackrel{f_3}{\rightarrow} & Y_3 \\ \downarrow & & \downarrow \\ X_2 & \stackrel{f_2}{\rightarrow} & Y_2 \end{array} \tag{15.24}$$

satisfying the conclusions of Lemma 14.10, so that the strict transform of C_2 is a resolving curve for f_3 at q_1. We have that f_3 is τ-prepared.

The vertical arrows of (15.24) are products of blow ups of 2-curves above V_q. Let

$$\begin{array}{ccc} X_4 & \stackrel{f_4}{\rightarrow} & Y_4 \\ \Phi_4 \downarrow & & \downarrow \Psi_4 \\ X_3 & \stackrel{f_3}{\rightarrow} & Y_3 \end{array}$$

be the pre-τ-quasi-well prepared diagram of the conclusions of Lemma 14.5, where Ψ_4 is the blow up of the strict transform of C_2. f_4 is τ-prepared.

Suppose that C_2 is a 2-curve. Then we construct (from Lemma 14.1)

$$\begin{array}{ccc} X_4 & \rightarrow & Y_4 \\ \downarrow & & \downarrow \\ X_2 & \rightarrow & Y_2 \end{array} \tag{15.25}$$

as a τ-quasi-well prepared diagram for R^2 and the blow up of C_2.

As commented after the construction of (15.23), we may assume that the diagram (15.25) restricts to the diagram

$$\begin{array}{ccc} W_2 & \rightarrow & V_2 \\ \downarrow & & \downarrow \\ W_1 & \rightarrow & V_1 \end{array}$$

of (15.2) above V_q.

Since (15.2) is finite, after finitely many iterations, we achieve a pre-τ-quasi-well prepared diagram

$$\begin{array}{ccc} X_5 & \stackrel{f_5}{\rightarrow} & Y_5 \\ \Phi_5 \downarrow & & \downarrow \Psi_5 \\ X_1 & \rightarrow & Y_1 \end{array}$$

which restricts to the diagram (15.23) above V_q, Ψ_5 is an isomorphism over $\Lambda(Y_1) - \{q\}$ and Φ_5 is a sequence of blow ups of 2-curves over $\Lambda(Y_1) - \{q\}$. Further, f_5 is τ-prepared.

By Lemma 10.3, and Remark 14.2, 3 of Lemma 14.1 and 5 of Lemma 14.5, there exists a sequence of blow ups of 2-curves $\Phi_6 : X_6 \rightarrow X_5$ such that $f_6 : X_6 \rightarrow Y_5$ is pre-τ-quasi-well prepared, τ-prepared, and (with the notation introduced with (15.23)), there exist algebraic permissible parameters $u_q, v_q, w_q \in \mathcal{O}_{Y_1,q}$ and $w_{qi} =$

$w_q - \phi_i(u_q, v_q)$ such that for $\overline{q}$ in the finite set

$$\Sigma = (\Psi_5 \circ \Psi_6)^{-1}(q) \cap G_{Y_5}(f_6, \tau) - \Theta(f_6, Y_5),$$

we have algebraic permissible parameters $u_{\overline{q}}, v_{\overline{q}}, w_q$ at $\overline{q}$, and series

$$\overline{\phi}_i(u_{\overline{q}}, v_{\overline{q}}) = \phi_i(u_q, v_q)$$

such that $u_{\overline{q}}, v_{\overline{q}}, w_{qi} = w_q - \overline{\phi}_i(u_{\overline{q}}, v_{\overline{q}})$ are super parameters for all i, and for

$$p \in V_{\overline{q}i} = f_6^{-1}(\overline{q}) \cap (\Phi_5 \circ \Phi_6)^{-1}(A_i) \cap G_{X_6}(f_6, \tau),$$

$w_{\overline{q}i} = w_q - \overline{\phi}_i(u_{\overline{q}}, v_{\overline{q}})$ is good for f_6 at p, and $w_{\overline{q}i} = 0$ is supported on D_{X_6} at p if $\tau > 1$.

We define new primitive relations $R^6_{\overline{q},i}$ for $\overline{q} \in \Sigma$ and $1 \le i \le n(q)$.

Let $U(R^6_{\overline{q},i}) = \{\overline{q}\}$, $T(R^6_{\overline{q},i}) = V_{\overline{q}i}$.

Suppose that $\tau > 1$.

If $p \in V_{\overline{q}i}$ is a 1-point, then we have an expression

$$u_{\overline{q}} = x^a, v_{\overline{q}} = y, w_{\overline{q}i} = x^c\gamma$$

where x, y, z are regular parameters in $\hat{\mathcal{O}}_{X_6,p}$, $\gamma \in \hat{\mathcal{O}}_{X_6,p}$ is a unit series and $a \not| c$.

Let

$$a' = \frac{a}{\gcd(a,c)} > 1, \ \ c' = \frac{c}{\gcd(a,c)}.$$

We define $R^6_{\overline{q},i}(p) = w_{\overline{q}i}^{a'} - \gamma(0,0,0)^{a'} u_{\overline{q}}^{c'}$.

If $p \in V_i$ is a 2-point, then we have an expression

$$u_{\overline{q}} = (x^a y^b)^k, v_{\overline{q}} = z, w_{\overline{q}i} = (x^a y^b)^l \gamma$$

where x, y, z are regular parameters in $\hat{\mathcal{O}}_{X_6,p}$, $\gamma \in \hat{\mathcal{O}}_{X_6,p}$ is a unit series and $k \not| l$.

Let

$$k' = \frac{k}{\gcd(k,l)} > 1, \ \ l' = \frac{l}{\gcd(k,l)}.$$

We define $R^6_{\overline{q},i}(p) = w_{\overline{q}i}^{k'} - \gamma(0,0,0)^{k'} u_{\overline{q}}^{l'}$.

Suppose that $\tau = 1$. We then define $R^6_{\overline{q},i}(p) = w_{\overline{q}i}$. By Remark 11.3, $w_{\overline{q}i}$ is a monomial form at p.

We can extend the transform R^6 of R^5 on X_6 to include these new primitive relations $R^6_{\overline{q},i}$.

We now restrict R^6 to remove the points of $\Theta(f_6, Y_5)$ from $U(R^6)$. Let

$$\Lambda(Y_6) = G_{Y_5}(f_6, \tau) - \Theta(f_6, Y_5) - U(R^6).$$

By our construction, $\Lambda(Y_6) \subset \Psi_5^{-1}(\Lambda(Y_1) - \{q\})$, and over $\Lambda(Y_1) - \{q\}$, the vertical arrows of the pre-τ-quasi-well prepared diagram

$$\begin{array}{ccc} X_6 & \stackrel{f_6}{\to} & Y_5 \\ \Phi_5 \circ \Phi_6 \downarrow & & \downarrow \Psi_5 \\ X_1 & \stackrel{f_1}{\to} & Y_1 \end{array}$$

are blow ups of 2-curves.

By induction on $|\Lambda(Y_1)|$, we may iterate the above procedure to construct a commutative diagram

$$\begin{array}{ccc} X_7 & \stackrel{f_7}{\to} & Y_7 \\ \downarrow & & \downarrow \\ X & \to & Y \end{array}$$

such that f_7 is pre-τ-quasi-well prepared with $U(R^7) = G_{Y_7}(f_7,\tau) - \Theta(f_7, Y_7)$, and f_7 is τ-prepared.

Now by Theorem 15.3 there exists a pre-τ-quasi-well diagram for R^7

$$\begin{array}{ccc} X_8 & \overset{f_8}{\to} & Y_8 \\ \downarrow & & \downarrow \\ X_7 & \to & Y_7 \end{array}$$

such that f_8 is τ-quasi-well prepared for the transform of R^7. □

LEMMA 15.5. *Suppose that $\tau \geq 1$, $f : X \to Y$ is τ-quasi-well prepared with relation R. Further suppose there exists a τ-quasi-well prepared diagram for R*

$$\begin{array}{ccc} \tilde X & \overset{\tilde f}{\to} & \tilde Y \\ \tilde\Phi \downarrow & & \downarrow \tilde\Psi \\ X & \overset{f}{\to} & Y, \end{array} \tag{15.26}$$

where $\tilde R$ is the transform of R on $\tilde X$, such that if $q_1 \in U(\tilde R)$ is on a component E of $D_{\tilde Y}$ such that $\tilde\Psi(E)$ is not a point, then $T(\tilde R) \cap \tilde f^{-1}(q_1) = \emptyset$. Then there exists a commutative diagram

$$\begin{array}{ccc} X_1 & \overset{f_1}{\to} & Y_1 \\ \Phi \downarrow & & \downarrow \Psi \\ \tilde X & \overset{\tilde f}{\to} & \tilde Y \end{array}$$

such that Φ, Ψ are products of blow ups of possible centers, and f_1 is τ-quasi-well prepared with relation R^1 and pre-algebraic structure. Further, R^1 is algebraic. (R^1 will in general not be the transform of $\tilde R$.)

PROOF. Given a diagram (15.26), we will define a new relation $\tilde R'$ on $\tilde X$ for $\tilde f$. This is accomplished as follows. We have that $|U(\tilde R) - \Theta(\tilde f, \tilde Y)| < \infty$ by Remark 11.12. Suppose that $q_1 \in U(\tilde R) - \Theta(\tilde f, \tilde Y)$ is such that $\tilde f^{-1}(q_1) \cap T(\tilde R) \neq \emptyset$. Let

$$J_{q_1} = \{i \mid T(\tilde R_i) \cap \tilde f^{-1}(q_1) \neq \emptyset\}.$$

Let $q = \tilde\Psi(q_1)$. For $j \in J_{q_1}$, let

$$u = u_{\overline{R}_j(q)}, v = v_{\overline{R}_j(q)}, w_j = w_{\overline{R}_j(q)}.$$

Let

$$u_1 = u_{\overline{\tilde R_j}(q_1)}, v_1 = v_{\overline{\tilde R_j}(q_1)}, w_{j,1} = w_{\overline{\tilde R_j}(q_1)}.$$

Since $\tilde\Psi$ is a composition of admissible blow ups for the transforms of the pre-relations $\overline{R}_i$ on Y, by the description of admissible blow ups (12.7) - (12.9) following Definition 12.2, we have in $\hat{\mathcal{O}}_{Y_1,q}$, one of the following relations: q_1 a 2-point

$$u = u_1^{\tilde a} v_1^{\tilde b}, v = u_1^{\tilde c} v_1^{\tilde d}, w_j = u_1^{\tilde e} v_1^{\tilde f} w_{j,1} \tag{15.27}$$

with $\tilde a \tilde d - \tilde b \tilde c = \pm 1$, or q_1 a 1-point,

$$u = u_1, v = u_1^{\tilde c} v_1, w_j = u_1^{\tilde e} w_{j,1} \tag{15.28}$$

or q_1 a 1-point

$$u = u_1^{\tilde a} \gamma_1(u_1, v_1), v = u_1^{\tilde b} \gamma_2(u_1, v_1), w_j = u_1^{\tilde c} \gamma_3(u_1, v_1) w_{j,1} \tag{15.29}$$

where $\gamma_1,\gamma_2,\gamma_3$ are unit series and $\gamma_1,\gamma_2,\gamma_3\in\mathcal{O}_{Y_1,q}$ or q_1 a 2-point

$$(15.30)\qquad u=(u_1^{\tilde a}v_1^{\tilde b})^{\tilde t}\gamma_1(u_1,v_1), v_1=(u_1^{\tilde c}v_1^{\tilde d})^{\tilde k}\gamma_2(u_1,v_1), w_j=u_1^{\tilde e}v_1^{\tilde f}\gamma_3(u_1,v_1)w_{j,1}$$

where $\gamma_1,\gamma_2,\gamma_3$ are unit series, and $\gamma_1,\gamma_2,\gamma_3\in\mathcal{O}_{Y_1,q}$, $\tilde a,\tilde b>0$. Since (15.26) is a τ-quasi-well prepared diagram, the exponents appearing in (15.27) - (15.30) are independent of $j\in J_{q_1}$.

By assumption, if q_1 is on a component E of $D_{\tilde Y}$ we must have $\tilde\Psi(E)$ is a point. Thus in (15.27) we have $\tilde a,\tilde b,\tilde c,\tilde d,\tilde e,\tilde f$ all nonzero. If (15.28) holds, we have $\tilde c,\tilde e>0$. In (15.29) we have $\tilde a,\tilde b,\tilde c>0$. In (15.30) we have $\tilde t,\tilde k,\tilde e,\tilde f>0$.

Suppose that $p_1\in T(\tilde R_j)\cap\tilde f^{-1}(q_1)$ and $\tau>1$. Let $p=\tilde\Phi(p_1)\in T(R_j)$. On X_1, $w_{j,1}=0$ is a divisor supported on D_{X_1} at p_1. In (15.27), (15.29), (15.30) we see that $u=0$ is a local equation of D_{X_1} at p_1, and $v=0$ is a local equation of D_{X_1} at p_1. In (15.28), we have that $u=0$ is a local equation of D_{X_1} at p_1, $w_j=0$ is a local equation of D_{X_1} at p_1, and $v=0$ is a local equation of a divisor that contains D_{X_1} at p_1. Thus in all cases, there exists a natural number r such that w_j divides u^r and v^r in $\hat{\mathcal{O}}_{\tilde X,p_1}$.

Define

$$\eta=\eta(q_1)=\begin{cases}\max\{2r^2,\tilde e,\tilde f\} & \text{if (15.27) or (15.30) holds}\\ \max\{2r^2,\tilde c\} & \text{if (15.29) holds}\\ \max\{2r^2,\tilde e\} & \text{if (15.28) holds}\end{cases}$$

where the maximum is over $j\in J_{q_1}$ and $p_1\in T(\tilde R_j)\cap\tilde f^{-1}(q_1)$.

Fix $j\in J_{q_1}$. There exists $\sigma(u,v,w_j)\in\mathbf{k}[[u,v,w_j]]=\hat{\mathcal{O}}_{Y,q}$ such that the order of the series σ is greater than η and $w_j+\sigma\in\mathcal{O}_{Y,q}$. Let

$$w_j^*=w_j+\sigma(u,v,w_j).$$

For $p_1\in T(\tilde R_j)\cap\tilde f^{-1}(q_1)$, we have

$$w_j^*=w_j\gamma_{p_1j}$$

where $\gamma_{p_1j}\in\hat{\mathcal{O}}_{X_1,p_1}$ is a unit series. If (15.27) holds at q_1, set

$$(15.31)\quad w_{q_1,j}=\frac{w_j^*}{u_1^{\tilde e}v_1^{\tilde f}}=w_{j,1}+\frac{\sigma(u_1^{\tilde a}v_1^{\tilde b},u_1^{\tilde c}v_1^{\tilde d},u_1^{\tilde e}v_1^{\tilde f}w_{j,1})}{u_1^{\tilde e}v_1^{\tilde f}}\in\hat{\mathcal{O}}_{\tilde Y,q_1}\cap\mathbf{k}(Y)=\mathcal{O}_{\tilde Y,q_1},$$

since $u_1,v_1\in\mathcal{O}_{\tilde Y,q_1}$ (this is part of the definition of a relation).

We further have

$$w_{q_1,j}=w_{j,1}\gamma_{p_1,j}.$$

There is a similar argument if (15.28), (15.29) or (15.30) holds.

For $k\in J_{q_1}$, there exists $\lambda_{jk}(u,v)\in\mathbf{k}[[u,v]]$ such that $w_k=w_j+\lambda_{jk}(u,v)$. Write

$$\lambda_{jk}(u,v)=\alpha_k(u,v)+h_k(u,v)$$

where $\alpha_k(u,v)$ is a polynomial, and $h_k(u,v)$ is a series of order greater than η. Set

$$w_k^*=w_j+\sigma(u,v,w_j)+\lambda_{jk}(u,v)-h_k(u,v)\in\mathcal{O}_{Y,q}.$$

$$\begin{aligned}w_k^*&=w_k+\sigma(u,v,w_k-\lambda_{jk}(u,v))-h_k(u,v)\\&=w_k+\overline{\sigma}_k(u,v,w_k)\end{aligned}$$

where $\overline{\sigma}_k$ is a series of order greater than η.

Suppose that (15.27) holds at q_1. Set

$$w_{q_1,k} = \frac{w_k^*}{u_1^{\tilde{e}} v_1^{\tilde{f}}}.$$

From (15.27) we see that $u_1, v_1, w_{q_1,k}$ are permissible parameters at q_1 and $w_{q_1,k} \in \mathcal{O}_{\tilde{Y},q_1}$. We further have that

$$w_{q_1,k} = w_{k,1}\gamma_{p_1 k} \tag{15.32}$$

for some unit series $\gamma_{p_1 k} \in \hat{\mathcal{O}}_{\tilde{X}_1,p_1}$.

We have that for $k \in J_{q_1}$,

$$w_{q_1,k} = w_{k,1} + \frac{\overline{\sigma}_k(u_1^{\tilde{a}} v_1^{\tilde{b}}, u_1^{\tilde{c}} v_1^{\tilde{d}}, u_1^{\tilde{e}} v_1^{\tilde{f}} w_{k,1})}{u_1^{\tilde{e}} v_1^{\tilde{f}}} \tag{15.33}$$

with

$$\frac{\overline{\sigma}_k}{u_1^{\tilde{e}} v_1^{\tilde{f}}} \in \hat{\mathcal{O}}_{\tilde{Y},q_1}.$$

We further have

$$\begin{aligned} w_{q_1,k} - w_{q_1,j} &= \tfrac{w_k^* - w_j^*}{u_1^{\tilde{e}} v_1^{\tilde{f}}} \\ &= \tfrac{\lambda_{jk}(u,v) - h_k(u,v)}{u_1^{\tilde{e}} v_1^{\tilde{f}}} \in \mathbf{k}((u_1, v_1)) \cap \mathbf{k}[[u_1, v_1, w_{j,1}]] = \mathbf{k}[[u_1, v_1]]. \end{aligned}$$

There is a similar argument if (15.28), (15.29) or (15.30) holds. In these cases (15.31) becomes

$$w_{q_1 j} = \frac{w_j^*}{u_1^{\tilde{e}}}, w_{q_1 j} = \frac{w_j^*}{u_1^{\tilde{c}}}, w_{q_1 j} = \frac{w_j^*}{u_1^{\tilde{e}} v_1^{\tilde{f}}}$$

respectively.

In all these cases, an equation (15.32) holds.

In case (15.28), a variant of equation (15.33) holds.

Suppose that (15.30) holds and $k \in J_{q_1}$. Then there exist series $\overline{\sigma}_k(u, v, w_k)$ such that ord $\overline{\sigma}_k > \eta$ and such that

$$w_k^* = w_k + \overline{\sigma}_k(u, v, w_k).$$

Thus

$$w_{q_1,k} = \gamma_3(u_1, v_1) w_{k,1} + \frac{\overline{\sigma}_k((u_1^{\tilde{a}} v_1^{\tilde{b}})^{\tilde{t}} \gamma_1(u_1, v_1), (u_1^{\tilde{c}} v_1^{\tilde{d}})^{\tilde{k}} \gamma_2(u_1, v_1), u_1^{\tilde{e}} v_1^{\tilde{f}} \gamma_3(u_1, v_1) w_{j,1})}{u_1^{\tilde{e}} v_1^{\tilde{f}}} \tag{15.34}$$

with

$$\frac{\overline{\sigma}_k}{u_1^{\tilde{e}} v_1^{\tilde{f}}} \in \hat{\mathcal{O}}_{\tilde{Y},q_1}.$$

There is a similar expression if (15.29) holds.

Now suppose that $\tau = 1$. We make the same argument as the $\tau > 1$ case if $w_{j,1}$ satisfies a form 3 or 4 of Definition 4.4 or a form 1, 2, 5 with $\beta \neq 0$. A similar, but slightly different argument is required if w_{j1} satisfies 1, 2 or 5 of Definition 4.4, with $\beta = 0$.

We now define the new relation $\tilde{R}'$ on $\tilde{X}$ for $\tilde{f}$. Set $T(\tilde{R}') = T(\tilde{R}) - \tilde{f}^{-1}(\Theta(\tilde{f}, \tilde{Y}))$, $U(\tilde{R}') = \tilde{f}(T(\tilde{R})) - \Theta(\tilde{f}, \tilde{Y})$. $U(\tilde{R}')$ is a finite set by Remark 11.12. For $q_1 \in U(\tilde{R}')$, we define primitive relations $R_{q_1,k}$ as follows. Set $U(R_{q_1,k}) = \{q_1\}$, $T(R_{q_1,k}) = T(\tilde{R}_k)\cap \tilde{f}^{-1}(q_1)$.

For $p_1 \in T(R_{q_1,k})$ define

$$u_{R_{q_1,k}(p_1)} = u_{\tilde{R}_k(p_1)}, v_{R_{q_1,k}(p_1)} = v_{\tilde{R}_k(p_1)}, w_{R_{q_1,k}(p_1)} = w_{q_1,k}.$$

If $\tau > 1$ and q_1 is a 2-point (from 15.32), we define $R_{q_1,k}(p_1)$ by

$$a_{R_{q_1,k}}(p_1) = a_{\tilde{R}_k}(p_1), b_{R_{q_1,k}}(p_1) = b_{\tilde{R}_k}(p_1), e_{R_{q_1,k}}(p_1) = e_{\tilde{R}_k}(p_1)$$

and

$$\lambda_{R_{q_1,k}}(p_1) = \lambda_{\tilde{R}_k}(p_1)\gamma_{p_1k}(0,0,0)^{e_{\tilde{R}_k}(p_1)}.$$

If $\tau = 1$, and q_1 is a 2-point, we define

$$a_{R_{q_1,k}}(p_1) = b_{R_{q_1,k}}(p_1) = \infty.$$

If q_1 is a 1-point, we define $R_{q_1,k}(p_1)$ in an analogous way.

From the above calculations, we see that $f : \tilde{X} \to \tilde{Y}$ with the relation $\tilde{R}'$ satisfies 1 - 4 of the conditions of Definition 13.1 of a pre-τ-quasi-well prepared morphism.

Recall that all exponents are positive in (15.27) - (15.30), and thus in (15.33) and (15.34). Thus we can choose a possibly larger $\eta(q_1)$ so that $\mathrm{ord}(\sigma_k)$ is sufficiently large in (15.33) and (15.34) that $\tilde{R}'$ satisfies 5 of the conditions of Definition 13.1 (as well as 1 – 4). Thus $\tilde{f}$ is pre-τ-quasi-well prepared with respect to $\tilde{R}'$. Further, f is τ-prepared.

For $q_i \in U(R_{q_i,k})$, let $\Omega(R_{q_i,k})$ be an affine neighborhood of q_i on the surface with local equation $w_{q_i,k} = 0$ at q_i, such that

1. $\Omega(R_{q_i,k})$ is nonsingular and makes SNCs with $D_{\tilde{Y}}$.
2. $\Omega(R_{q_i,k})\cap U(\tilde{R}') = \{q_i\}$

We now restrict $\tilde{R}'$ so that $U(\tilde{R}') = G_{\tilde{Y}}(\tilde{f},\tau) - \Theta(\tilde{f},\tilde{Y})$, $\tilde{f}$ is pre-τ-quasi-well prepared with relation $\tilde{R}'$ and $\tilde{R}'$ is algebraic.

By Theorem 15.3 there exists a pre-τ-quasi-well prepared diagram for $\tilde{R}'$

$$\begin{array}{ccc} X' & \xrightarrow{f'} & Y' \\ \Phi' \downarrow & & \downarrow \Psi' \\ \tilde{X} & \to & \tilde{Y} \end{array}$$

such that f' is τ-quasi-well prepared for the transform R' of $\tilde{R}'$. By our construction, R' has pre-algebraic structure. Further, R' is algebraic. Thus the conclusions of Lemma 15.5 hold. $\square$

REMARK 15.6. *Suppose that $f : X \to Y$ is τ-quasi-well prepared and $q \in U(\overline{R}_i)$ is a 1-point. Let*

$$u = u_{\overline{R}_i(q)}, v = v_{\overline{R}_i(q)}, w_i = w_{\overline{R}_i(q)}.$$

Then f is not toroidal above q if and only if q is contained in the nonfinite locus of f.

Suppose that f is not toroidal above q. Then the germ of the nonfinite locus of f at q is a nonsingular (algebraic) curve, and $u = w_i = 0$ are (formal) local equations of the nonfinite locus of f at q.

PROOF. Suppose that f is not toroidal above q. u, v, w_i are super parameters at p for all $p \in f^{-1}(q)$. From consideration of the local forms 5 and 6 of Definition 10.1 of super parameters at a 1-point q, we see that $u = w_i = 0$ are local equations of (a formal branch of) the fundamental locus of f at q. Since the fundamental locus of f is algebraic, we obtain the conclusions of the remark. □

THEOREM 15.7. *Suppose that $\tau \geq 1$, $f : X \to Y$ is τ-quasi-well prepared with relation R. Then there exists a τ-quasi-well prepared diagram*

$$\begin{array}{ccc} \tilde{X} & \stackrel{\tilde{f}}{\to} & \tilde{Y} \\ \tilde{\Phi} \downarrow & & \downarrow \tilde{\Psi} \\ X & \stackrel{f}{\to} & Y \end{array}$$

where $\tilde{R}$ is the transform of R on $\tilde{X}$ such that if $q_1 \in U(\tilde{R})$ is on a component E of $D_{\tilde{Y}}$ such that $\tilde{\Psi}(E)$ is not a point, then $T(\tilde{R}) \cap \tilde{f}^{-1}(q_1) = \emptyset$.

PROOF. **Step 1.** Let A_0 be the set of 2-points $q \in Y$ such that $q \in U(\overline{R}_i)$ for some $\overline{R}_i$ associated to R and $f^{-1}(q) \cap T(R_i) \neq \emptyset$. A_0 is a finite set since f is τ-prepared.

For $q \in A_0 \cap U(\overline{R}_i)$, set

$$u = u_{\overline{R}_i(q)}, v = v_{\overline{R}_i(q)}, w_i = w_{\overline{R}_i(q)}. \tag{15.35}$$

Let $\Psi_1 : Y_1 \to Y$ be the blowup of all $q \in A_0$, and let

$$\begin{array}{ccc} X_1 & \stackrel{f_1}{\to} & Y_1 \\ \Phi_1 \downarrow & & \downarrow \Psi_1 \\ X & \stackrel{f}{\to} & Y \end{array}$$

be a τ-quasi-well prepared diagram of R and Ψ_1. Such a diagram exits by Lemma 14.7. Suppose that $q \in A_0 \cap U(\overline{R}_i)$ and $q_1 \in \Psi_1^{-1}(q) \cap U(\overline{R}_i^1)$ is a 1-point lying on a component E of D_{Y_1} and $f_1^{-1}(q_1) \cap T(\overline{R}_i^1) \neq \emptyset$. Then q_1 has regular parameters $u_1, v_1, w_{i,1}$ with

$$u = u_1, v = u_1(v_1 + \alpha), w_i = u_1 w_{i,1}$$

with $0 \neq \alpha \in \mathbf{k}$, which implies that E has local equation $u_1 = 0$, so that $\Psi_1(E)$ is a point.

Let A_1 be the set of all 2-points $q_1 \in Y_1$ such that for some i, $q_1 \in U(\overline{R}_i^1)$, $f_1^{-1}(q_1) \cap T(R_i^1) \neq \emptyset$ and q_1 is on a component E of D_{Y_1} such that $\Psi_1(E)$ is not a point. We have $A_1 \subset \Psi_1^{-1}(A_0)$. Let $\Psi_2 : Y_2 \to Y_1$ be the blowup of all $q_1 \in A_1$, and let (by Lemma 14.7)

$$\begin{array}{ccc} X_2 & \stackrel{f_2}{\to} & Y_2 \\ \Phi_2 \downarrow & & \downarrow \Psi_2 \\ X_1 & \stackrel{f_1}{\to} & Y_1 \end{array}$$

be a τ-quasi-well prepared diagram of R^1 and Ψ_2. Continue in this way to construct (for arbitrary n) a sequence of n blow ups of sets of 2-points $\Psi_{k+1} : Y_{k+1} \to Y_k$ for $0 \leq k \leq n-1$ with τ-quasi-well prepared diagrams

$$\begin{array}{ccc} X_{k+1} & \stackrel{f_{k+1}}{\to} & Y_{k+1} \\ \Phi_{k+1} \downarrow & & \downarrow \Psi_{k+1} \\ X_k & \stackrel{f_k}{\to} & Y_k \end{array}$$

of R^k and Ψ_{k+1}. We have a resulting τ-quasi-well prepared diagram of R

$$\begin{array}{ccc} X_n & \overset{f_n}{\to} & Y_n \\ \Phi\downarrow & & \downarrow\Psi \\ X & \overset{f}{\to} & Y. \end{array} \tag{15.36}$$

Suppose that $q_n \in Y_n$ is a 2-point such that q_n is on a component E of D_{Y_n} such that $\Psi(E)$ is not a point, and $q_n \in U(\overline{R}_i^n)$, $f_n^{-1}(q_n) \cap T(R_i^n) \neq \emptyset$ for some i. We have permissible parameters

$$u_1 = u_{\overline{R}_i^n(q_n)}, v_1 = v_{\overline{R}_i^n(q_n)}, w_{i,1} = w_{\overline{R}_i^n(q_n)} \tag{15.37}$$

at q_n such that for $\Psi(q_n) = q$ and with notation of (15.35),

$$\begin{array}{ll} u & = u_1 \\ v & = u_1^n v_1 \\ w_i & = u_1^n w_{i,1} \end{array} \tag{15.38}$$

or

$$\begin{array}{ll} u & = u_1 v_1^n \\ v & = v_1 \\ w_i & = v_1^n w_{i,1}. \end{array}$$

Suppose that $p \in T(R_i) \cap f^{-1}(q)$ is a 1-point. First suppose that $\tau > 1$. There are permissible parameters x, y, z at p, $0 \neq \alpha \in \mathbf{k}$, and a unit $\gamma \in \hat{\mathcal{O}}_{X,p}$, such that we have

$$u = x^a, v = x^b(\alpha + y), w_i = x^c \gamma.$$

From the construction of (15.36) and the algorithm of Lemma 14.7, we have that Φ is an isomorphism at points of $\Phi^{-1}(p) \cap f_n^{-1}(q_n)$. Thus for $n \geq \max\{\frac{a}{b}, \frac{b}{a}\}$, $f_n^{-1}(q_n) \cap \Phi^{-1}(p) = \emptyset$.

Suppose that $\tau = 1$ (and $p \in T(R_i) \cap f^{-1}(q)$ is a 1-point). Then there exist permissible parameters x, y, z at p, $\beta \in \mathbf{k}$, $c > 0$ and $0 \neq \alpha \in \mathbf{k}$ such that

$$u = x^a, v = x^b(\alpha + y), w = x^c(\beta + z).$$

If $\beta \neq 0$, then Φ is an isomorphism at points of $\Phi^{-1}(p) \cap f_n^{-1}(q_n)$. For $n \geq \max\{\frac{a}{b}, \frac{b}{a}\}$, we have $f_n^{-1}(q_n) \cap \Phi^{-1}(p) = \emptyset$.

If $\beta = 0$, then Φ is a product of blow ups of sections over the curve $x = z = 0$ at points of $f_n^{-1}(q_n)$, and in this case also, $\Phi^{-1}(p) \cap f_n^{-1}(q_n) = \emptyset$ for $n \geq \max\{\frac{a}{b}, \frac{b}{a}\}$.

Suppose that $p \in T(R_i) \cap f^{-1}(q)$ is a 2-point. First suppose that $\tau > 1$. There are permissible parameters x, y, z at p and $0 \neq \alpha \in \mathbf{k}$ such that we have one of the forms

$$u = (x^a y^b)^k, v = (x^a y^b)^t(\alpha + z), w_i = (x^a y^b)^l \gamma, \tag{15.39}$$

where $0 \neq \alpha \in \mathbf{k}$, $\gcd(a, b) = 1$, $\gamma \in \hat{\mathcal{O}}_{X,p}$ is a unit,

or we have

$$u = x^a y^b, v = x^c y^d, w_i = x^e y^f \gamma \tag{15.40}$$

where $ad - bc \neq 0$, $\gamma \in \hat{\mathcal{O}}_{X,p}$ is a unit.

Suppose that (15.39) holds. From the construction of (15.36) and the algorithm of Lemma 14.7, we have that Φ is a sequence of blow ups of 2-curves at points of $\Phi^{-1}(p) \cap f_n^{-1}(q_n)$, so that for $n \geq \max\{\frac{k}{t}, \frac{t}{k}\}$, $f_n^{-1}(q_n) \cap \Phi^{-1}(p) = \emptyset$,

Suppose that (15.40) holds. If $f_n^{-1}(q_n)\cap\Phi^{-1}(p)\cap T(R_i^n)\neq\emptyset$ for all n, we will show that, after possibly interchanging u,v and x,y, (15.40) must be

$$u=x^a, v=x^cy^d, w_i=x^ey^f\gamma, \tag{15.41}$$

with $d,f>0$, and (15.38) holds.

We see this as follows.

Suppose that $p_n\in f_n^{-1}(q_n)\cap T(R_i^n)$ and $\Phi(p_n)=p$.

Let ν be a valuation of $\mathbf{k}(X)$ whose center on X_n is p_n. We identify ν with an extension of ν to the quotient field of $\hat{\mathcal{O}}_{X_n,p_n}$ which dominates $\hat{\mathcal{O}}_{X_n,p_n}$. q_n has permissible parameters (15.37). After possibly interchanging u and v, we have a relation (15.38), so that $\nu(v)>n\nu(u)$, and $\nu(w_i)>n\nu(u)$. We can reindex x,y,z so that $0<\nu(x)\le\nu(y)$. Then

$$(c+d-nb)\nu(y)\ge(d-nb)\nu(y)+(c-na)\nu(x)>0,$$

and

$$(e+f-nb)\nu(y)\ge(f-nb)\nu(y)+(e-na)\nu(x)>0. \tag{15.42}$$

Thus if $b\neq0$, and $n>c+d$, we have a contradiction.

Taking $n>c+d$ for all c,d in local forms (15.40) for 2-points $p\in T(R)$, we achieve that $b=0$ in all local forms (15.40) which are the images of 2-points $p_n\in T(R^n)$ which map to a point q_n of Y_n which is on a component E of D_{Y_n} such that $\Psi(E)$ is not a point. $d>0$ since $ad-bc\neq0$.

We have $f>0$ if $n>>0$. In fact, if $f=0$ in (15.40), we then have $e>0$, and for $n>\frac{a}{e}$, we have a contradiction to (15.42).

Suppose that $\tau=1$ (and $p\in T(R_i)\cap f^{-1}(q)$ is a 2-point). Then there exist permissible parameters x,y,z at p and $0\neq\alpha\in\mathbf{k}$ such that we have one of the forms:

$$u=(x^ay^b)^k, v=(x^ay^b)^t(\alpha+z), w_i=x^cy^d \tag{15.43}$$

with $0\neq\alpha\in\mathbf{k}$, $ad-bc\neq0$, $\gcd(a,b)=1$, or we have

$$u=x^ay^b, v=x^cy^d, w_i=x^ey^f(\beta+z) \tag{15.44}$$

with $ad-bc\neq0$, $\beta\in\mathbf{k}$.

Suppose that (15.43) holds. From the construction of (15.36), we have that Φ is a sequence of blow ups of 2-curves at points of $\Phi^{-1}(p)\cap f_n^{-1}(q_n)$, so that for $n\ge\max\{\frac{k}{t},\frac{t}{k}\}$, $f_n^{-1}(q_n)\cap\Phi^{-1}(p)=\emptyset$.

Suppose that (15.44) holds. If $\beta\neq0$, the analysis of (15.40) shows that if $f_n^{-1}(q_n)\cap\Phi^{-1}(p)\cap T(R_i^n)\neq\emptyset$ for all n, then after possibly interchanging u,v and interchanging x,y (15.44) has the form (15.41), with $\gamma=\beta+z$, $d,f>0$ and (15.38) holds.

Suppose that $\beta=0$ (in (15.44)), and $f_n^{-1}(q_n)\cap\Phi^{-1}(p)\cap T(R_i^n)\neq\emptyset$ for all n. After possibly interchanging u and v, we may assume that (15.38) holds. Let ν be a valuation of $\mathbf{k}(X)$ whose center on X_n is p_n. We identify ν with an extension of ν to the quotient field of $\hat{\mathcal{O}}_{X_n,p_n}$ which dominates $\hat{\mathcal{O}}_{X_n,p_n}$. Then we see that

$$\nu(v)>n\nu(u) \text{ and } \nu(w_i)>n\nu(u) \tag{15.45}$$

for all n.

Thus for n sufficiently large, after possibly interchanging x and y, (15.44) must be

$$u = x^a, v = x^c y^d, w_i = x^e y^f z \tag{15.46}$$

with $d > 0$.

We will now show that we can take n sufficiently large that $f > 0$ in (15.46).

Suppose that $f = 0$ in (15.46). Then we have

$$\nu(y) > n\nu(x) \text{ and } \nu(z) > n\nu(x) \tag{15.47}$$

for all n.

In the algorithm of Lemma 14.7, we see that $\Phi_1 : X_1 \to X$ can be factored by morphisms

$$X_1 = Z_m \to \cdots \to Z_2 \to Z_1 \to X$$

where $Z_1 \to X$ is a sequence of blow ups of 2-curves and 3-points, and each $Z_{i+1} \to Z_i$ is the blow up of a possible curve containing a 1-point, which is in the locus where $\mathcal{I}_q\mathcal{O}_{Z_i}$ is not invertible.

By (15.47), we see that there exist permissible parameters $\tilde{x}_1, \tilde{y}_1, \tilde{z}_1$ at the center $\tilde{p}_1$ of ν on Z_1 such that

$$x = \tilde{x}_1, y = \tilde{x}_1^g \tilde{y}_1, z = \tilde{z}_1.$$

We have

$$u = \tilde{x}_1^a, v = \tilde{x}_1^{c+gd}\tilde{y}_1^d, w_i = \tilde{x}^e \tilde{z}_1$$

with $a \le c + gd$, since from the construction of $Z_1 \to X_1$, we have that $(u, v)\mathcal{O}_{Z_1,\tilde{p}_1}$ is invertible.

If $e \ge a$, then $\mathcal{I}_q\mathcal{O}_{Z_1,\tilde{p}_1}$ is invertible, and $X_1 \to Z_1$ is an isomorphism above $\tilde{p}_1$. Then at $p_1 \in X_1$, there are permissible parameters $\hat{x}_1, \hat{y}_1, \hat{z}_1$ such that

$$u_{\overline{R}_i^1(q_1)} = \hat{x}_1^a, v_{\overline{R}_i^1(q_1)} = \hat{x}_1^{c+gd-a}\hat{y}_1^d, w_{\overline{R}_i^1(q_1)} = \hat{x}_1^{e-a}\hat{z}_1.$$

If $e - a = 0$, we have that f_1 is toroidal at p_1 (so that $\tau_{f_1}(p_1) = -\infty$).

If $e < a$, we have that $X_1 \to Z_1$ is not an isomorphism above $\tilde{p}_1$. Without loss of generality, we may assume that each $Z_{i+1} \to Z_i$ is not an isomorphism at the center of ν.

We have that $\tilde{x}_1 = \tilde{z}_1 = 0$ are (formal) local equations at $\tilde{p}_1$ of the curve blown up in $Z_2 \to Z_1$.

Let $\tilde{p}_2$ be the center of ν on $\tilde{Z}_2$. By (15.47), we have that there are permissible parameters $\tilde{x}_2, \tilde{y}_2, \tilde{z}_2$ at $\tilde{p}_2$ such that $\tilde{x}_1 = \tilde{x}_2$, $\tilde{y}_1 = \tilde{y}_2$, $\tilde{z}_1 = \tilde{x}_2\tilde{z}_2$.

We have

$$u = \tilde{x}_2^a, v = \tilde{x}_2^{c+gd}\tilde{y}_2^d, w_i = \tilde{x}_2^{e+1}\tilde{z}_2.$$

We see that in $X_1 = Z_m$, there are regular parameters $\hat{x}_1, \hat{y}_1, \hat{z}_1$ such that

$$u_{\overline{R}_i^1(q_1)} = \hat{x}_1^a, v_{\overline{R}_i^1(q_1)} = \hat{x}_1^{c+gd-a}\hat{y}_1^d, w_{\overline{R}_i^1(q_1)} = \hat{z}_2.$$

thus f_1 is toroidal at p_1.

Iterating this analysis for the morphisms $\Psi_2, \ldots, \Psi_n$, we see that for $n >> 0$, f_n is toroidal at p_n. In fact, if we take $n \ge \frac{e}{a}$ in (15.36), we see that $f_n^{-1}(q_n) \cap T(R_i^n) \cap \Phi^{-1}(p) = \emptyset$.

We can thus take n sufficiently large so that $f > 0$ in all local forms (15.46), which are the images of 2-points $p_n \in T(R^n)$ which map to a point q_n of Y_n which is on a component E of D_{Y_n} such that $\Psi(E)$ is not a point.

Suppose that $p \in X$ is a 3-point such that $p \in T(R_i) \cap f^{-1}(q)$. Then there are permissible parameters x, y, z for u, v, w_i at p such that

$$
\begin{aligned}
u &= x^a y^b z^c \\
v &= x^d y^e z^f \\
w_i &= x^g y^h z^i \gamma
\end{aligned} \tag{15.48}
$$

where γ is a unit series.

We will show that we can choose n sufficiently large in the diagram (15.36), so that if $p_n \in X_n$ is such that $p_n \in \Phi^{-1}(p) \cap T(R_i^n)$ and $q_n = f_n(p_n)$ is on a component E of D_{Y_n} such that $\Psi(E)$ is not a point, then (15.48) must have one of the following forms (after possibly interchanging u, v and x, y, z):

$$
\begin{aligned}
u &= x^a y^b \\
v &= x^d y^e z^f \\
w_i &= x^g y^h z^i \gamma
\end{aligned} \tag{15.49}
$$

where $b \neq 0$, $f \neq 0$, $i \neq 0$ and (15.38) holds, or

$$
\begin{aligned}
u &= x^a \\
v &= x^d y^e z^f \\
w_i &= x^g y^h z^i \gamma
\end{aligned} \tag{15.50}
$$

with e and $f \neq 0$, h or $i \neq 0$ and (15.38) holds.

We will now prove this statement.

Let ν be any valuation of $\mathbf{k}(X)$ which has center p_n on X_n. We identify ν with an extension of ν to the quotient field of $\hat{\mathcal{O}}_{X_n,p_n}$ which dominates $\hat{\mathcal{O}}_{X_n,p_n}$.

q_n has permissible parameters (15.37).

After possibly interchanging u and v, we have a relation (15.38), so that $\nu(v) > n\nu(u)$. We can reindex x, y, z so that

$$0 < \nu(x) \leq \nu(y) \leq \nu(z).$$

Then

$$(f + e + d - nc)\nu(z) \geq (f - nc)\nu(z) + (e - nb)\nu(y) + (d - na)\nu(x) > 0.$$

If $c \neq 0$, and $n > f + e + d$, we have a contradiction. Thus taking $n > f + e + d$ for all d, e, f in local forms (15.48) for 3-points $p \in T(R)$, we achieve that $c = 0$ in all local forms (15.48) which are the images of 3-points $p_n \in T(R^n)$ which map to a point q_n of Y_n which is on a component E of D_{Y_n} such that $\Psi(E)$ is not a point.

If $i = 0$ (and $c = 0$) in (15.48) we have

$$(h + g - nb)\nu(y) \geq (h - nb)\nu(y) + (g - na)\nu(x) > 0$$

so that if $b \neq 0$ and $n > h + g$ we have a contradiction. Thus, by taking $n \gg 0$ in (15.36), we see that if $b \neq 0$, then a form (15.49) must hold at p (since $uv = 0$ is a local equation of D_X at p implies $f \neq 0$). If $b = c = 0$ in (15.48), then a similar calculation shows that a form (15.50) must hold at p (for $n \gg 0$).

We observe that in (15.49) we have

$$(z) \cap \hat{\mathcal{O}}_{Y,q} = (v, w_i). \tag{15.51}$$

Suppose that (15.50) holds. If $i \neq 0$ then

$$(z) \cap \hat{\mathcal{O}}_{Y,q} = (v, w_i). \tag{15.52}$$

If $h \neq 0$, then

$$(y) \cap \hat{\mathcal{O}}_{Y,q} = (v, w_i). \tag{15.53}$$

Suppose that (15.41) or (15.46) hold. Then

$$(y) \cap \hat{\mathcal{O}}_{Y,q} = (v, w_i). \tag{15.54}$$

We will show that in (15.49), $v = w_i = 0$ is a formal branch of an algebraic curve C in the nonfinite locus of $f : X \to Y$. Let $R = \mathcal{O}_{Y,q}$, $S = \mathcal{O}_{X,p}$.

Since p is a 3-point, there exist regular parameters $\overline{x}, \overline{y}, \overline{z}$ in $\mathcal{O}_{X,p}$ and units $\lambda_1, \lambda_2, \lambda_3 \in \hat{\mathcal{O}}_{X,p}$ such that $\overline{x} = x\lambda_1$, $\overline{y} = y\lambda_2$, $\overline{z} = z\lambda_3$.

$\overline{z} = 0$ is a local equation for a component of D_X. We have that $v \in (\overline{z}) \cap R$ and $u \notin (\overline{z}) \cap R$ so that $(\overline{z}) \cap R = (v)$ or $(\overline{z}) \cap R = a$ where a is a height two prime containing v. We have $(z\hat{S}) \cap \hat{R} = (v, w_i)$. Suppose that $(\overline{z}) \cap R = (v)$. We then have an induced morphism

$$\hat{R}/(v) \to \hat{S}/(z)$$

which is an inclusion by the Zariski Subspace Theorem (Theorem 10.14 [**Ab3**]). This is impossible, so that a is a height 2 prime in R, and defines a curve C, which is necessarily in the nonfinite locus of f since $\overline{z} = 0$ is a local equation at p of a component of D_X which dominates C. A similar argument shows that in (15.50), (15.41) and (15.46), $v = w_i = 0$ is a formal branch of an algebraic curve C in the nonfinite locus of f.

Step 2. Let C be the reduced curve in Y whose components are the curves in the nonfinite locus of f which are not 2-curves. Let $\overline{C}$ be the reduced curve in Y_n which is the strict transform of C. The components of $\overline{C}$ are then in the nonfinite locus of f_n. By Theorem 15.2, we can perform a sequence of blow ups of prepared 1-points, prepared 2-points, 2-curves and resolving curves $\Psi' : Y' \to Y_n$ so that we can construct a τ-quasi-well prepared diagram of Ψ' and R^n

$$\begin{array}{ccc} X' & \overset{f'}{\to} & Y' \\ \Phi' \downarrow & & \downarrow \Psi' \\ X_n & \to & Y_n \end{array} \tag{15.55}$$

where R' is the transform of R^n on X', such that the strict transform $\tilde{C}$ of $\overline{C}$ on Y' is nonsingular, and makes SNCs with $D_{Y'}$. If $q' \in U(R'_i) \cap \tilde{C}$ for some i then the germ at q' of $\tilde{C}$ is contained in $S_{R'_i}(q')$, and the (disjoint) components of $\tilde{C}$ are permissible centers for R'.

Let $\Psi(1) : Y(1) \to Y'$ be the blow up of $\tilde{C}$. By Theorem 15.2, we have a τ-quasi-well prepared diagram of $\Psi(1)$ and R'

$$\begin{array}{ccc} X(1) & \overset{f(1)}{\to} & Y(1) \\ \Phi(1) \downarrow & & \downarrow \Psi(1) \\ X' & \overset{f'}{\to} & Y'. \end{array} \tag{15.56}$$

Let $R(1)$ be the transform of R' on $X(1)$.

Suppose that $q \in U(R) \subset Y$ is a 2-point. Suppose that $\tilde{q} \in (\Psi \circ \Psi')^{-1}(q)$ and $(f')^{-1}(\tilde{q}) \cap T(R'_i) \neq \emptyset$. Then $q \in A_0$.

Let

$$\tilde{u} = u_{\overline{R}'_i(\tilde{q})}, \tilde{v} = v_{\overline{R}'_i(\tilde{q})}, \tilde{w}_i = w_{\overline{R}'_i(\tilde{q})},$$

$$u = u_{\overline{R}(q)}, v = v_{\overline{R}(q)}, w_i = w_{\overline{R}_i(q)}.$$

We have (since $q \in A_0$ and $\Psi \circ \Psi'$ is a nontrivial sequence of blow ups of points, followed by a sequence of blow ups of points and 2-curves at $\tilde{q}$) one of the following forms (by (12.7) - (12.9):

$$u = \tilde{u}^a\tilde{v}^b, v = \tilde{u}^c\tilde{v}^d, w_i = \tilde{u}^l\tilde{v}^m\tilde{w}_i \tag{15.57}$$

with $ad - bc = \pm 1$ and $l > 0$ if $c > 0$, $m > 0$ if $b > 0$, or

$$u = (\tilde{u}^a\tilde{v}^b)^t\gamma_1(\tilde{u}, \tilde{v}), v = (\tilde{u}^a\tilde{v}^b)^k\gamma_2(\tilde{u}, \tilde{v}), w_i = \tilde{u}^e\tilde{v}^f\tilde{w}_i\gamma_3(\tilde{u}, \tilde{v}) \tag{15.58}$$

where $\gamma_1, \gamma_2, \gamma_3$ are unit series, $a, b, t, k, e, f > 0$, $\gcd(a, b) = 1$, or

$$u = \tilde{u}^a\gamma_1(\tilde{u}, \tilde{v}), v = \tilde{u}^b\gamma_2(\tilde{u}, \tilde{v}), w_i = \tilde{u}^c\tilde{w}_i\gamma_3(\tilde{u}, \tilde{v}) \tag{15.59}$$

where $\gamma_1, \gamma_2, \gamma_3$ are unit series, $a, b, c > 0$.

We have that $(\Psi \circ \Psi')(E) = q$ for all components E of $D_{Y'}$ containing $\tilde{q}$ (which implies $\tilde{q} \notin \tilde{C}$) unless $\tilde{q}$ is a 2-point, and we have an expression

$$\begin{aligned} u &= \tilde{u} \\ v &= \tilde{u}^c\tilde{v} \\ w_i &= \tilde{u}^l\tilde{v}^m\tilde{w}_i \end{aligned} \tag{15.60}$$

with $c, l > 0$ and $\tilde{v} = 0$ is a local equation of E, or

$$\begin{aligned} u &= \tilde{u}\tilde{v}^b \\ v &= \tilde{v} \\ w_i &= \tilde{u}^l\tilde{v}^m\tilde{w}_i \end{aligned}$$

with $b, m > 0$ and $\tilde{u} = 0$ is a local equation of E.

Let $q^* = \Psi'(\tilde{q})$. q^* is a 2-point and $f^{-1}(q^*) \cap T(R_i') \neq \emptyset$. After possibly interchanging u and v we have that a form (15.38) holds at q_n and q, and thus by (15.51) - (15.54) that $v = w_i = 0$ are local equations of a formal component of C at q. (15.60) thus holds at $\tilde{q}$, and since Ψ' is a sequence of blow ups of points and 2-curves at $\tilde{q}$, $m = 0$ in (15.60). We thus have an expression

$$u = \tilde{u}, v = \tilde{u}^e\tilde{v}, w_i = \tilde{u}^f\tilde{w}_i \tag{15.61}$$

for some $e, f > 0$. $\tilde{v} = 0$ is a local equation of the strict transform of D_Y at $\tilde{q}$, and $\tilde{v} = \tilde{w}_i = 0$ are local equations of $\tilde{C}$ at $\tilde{q}$ since $\tilde{C}$ is nonsingular.

Suppose that $\overline{q} \in (\Psi \circ \Psi' \circ \Psi(1))^{-1}(q)$ and $f(1)^{-1}(\overline{q}) \cap T(R_i(1)) \neq \emptyset$. Let $\tilde{q} = \Psi(1)(\overline{q})$. then $(f')^{-1}(\tilde{q}) \cap T(R_i') \neq \emptyset$. If $\tilde{q} \notin \tilde{C}$ (so that $\overline{q} = \tilde{q}$) then we have seen that $(\Psi \circ \Psi')(E) = q$ for all components E of $D_{Y'}$ containing $\tilde{q}$. Suppose that $\tilde{q} \in \tilde{C}$. Then an expression (15.61) holds at $\tilde{q}$.

$\Psi(1)$ is the blow up of $\tilde{v} = \tilde{w}_i = 0$ above $\tilde{q}$. Since $\overline{q} \in U(R_i(1))$, we must have

$$\tilde{u} = u_{\overline{R}_i(1)(\overline{q})}, \tilde{v} = v_{\overline{R}_i(1)(\overline{q})}, \tilde{w}_i = v_{\overline{R}_i(1)(\overline{q})}w_{\overline{R}_i(1)(\overline{q})}.$$

Substituting into (15.61), we have

$$u = u_{\overline{R}_i(1)(\overline{q})}, v = u^e_{\overline{R}_i(1)(\overline{q})}v_{\overline{R}_i(1)(\overline{q})}, w_i = u^f_{\overline{R}_i(1)(\overline{q})}v_{\overline{R}_i(1)(\overline{q})}w_{\overline{R}_i(1)(\overline{q})} \tag{15.62}$$

with $e, f \geq 1$. We have that q is a 2-point and $\tilde{q}$ is a 2-point.

Suppose that $q \in U(R) \subset Y$ is a 1-point. Then Ψ is an isomorphism over q. Suppose that $\tilde{q} \in (\Psi \circ \Psi')^{-1}(q)$ and $(f')^{-1}(\tilde{q}) \cap T(R_i') \neq \emptyset$. Let

$$\tilde{u} = u_{\overline{R}_i'(\tilde{q})}, \tilde{v} = v_{\overline{R}_i'(\tilde{q})}, \tilde{w}_i = w_{\overline{R}_i'(\tilde{q})},$$

$$u = u_{\overline{R}_i(q)}, v = v_{\overline{R}_i(q)}, w_i = w_{\overline{R}_i(q)}.$$

We have that $u = w_i = 0$ are local equations of C since $T(R_i) \cap f^{-1}(q) \neq \emptyset$, by Remark 15.6. Then Ψ' is either an isomorphism at $\tilde{q}$, or factors at $\tilde{q}$ as the blow up of q, followed by a sequence of blow ups of 2-points and 2-curves.

First suppose that Ψ' is not an isomorphism at $\tilde{q}$. We have one of the following forms (by (12.7) - (12.9):

$$u = \tilde{u}^a\tilde{v}^b, v = \tilde{u}^c\tilde{v}^d, w_i = \tilde{u}^l\tilde{v}^m\tilde{w}_i \tag{15.63}$$

with $ad - bc = \pm 1$ and $l > 0$ if $c > 0$, $m > 0$ if $b > 0$, or

$$u = (\tilde{u}^a\tilde{v}^b)^t\gamma_1(\tilde{u}, \tilde{v}), v = (\tilde{u}^a\tilde{v}^b)^k\gamma_2(\tilde{u}, \tilde{v}), w_i = \tilde{u}^e\tilde{v}^f\tilde{w}_i\gamma_3(\tilde{u}, \tilde{v}) \tag{15.64}$$

where $\gamma_1, \gamma_2, \gamma_3$ are unit series, $a, b, t, k, e, f > 0$, $\gcd(a, b) = 1$, or

$$u = \tilde{u}^a\gamma_1(\tilde{u}, \tilde{v}), v = \tilde{u}^b\gamma_2(\tilde{u}, \tilde{v}), w_i = \tilde{u}^c\tilde{w}_i\gamma_3(\tilde{u}, \tilde{v}) \tag{15.65}$$

where $\gamma_1, \gamma_2, \gamma_3$ are unit series, $a, b, c > 0$.

We have that $\Psi'(E) = q$ for all components E of $D_{Y'}$ containing $\tilde{q}$, (which implies $\tilde{q} \notin \tilde{C}$) unless we have an expression

$$\begin{aligned} u &= \tilde{u}\tilde{v}^b \\ v &= \tilde{v} \\ w_i &= \tilde{u}^l\tilde{v}^m\tilde{w}_i \end{aligned} \tag{15.66}$$

with $b, m > 0$.

Since Ψ' is an isomorphism over a generic point of every component of C, we have that $\tilde{q}$ is a 2-point and

$$u = \tilde{u}\tilde{v}^m, v = \tilde{v}, w_i = \tilde{v}^n\tilde{w}_i \tag{15.67}$$

with $m, n \geq 1$. $\tilde{u} = \tilde{w}_i = 0$ are local equations of $\tilde{C}$ at $\tilde{q}$.

Suppose that $\overline{q} \in (\Psi \circ \Psi' \circ \Psi(1))^{-1}(q)$ and $f(1)^{-1}(\overline{q}) \cap T(R_i(1)) \neq \emptyset$. Let $\tilde{q} = \Psi(1)(\overline{q})$. then $(f')^{-1}(\tilde{q}) \cap T(R_i') \neq \emptyset$. If $\tilde{q} \notin \tilde{C}$ (so that $\overline{q} = \tilde{q}$) then we have seen that $(\Psi \circ \Psi')(E) = q$ for all components E of $D_{Y'}$ containing $\tilde{q}$. Suppose that $\tilde{q} \in \tilde{C}$. Then an expression (15.67) holds at $\tilde{q}$. Then $\overline{q}$ is a 2-point with

$$\tilde{u} = u_{\overline{R}_i(1)(\overline{q})}, \tilde{v} = v_{\overline{R}_i(1)(\overline{q})}, \tilde{w}_i = u_{\overline{R}_i(1)(\overline{q})}w_{\overline{R}_i(1)(\overline{q})},$$

and thus

$$u = u_{\overline{R}_i(1)(\overline{q})}v^m_{\overline{R}_i(1)(\overline{q})}, v = v_{\overline{R}_i(1)(\overline{q})}, w_i = u_{\overline{R}_i(1)(\overline{q})}v^n_{\overline{R}_i(1)(\overline{q})}w_{\overline{R}_i(1)(\overline{q})} \tag{15.68}$$

with $m, n \geq 1$. We have that q is a 1-point and $\tilde{q}$ is a 2-point.

Now suppose that Ψ' is an isomorphism over q ($q \in U(R)$ is a 1-point), and $\overline{q} \in \Psi(1)^{-1}(q)$ is such that $f(1)^{-1}(q) \cap T(R_i(1)) \neq \emptyset$. then $\overline{q}$ is a 1-point, and (by Remark 15.6)

$$u = u_{\overline{R}_i(1)(\overline{q})}, v = v_{\overline{R}_i(1)(\overline{q})}, w_i = u_{\overline{R}_i(1)(\overline{q})}w_{\overline{R}_i(1)(\overline{q})}. \tag{15.69}$$

We have that q is a 1-point and $\tilde{q}$ is a 1-point.

Step 3. We now apply steps 1 and 2 of the proof to $f(1): X(1) \to Y(1)$ and $R(1)$. We construct a τ-quasi-well prepared diagram

$$\begin{array}{ccc} X(2) & \overset{f(2)}{\to} & Y(2)\\ \overline{\Phi}(2)\downarrow & & \downarrow \overline{\Psi}(2)\\ X(1) & \overset{f(1)}{\to} & Y(1), \end{array}$$

where $R(2)$ is the transform of $R(1)$ on $X(2)$. Suppose that $q_2 \in U(R_i(2)) \subset Y(2)$ is on a component E_2 of $D_{Y(2)}$ such that $\Psi \circ \Psi' \circ \Psi(1) \circ \overline{\Psi}(2)(E_2)$ is not a point of Y and there exists a point $p_2 \in f(2)^{-1}(q_2) \cap T(R_i(2)) \subset X(2)$. Let $q_1 = \overline{\Psi}(2)(q_2)$. $\overline{\Psi}_2(2)(E_2)$ is necessarily not a point of $Y(1)$.

Suppose that q_1 is a 2-point. Then we have an expression (analogous to (15.62)):

$$(15.70)\qquad \begin{aligned} u_{\overline{R}_i(1)(q_1)} &= u_{\overline{R}_i(2)(q_2)}\\ v_{\overline{R}_i(1)(q_1)} &= u^e_{\overline{R}_i(2)(q_2)} v_{\overline{R}_i(2)(q_2)}\\ w_{\overline{R}_i(1)(q_1)} &= u^f_{\overline{R}_i(2)(q_2)} v_{\overline{R}_i(2)(q_2)} w_{\overline{R}_i(2)(q_2)} \end{aligned}$$

or

$$(15.71)\qquad \begin{aligned} u_{\overline{R}_i(1)(q_1)} &= u_{\overline{R}_i(2)(q_2)} v^e_{\overline{R}_i(2)(q_2)}\\ v_{\overline{R}_i(1)(q_1)} &= v_{\overline{R}_i(2)(q_2)}\\ w_{\overline{R}_i(1)(q_1)} &= u_{\overline{R}_i(2)(q_2)} v^f_{\overline{R}_i(2)(q_2)} w_{\overline{R}_i(2)(q_2)} \end{aligned}$$

with $e, f \geq 1$.

Let $q = (\Psi \circ \Psi' \circ \Psi(1))(q_1)$. Since q_1 lies on a component E_1 of $D_{Y(1)}$ which does not contract to a point of Y, and $f(1)^{-1}(q_1) \cap T(R_i(1)) \neq \emptyset$, a form (15.62) or (15.68) holds for

$$u = u_{\overline{R}_i(q)}, v = v_{\overline{R}_i(q)}, w_i = w_{\overline{R}_i(q)}$$

and

$$u_{\overline{R}_i(1)(q_1)}, v_{\overline{R}_i(1)(q_1)}, w_{\overline{R}_i(1)(q_1)}.$$

Suppose that (15.62) holds at q_1. Substituting (15.70) and (15.71) into (15.62), we see that since q_2 is on a component E_2 of $D_{Y(2)}$ which does not contract to q, then we have that (15.70) holds, and an expression

$$(15.72)\qquad \begin{aligned} u = u_{R_i(q)} &= u_{\overline{R}_i(2)(q_2)}\\ v = v_{R_i(q)} &= u^{e_2}_{\overline{R}_i(2)(q_2)} v_{\overline{R}_i(2)(q_2)}\\ w = w_{R_i(q)} &= u^{f_2}_{\overline{R}_i(2)(q_2)} v^2_{\overline{R}_i(2)(q_2)} w_{\overline{R}_i(2)(q_2)} \end{aligned}$$

with $e_2, f_2 \geq 2$.

Suppose that (15.68) holds at q_1. Substituting (15.70) or (15.71) into (15.68), we see that since q_2 is on a component of $D_{Y(2)}$ which does not contract to q, we have that (15.71) holds we have and an expression

$$(15.73)\qquad u = u_{\overline{R}_i(2)(q_2)} v^{e_2}_{\overline{R}_i(2)(q_2)}, v = v_{\overline{R}_i(2)(q_2)}, w_i = u^2_{\overline{R}_i(2)(q_2)} v^{f_2}_{\overline{R}_i(2)(q_2)} w_{\overline{R}_i(2)(q_2)}$$

with $e_2, f_2 \geq 2$.

Suppose that q_1 is a 1-point. Then we have an expression (analogous to (15.68))

$$(15.74)\qquad \begin{aligned} u_{\overline{R}_i(1)(q_1)} &= u_{\overline{R}_i(2)(q_2)} v^m_{\overline{R}_i(2)(q_2)},\\ v_{\overline{R}_i(1)(q_1)} &= v_{\overline{R}_i(2)(q_2)},\\ w_{\overline{R}_i(1)(q_1)} &= u_{\overline{R}_i(2)(q_2)} v^n_{\overline{R}_i(2)(q_2)} w_{\overline{R}_i(2)(q_2)} \end{aligned}$$

with $m, n \geq 1$, or (analogous to (15.69))
(15.75)

$$u_{\overline{R}_i(1)(q_1)} = u_{\overline{R}_i(2)(q_2)}, v_{\overline{R}_i(1)(q_1)} = v_{\overline{R}_i(2)(q_2)}, w_{\overline{R}_i(1)(q_1)} = u_{\overline{R}_i(2)(q_2)} w_{\overline{R}_i(2)(q_2)}.$$

Since q_1 lies on a component E_1 of $D_{Y(1)}$ which does not contract to a point of Y, and $f(1)^{-1}(q_1) \cap T(R_i(1)) \neq \emptyset$, a form (15.69) holds at q_1 for

$$u = u_{\overline{R}_i(q)}, v = v_{\overline{R}_i(q)}, w_i = w_{\overline{R}_i}(q)$$

and

$$u_{\overline{R}_1(1)(q_1)}, v_{\overline{R}_i(1)(q_1)}, w_{\overline{R}_i(1)(q_1)}.$$

Substituting (15.74) and (15.75) into (15.69), we see that since q_1 is on a component E_1 of $D_{Y(1)}$ which does not contract to q, we have

$$(15.76) \quad u = u_{\overline{R}_i(2)(q_2)} v^m_{\overline{R}_i(2)(q_2)}, v = v_{\overline{R}_i(2)(q_2)}, w_i = u^2_{\overline{R}_i(2)(q_2)} v^n_{\overline{R}_i(2)(q_2)} w_{\overline{R}_i(2)(q_2)}$$

with $m, n \geq 0$.

Iterating steps 1 and 2, we construct a sequence of τ-quasi-well prepared diagrams

$$(15.77) \quad \begin{array}{ccc} \vdots & & \vdots \\ \downarrow & & \downarrow \\ X(n) & \stackrel{f(n)}{\rightarrow} & Y(n) \\ \overline{\Phi}(n)\downarrow & & \downarrow \overline{\Psi}(n) \\ X(n-1) & \stackrel{f(n-1)}{\rightarrow} & Y(n-1) \\ \downarrow & & \downarrow \\ \vdots & & \vdots \\ \downarrow & & \downarrow \\ X(1) & \stackrel{f(1)}{\rightarrow} & Y(1) \\ \overline{\Phi}(1)\downarrow & & \downarrow \overline{\Psi}(1) \\ X & \stackrel{f}{\rightarrow} & Y. \end{array}$$

We continue this algorithm as long as there exists $q_n \in U(R_i(n))$ for some i such that q_n is on a component E of $D_{Y(n)}$ which does not contract to a point of Y, and $f(n)^{-1}(q_n) \cap T(R_i(n)) \neq \emptyset$.

Step 4. Suppose that the algorithm never terminates.

Let ν be a 0-dimensional valuation of $\mathbf{k}(X)$. We will say that ν is resolved on $X(n)$ if the center of ν on $X(n)$ is at a point p_n of $X(n)$ such that either $p_n \notin T(R(n))$ or $p_n \in T(R(n))$ and all components E of $D_{Y(n)}$ containing $q_n = f(n)(p_n)$ contract to a point of Y.

By our construction, if ν is resolved on $X(n)$, then ν is resolved on $X(m)$ for all $m \geq n$. Further, the set of ν in the Zariski-Riemann manifold $\Omega(X)$ of X ([**Z**]) which are resolved on X is an open subset of $\Omega(X)$.

Suppose that ν is a 0-dimensional valuation of $\mathbf{k}(X)$ such that ν is not resolved on $X(n)$ for all n. Let p_n be the center of ν on $X(n)$, q_n be the center of ν on $Y(n)$.

For all n, we identify ν with an extension of ν to the quotient field of $\hat{\mathcal{O}}_{X_n,p_n}$ which dominates $\hat{\mathcal{O}}_{X_n,p_n}$.

First suppose that the center of ν on Y is a 2-point.

There exists an i such that for all n, $q_n \in U(R_i(n))$ and $p_n \in f(n)^{-1}(q_n) \cap T(R_i(n))$. We have expressions (after possibly interchanging u and v)

$$
\begin{aligned}
u = u_{\overline{R}_i(q)} &= u_{\overline{R}_i(n)(q_n)} \\
v = v_{\overline{R}_i(q)} &= u^{e_n}_{\overline{R}_i(n)(q_n)} v_{\overline{R}_i(n)(q_n)} \\
w_i = w_{\overline{R}_i(q)} &= u^{f_n}_{\overline{R}_i(n)(q_n)} v^{n}_{\overline{R}_i(n)(q_n)} w_{\overline{R}_i(n)(q_n)}
\end{aligned}
\tag{15.78}
$$

with $e_n, f_n \geq n$ for all n.

From (15.78), we see that

$$\nu(v) > n\nu(u) \text{ for all } n \in \mathbf{N}. \tag{15.79}$$

Thus ν is a composite valuation, and there exists a prime ideal P of the valuation ring V of ν such that $v \in P$, $u \notin P$. Let ν_1 be a valuation whose valuation ring is V_P. We have $\nu_1(u) = 0$, $\nu_1(v) > 0$. From (15.78) we see that

$$\nu_1(w_i) > n\nu_1(v) > 0 \tag{15.80}$$

for all $n \in \mathbf{N}$.

At $p = p_0 \in X$, we have a form (15.41), (15.46), (15.49) or (15.50). In (15.41) we have $\nu_1(x) = 0$ and $d > 0$, a contradiction to (15.80). In (15.49) we have $\nu_1(y) = 0$. $uv = 0$ is a local equation of D_X at p. Thus either $a > 0$ or $d > 0$. If $a > 0$ then $\nu_1(x) = 0$ and $\nu_1(z) > 0$, a contradiction to (15.80) since $f \neq 0$. If $d > 0$, we again have a contradiction to (15.80). In (15.50) we have a contradiction to (15.80), since $\nu_1(x) = 0$ and $e, f > 0$.

Suppose that (15.46) holds. In this case a more detailed analysis is required. By our construction and with the notation of steps 1 and 2, there exists a factorization

$$
\begin{array}{ccc}
X(1) & \overset{f(1)}{\rightarrow} & Y(1) \\
\Phi(1) \downarrow & & \downarrow \Psi(1) \\
X' & \overset{f'}{\rightarrow} & Y' \\
\Phi' \downarrow & & \downarrow \Psi' \\
X_n & \overset{f_n}{\rightarrow} & Y_n \\
\Phi \downarrow & & \downarrow \Psi \\
X & \overset{f}{\rightarrow} & Y.
\end{array}
$$

Recall that p_1 is the center of ν on $X(1)$, p is the center of ν on X, q_1 is the center of ν on $Y(1)$, and q is the center of ν on Y.

Let p' be the center of ν on X', q' be the center of ν on Y'.

At q', $Y' \rightarrow Y$ is a sequence of blow ups of prepared 2-points of type 1, and 2-curves, and at p',

$$
\begin{array}{ccc}
X' & \rightarrow & Y' \\
\downarrow & & \downarrow \\
X & \rightarrow & Y
\end{array}
$$

is obtained by iterating the constructions of Remark 14.3 and Lemma 14.7.

Let

$$u' = u_{\overline{R}'_i(q')}, v' = v_{\overline{R}'_i(q')}, w'_i = w_{\overline{R}'_i(q')}.$$

By equations (15.79), (15.80) and (15.46) we see that

$$\nu(y) > n\nu(x) \text{ and } \nu_1(x) = 0, \ \ \nu_1(z) > n\nu_1(y) \tag{15.81}$$

for all $n \in \mathbf{N}$.

We see (by a variant of the analysis of (15.46) in Step 1, using (15.81)), that there exist permissible parameters x', y', z' at p' such that

$$u' = (x')^a, v' = (x')^{c'}(y')^d, w_i' = (x')^{e'}(y')^f z'$$

where a, d, f are the constants of (15.46) and $c', e' \in \mathbf{N}$. $\Psi(1)$ is the blow up of the curve $\tilde{C}$ which has local equations $v' = w_i' = 0$ at q'. The construction of $X(1) \rightarrow X'$ at p_1 (from Remark 14.12) is analogous to the analysis of Lemma 14.11. There exists a factorization

$$X(1) = W_m \rightarrow \cdots \rightarrow W_2 \rightarrow W_1 \rightarrow X'$$

where $W_1 \rightarrow X'$ is a sequence of blow ups of 2-curves and 3-points, and each $W_{i+1} \rightarrow W_i$ is a curve containing a 1-point in the locus where $\mathcal{I}_{\tilde{C}}\mathcal{O}_{W_i}$ is not invertible.

Let $\tilde{p}_i$ be the center of ν on W_i. By the analysis of the proof of Lemma 14.11, and (15.81), we see that there exist permissible parameters $\tilde{x}_1, \tilde{y}_1, \tilde{z}_1$ at $\tilde{p}_1$ such that

$$u' = \tilde{x}_1^a, v' = \tilde{x}_1^{c_1}\tilde{y}_1^d, w_i' = \tilde{x}_1^{e_1}\tilde{y}_1^f\tilde{z}_1$$

with $(c_1, d) \leq (e_1, f)$, or $(c_1, d) > (e_1, f)$.

If $(c_1, d) \leq (e_1, f)$, then $X(1) \rightarrow W_1$ is an isomorphism at p_1, and we have

$$u' = u_{\overline{R}_i(1)(q_1)}, v' = v_{\overline{R}_i(1)(q_1)}, w_i' = w_{\overline{R}_i(1)(q_1)}v_{\overline{R}_i(1)(q_1)}.$$

We have

$$u_{\overline{R}_i(1)(q_1)} = \tilde{x}_1^a, v_{\overline{R}_i(1)(q_1)} = \tilde{x}_1^{c_1}\tilde{y}_1^d, w_{\overline{R}_i(1)(q_1)} = \tilde{x}_1^{e_1-c_1}\tilde{y}_1^{f-d}\tilde{z}_1. \tag{15.82}$$

Suppose that $(c_1, d) > (e_1, f)$. We may assume without loss of generality that each $W_{i+1} \rightarrow W_i$ is not an isomorphism at the center of ν.

$W_2 \rightarrow W_1$ is the blow up of a curve which either has local equations

$$\tilde{x}_1 = \tilde{z}_1 = 0 \tag{15.83}$$

(and $c_1 > e_1$), or

$$\tilde{y}_1 = \tilde{z}_1 = 0 \tag{15.84}$$

(and $d > f$).

By (15.81), there exist permissible parameters $\tilde{x}_2, \tilde{y}_2, \tilde{z}_2$ at $\tilde{p}_2$ such that

$$\tilde{x}_1 = \tilde{x}_2, \tilde{y}_1 = \tilde{y}_2, \tilde{z}_1 = \tilde{x}_2\tilde{z}_2$$

if (15.83) holds,

$$\tilde{x}_1 = \tilde{x}_2, \tilde{y}_1 = \tilde{y}_2, \tilde{z}_1 = \tilde{y}_2\tilde{z}_2$$

if (15.84) holds.

We then have that

$$u' = \tilde{x}_2^a, v' = \tilde{x}_2^{c_1}\tilde{y}_2^d, w_i' = \tilde{x}_2^{e_1+1}\tilde{y}_2^f\tilde{z}_1$$

or

$$u' = \tilde{x}_2^a, v' = \tilde{x}_2^{c_1}\tilde{y}_2^d, w_i' = \tilde{x}_2^{e_1}\tilde{y}_2^{f+1}\tilde{z}_1.$$

By iteration of this analysis for local equations of $W_{i+1} \rightarrow W_i$, we see that at $p_1 = \tilde{p}_m$, we have permissible parameters $\tilde{x}_m, \tilde{y}_m, \tilde{z}_m$ such that

$$u' = \tilde{x}_m^a, v' = \tilde{x}_m^{c_1}\tilde{y}_m^d, w_i' = \tilde{x}_m^{c_1}\tilde{y}_m^d\tilde{z}_m.$$

We have

$$(15.85) \qquad u_{\overline{R}_i(1)(q_1)} = \tilde{x}_m^a, v_{\overline{R}_i(1)(q_1)} = \tilde{x}_m^{c_1}\tilde{y}_m^d, w_{\overline{R}_i(1)(q_1)} = \tilde{z}_m.$$

We see that ν is resolved on $X(1)$ if (15.85) holds, since (15.85) is a toroidal form, so we must have that (15.82) holds at p_1. Observe that we must have a reduction $f_1 = f - d < f$ in (15.82) from (15.46).

Iterating this analysis, we see that we must reach the case (15.85) after a finite number of iterations of step 2. This is a contradiction to the assumption that ν is never resolved on $X(n)$.

Now suppose that the center of ν on Y is a 1-point. Then there exists an i such that for all n, $q_n \in U(R_i(n))$ and $p_n \in f(n)^{-1}(q_n) \cap T(R_i(n))$, and either there exists n_0 such that q_n is a 1-point for $n < n_0$ and q_n is a 2-point for $n \ge n_0$ or q_n is a 1-point for all n.

$q = q_0 \in Y$ is a 1-point and $p = p_0 \in f^{-1}(q) \cap T(\overline{R}_i)$. Let

$$u = u_{\overline{R}_i(q)}, v = v_{\overline{R}_i(q)}, w_i = w_{\overline{R}_i(q)}.$$

For all n, let

$$u(n) = u_{\overline{R}_i(n)(q_n)}, v(n) = v_{\overline{R}_i(n)(q_n)}, w(n) = w_{\overline{R}_i(n)(q_n)}.$$

If $\tau > 1$, there exist permissible parameters x, y, z at p such that one of the following forms hold: p a 1-point

$$(15.86) \qquad u = x^a, v = y, w_i = x^b(\gamma(x,y) + x^{c-b}z)$$

where γ is a unit series or p a 2-point

$$(15.87) \qquad u = (x^ay^b)^k, v = z, w_i = (x^ay^b)^t(\gamma(x^ay^b, z) + x^cy^d)$$

where γ is a unit series.

If $\tau = 1$, then there exist permissible parameters x, y, z at p such that either p is a 1-point and

$$(15.88) \qquad u = x^a, v = y, w_i = x^b(\beta + z)$$

with $\beta \in \mathbf{k}$, $b > 0$, or p is a 2-point and

$$(15.89) \qquad u = (x^ay^b)^k, v = z, w_i = x^cy^d$$

with $ad - bc \neq 0$.

Suppose that there exists n_0 such that q_n is a 2-point for all $n \ge n_0$ (and q_n is a 1-point for $n < n_0$). By iterating (15.69), we see that

$$(15.90) \qquad u = u(n_0-1), v = v(n_0-1), w_i = u(n_0-1)^{n_0-1}w(n_0-1).$$

After possibly interchanging $u(n_0-1)$ and $v(n_0-1)$, we have that $u(n_0-1), v(n_0-1), w(n_0-1)$ and $u(n_0), v(n_0), w(n_0)$ are related by an expression (15.68), and after possibly interchanging $u(n_0)$ and $v(n_0)$, we have that $u(n_0), v(n_0), w(n_0)$ and $u(n_0+1), v(n_0+1), w(n_0+1)$ are related by an expression (15.62). From then on, $u(n-1), v(n-1), w(n-1)$ are related to $u(n), v(n), w(n)$ by an expression (15.62).

Since we assume some component of $D_{Y(n)}$ containing q_n does not contract to q, we have an expression:

$$\begin{aligned} u &= u(n)v(n)^{e_n},\\ v &= v(n),\\ w_i &= u(n)^{f_n}v(n)^{g_n}w(n), \end{aligned}$$

and $e_n, f_n, g_n \ge n - n_0 + 1$.

We have $\nu(u) > n\nu(v) > 0$ for all $n \in \mathbf{N}$. Thus ν is a composite valuation, and there exists a prime ideal P of the valuation ring V of ν such that $u \in P$, $v \notin P$. Let ν_1 be a valuation whose valuation ring is V_P. We have $\nu_1(v) = 0$, $\nu_1(u) > 0$. We further have

$$\nu_1(w_i) > n\nu_1(u) > 0 \tag{15.91}$$

for all $n \in \mathbf{N}$.

Suppose that $\tau > 1$ (and q_n is a 2-point for $n \geq n_0$). At $p = p_0 \in X$, we have a form (15.86) or (15.87). In (15.86) we have

$$\nu_1(w_i) = \frac{b}{a}\nu_1(u)$$

and in (15.87) we have

$$\nu_1(w_i) = \frac{t}{k}\nu_1(u),$$

a contradiction (to (15.91).

Suppose that $\tau = 1$. If $\beta \neq 0$ in (15.88), then the analysis is the same as for the $\tau > 1$ case, so we may assume that $\beta = 0$ if (15.88) holds.

If (15.88) holds with $\beta = 0$, or if (15.89) holds, we finish the analysis in a similar way to the proof when q is a 2-point given above. We must end up with $f(n)$ being toroidal at p_n, which is impossible.

The final case is when q_n is a 1-point for all n. From (15.69) we see that

$$u = u(n), v = v(n), w_i = u(n)^n w(n)$$

so that

$$\nu(w_i) > n\nu(u) > 0 \tag{15.92}$$

for all $n \in \mathbf{N}$.

Suppose that $\tau > 1$ (and q_n is a 1-point for all n). At $p = p_0 \in X$ we have a form (15.86) or (15.87), which implies

$$\nu(w_i) = \frac{b}{a}\nu(u)$$

or

$$\nu(w_i) = \frac{t}{k}\nu(u),$$

a contradiction to (15.92).

When $\tau = 1$, the proof follows in a similar way to the proof when q is a 2-point and $\tau = 1$.

We have shown that for all 0-dimensional valuations ν of $\mathbf{k}(X)$, there exists n such that ν is resolved on $X(n)$.

By compactness of the Zariski-Riemann manifold [**Z**] there exists N such that all $\nu \in \Omega(X)$ are resolved on $X(N)$, a contradiction to our assumption that (15.77) is of infinite length. The diagram

$$\begin{array}{ccc} X(N) & \rightarrow & Y(N) \\ \downarrow & & \downarrow \\ X & \rightarrow & Y \end{array}$$

thus satisfies the conclusions of Theorem 15.7.

$\square$

THEOREM 15.8. *Suppose that $\tau \geq 1$, $f : X \to Y$ is τ-quasi-well prepared with relation R. Then there exists a commutative diagram*

$$\begin{array}{ccc} X_1 & \overset{f_1}{\to} & Y_1 \\ \Phi\downarrow & & \downarrow\Psi \\ X & \overset{f}{\to} & Y \end{array}$$

such that Φ and Ψ are products of blow ups of possible centers and f_1 is τ-quasi-well prepared with relation R^1 and pre-algebraic structure. Further, R has an algebraic structure. (R^1 will in general not be the transform of R.)

PROOF. By Theorem 15.7 there exists a τ-quasi-well prepared diagram (15.26) as in the hypothesis of Lemma 15.5. Then Lemma 15.5 implies that the conclusions of Theorem 15.8 hold. ☐

LEMMA 15.9. *Suppose that $\tau \geq 1$, $f : X \to Y$ is τ-quasi-well prepared with relation R and pre-algebraic structure (or τ-well prepared with relation R), $q \in U(R)$ is a 2-point and $p \in f^{-1}(q) \cap T(R_i)$ for some i. Suppose that E is a component of D_Y containing q. Let $C = \overline{E \cdot S_{\overline{R}_i}(q)}$.*

Let $\Psi_n : Y_n \to Y$ be obtained by blowing up q, then blowing up the point q_1 which is the intersection of the exceptional divisor over q and the strict transform of C on Y_1, and iterating this procedure n times, blowing up the intersection point of the last exceptional divisor with the strict transform of C. Let

$$\begin{array}{ccc} X_n & \overset{f_n}{\to} & Y_n \\ \Phi_n\downarrow & & \downarrow\Psi_n \\ X & \overset{f}{\to} & Y \end{array}$$

be a τ-quasi-well prepared (or τ-well prepared) diagram of R and Ψ_n obtained from Lemma 14.7 (so that Φ_n is an isomorphism above $f^{-1}(Y - \Sigma(Y))$).

Suppose that for all $n > 0$ there exists a point $p_n \in \Phi_n^{-1}(p) \cap T(R_i^n)$ such that $f_n(p_n) = q_n \in \Psi_n^{-1}(q) \cap C_n$, where C_n is the strict transform of C on Y_n. Then C is a component of the nonfinite locus of f.

PROOF. The proof follows from Step 1 of the proof of Theorem 15.7. Let

$$u = u_{\overline{R}_i(q)}, v = v_{\overline{R}_i(q)}, w_i = w_{\overline{R}_i(q)}.$$

After possibly interchanging u and v, $v = w_i = 0$ are local equations of C at q.

By our construction of Ψ_n, we have that

$$u_1 = u_{\overline{R}_i^n(q_n)}, v_1 = v_{\overline{R}_i^n(q_n)}, w_{i,1} = w_{\overline{R}_i^n(q_n)}$$

are defined by

$$u = u_1, v = u_1^n v_1, w_i = u_1^n w_{i,1}.$$

We must have an expression (15.41), (15.46), (15.49) or (15.50). It follows, as in the analysis of step 1 of Theorem 15.7 that C is in the nonfinite locus of f. ☐

THEOREM 15.10. *Suppose that $\tau \geq 1$ and $f : X \to Y$ is τ-quasi-well prepared with relation R and pre-algebraic structure. Further suppose that R has an algebraic* *structure. Then there exists a commutative diagram*

$$\begin{array}{ccc} X_1 & \overset{f_1}{\rightarrow} & Y_1 \\ \Phi\downarrow & & \downarrow\Psi \\ X & \overset{f}{\rightarrow} & Y \end{array}$$

such that Φ,Ψ are products of blow ups of possible centers and f_1 is τ-well prepared with relation R^1. (In general, R^1 is not the transform of R).

PROOF. 1 and 2 of Definition 13.5 of a τ-well prepared relation R hold by assumption.

For $i\neq j$, Let H_{ij} be the set of points $q\in U(R_i)\cap U(R_j)$ such that neither an expression (13.7) nor (13.8) of 3 of Definition 13.5 holds between $w_{\overline{R}_i(q)}$ and $w_{\overline{R}_j(q)}$. We see by 4 of Definition 13.1 (and (12.7) - (12.9)) that H_{ij} is a finite set.

Let $\tilde\Psi_1:\tilde Y_1\rightarrow Y$ be the blow up of the union of the sets

$$\{q\in H_{ij}\mid f^{-1}(H_{ij})\cap[T(R_i)\cup T(R_j)]\neq\emptyset\}.$$

By Lemmas 14.7 and 14.8, there exists a τ-quasi-well prepared diagram

$$\begin{array}{ccc} \tilde X_1 & \overset{\tilde f_1}{\rightarrow} & \tilde Y_1 \\ \tilde\Phi_1\downarrow & & \downarrow\tilde\Psi_1 \\ X & \overset{f}{\rightarrow} & Y. \end{array}$$

We define finite sets H^1_{ij} in the same way for the transform $\tilde R^1$ of R, and iterate to construct a τ-quasi-well prepared diagram

$$(15.93)\qquad \begin{array}{ccc} \tilde X_n & \overset{\tilde f_n}{\rightarrow} & \tilde Y_n \\ \tilde\Phi_n\downarrow & & \downarrow\tilde\Psi_n \\ \tilde X_{n-1} & \overset{\tilde f_{n-1}}{\rightarrow} & \tilde Y_{n-1} \\ \downarrow & & \downarrow \\ \vdots & & \vdots \\ \downarrow & & \downarrow \\ \tilde X_1 & \overset{\tilde f_1}{\rightarrow} & \tilde Y_1 \\ \tilde\Phi_1\downarrow & & \downarrow\tilde\Psi_1 \\ X & \overset{f}{\rightarrow} & Y, \end{array}$$

continuing as long as $\tilde f_n^{-1}(H^n_{ij})\cap[T(R^n_i)\cup T(R^n_j)]\neq\emptyset$ for some $i\neq j$.

Suppose that (15.93) doesn't terminate after a finite number of blow ups n. Then there exist $i\neq j$ and a valuation ν of the function field $\mathbf{k}(X)$ of X such that the center p_n of ν on $\tilde X_n$ and the center q_n of ν on $\tilde Y_n$ satisfy $q_n\in H^n_{ij}$, $p_n\in\tilde f_n^{-1}(H^n_{ij})\cap T(\tilde R^n_i)$ for all n.

For all n, we may identify ν with an extension of ν to the quotient field of $\hat{\mathcal{O}}_{\tilde X_n,p_n}$ which dominates $\hat{\mathcal{O}}_{\tilde X_n,p_n}$.

Let

$$u_n=u_{\tilde R^n_i(q_n)},v_n=v_{\tilde R^n_i(q_n)},w_{ni}=w_{\tilde R^n_i(q_n)},w_{nj}=w_{\tilde R^n_j(q_n)}.$$

We have

(15.94)

$$u_n=u_{n+1},v_n=u_{n+1}(v_{n+1}+\alpha_{n+1}),w_{ni}=u_{n+1}w_{n+1,i},w_{nj}=u_{n+1}w_{n+1,j}$$

with $\alpha_{n+1} \in \mathbf{k}$, or

$$u_n = u_{n+1}v_{n+1}, v_n = v_{n+1}, w_{ni} = v_{n+1}w_{n+1,i}, w_{nj} = v_{n+1}w_{n+1,j} \tag{15.95}$$

for all n.

The relation

$$w_{nj} = w_{ni} + \lambda_{ij}^n(u_n, v_n)$$

of 4 of Definition 13.1 transforms as

$$\lambda_{ij}^{n+1}(u_{n+1}, v_{n+1}) = \frac{\lambda_{ij}^n(u_{n+1}, u_{n+1}(v_{n+1} + \alpha_{n+1}))}{u_{n+1}}$$

if (15.94) holds, and transforms as

$$\lambda_{ij}^{n+1}(u_{n+1}, v_{n+1}) = \frac{\lambda_{ij}^n(u_{n+1}v_{n+1}, v_{n+1})}{v_{n+1}}$$

if (15.95) holds.

Let $\mathcal{F}_n$ be the germ at q_n of the divisor $\lambda_{ij}^n = 0$. By embedded resolution of plane curve singularities, there exists n_0 such that $D_{\tilde{X}_n} + \mathcal{F}_n$ is a SNC divisor at q_n for all $n \geq n_0$. If q_n is a 2-point for some $n \geq n_0$, we have that $q_n \notin H_{ij}^n$, a contradiction, so we have that q_n is a 1-point for all $n \geq n_0$. Thus a form (15.94) holds for all $n \geq n_0$. We have

$$w_{n_0 i} = u_{n_0}^{n-n_0} w_{ni}.$$

Thus

$$\nu(w_{n_0 i}) > n\nu(u_{n_0}) > 0 \tag{15.96}$$

for all positive n.

Suppose that $\tau > 1$. Since $q_{n_0} \in U(\tilde{R}_i^{n_0})$ is a 1-point, we have permissible parameters x, y, z at p_{n_0} such that

$$u_{n_0} = x^a, v_{n_0} = y, w_{n_0 i} = x^c \gamma \tag{15.97}$$

where $\gamma \in \hat{\mathcal{O}}_{\tilde{X}_{n_0}, p_{n_0}}$ is a unit series, or

$$u_{n_0} = (x^a y^b)^k, v_{n_0} = z, w_{n_0,i} = (x^a y^b)^l \gamma \tag{15.98}$$

where $\gamma \in \hat{\mathcal{O}}_{\tilde{X}_{n_0}, p_{n_0}}$ is a unit series. In either case, we have a contradiction to (15.96).

If $\tau = 1$, then we have one of the following forms at p_{n_0}.

$$u_{n_0} = x^a, v_{n_0} = y, w_{n_0 i} = x^c(z + \beta)$$

with $\beta \in \mathbf{k}$, or

$$u_{n_0} = (x^a y^b)^k, v_{n_0} = z, w_{n_0,i} = x^c y^d$$

with $a, b > 0$ and $ad - bc \neq 0$.

We obtain a contradiction to (15.96), as in the $\tau = 1$ case, unless we have the form

$$u_{n_0} = x^a, v_{n_0} = y, w_{n_0 i} = x^c z.$$

Now by an argument similar to the case $\tau = 1$ of Theorem 15.7, we obtain

$$\tau_{\tilde{f}_n}(p_n) = -\infty$$

for $n >> 0$, a contradiction.

Thus (15.93) terminates in a finite number of steps n. Let $(R^n)'$ be the restriction of $\tilde{R}^n$, defined by

$$U((R^n)'_i) = U(\tilde{R}^n_i) - \cup_j H_{ij},$$

where H_{ij} is defined by $\tilde{R}^n$, Let

$$\Omega((R^n)'_i) = \Omega(\tilde{R}^n_i) - \cup_j H_{ij}.$$

We have that $\tilde{f}_n : \tilde{X}_n \rightarrow \tilde{Y}_n$ with relation $(R^n)'$ satisfies 1, 2 and 3 of Definition 13.5 of a τ-well prepared morphism.

There exists a sequence of blow ups of 2-curves $\Psi_1 : Y_1 \rightarrow \tilde{Y}_n$ such that 4 (as well as 3) of Definition 13.5 hold for the transforms $\{\overline{R}^1_i\}$ of the $\{(\overline{R}^n)'_i\}$ on Y_1 at all 2-points of $U((\overline{R}^n)'_i)$, by Lemma 5.14 [**C6**]). By Lemma 14.1, there exists a τ-quasi-well prepared diagram

$$\begin{array}{ccc} X_1 & \overset{f_1}{\rightarrow} & Y_1 \\ \Phi_1 \downarrow & & \downarrow \Psi_1 \\ \tilde{X}_n & \overset{\tilde{f}_n}{\rightarrow} & \tilde{Y}_n \end{array}$$

of $(R^n)'_i$ and Ψ_1 where Φ_1 is a product of blow ups of 2-curves. Let R^1 be the transform of $(R^n)'$ on X_1. f_1 is τ-well prepared with relation R^1. □

THEOREM 15.11. *Suppose that $\tau \geq 1$ and $f : X \rightarrow Y$ is τ-well prepared with relation R. Then there exists a commutative diagram*

$$\begin{array}{ccc} X_1 & \overset{f_1}{\rightarrow} & Y_1 \\ \Phi \downarrow & & \downarrow \Psi \\ X & \overset{f}{\rightarrow} & Y \end{array}$$

such that Φ,Ψ are products of blow ups of possible centers and f_1 is τ-very-well prepared with relation R^1. (In general, R^1 is not the transform of R).

PROOF. Let $\Omega = G_Y(f,\tau) - \Theta(f,Y)$. Ω is a finite set by Remark 11.12.

Let R' be the restriction of R to Ω. R' has an algebraic structure determined by the algebraic structure of R.

By Theorem 15.3, there exists a τ-well prepared diagram

$$\begin{array}{ccc} X_2 & \overset{f_2}{\rightarrow} & Y_2 \\ \Phi_2 \downarrow & & \downarrow \Psi_2 \\ X & \overset{f}{\rightarrow} & Y \end{array}$$

for R where Φ_2 and Ψ_2 are products of blow ups of possible centers such that f_2 is τ-well prepared for the transform R^2 of R', and there exists an open subset $V \subset Y$ such that $Y - V$ is a finite set of points and f_2 is toroidal over $\Psi_2^{-1}(V)$. Further, $V \cap G_Y(f,\tau) = \Theta(f,Y)$.

Let R^2_i be a primitive relation associated to R^2. Then $U(R'_i) = \{q_i\}$ for some $q_i \in \Omega$. $\Omega(R'_i)$ is a neighborhood of q_i on a surface in Y, and $\Omega(R^2_i) \rightarrow \Omega(R'_i)$ is a projective birational map. Suppose that E is a component of D_{Y_2} such that $\gamma = E \cdot \Omega(R^2_i)$ dominates a curve (containing q_i) of $\Omega(R'_i)$. Then a general point η of γ is a 1-point over which f_2 is toroidal, since $\Psi_2(\eta) \in V$. Hence f_2 is finite over a general point of γ, and γ is not in the nonfinite locus of f_2. Further, by Remark 15.6, if $q \in \gamma \cap U(R^2_i)$, and $f_2^{-1}(q) \cap T(R^2_i) \neq \emptyset$, then q is a 2-point.

By Lemma 15.9, there exists a τ-well prepared diagram (for R^2)

$$\begin{array}{ccc} X_3 & \overset{f_3}{\to} & Y_3 \\ \Phi_3 \downarrow & & \downarrow \Psi_3 \\ X_2 & \overset{f_2}{\to} & Y_2 \end{array}$$

where Ψ_3 is a product of blow ups of prepared 2-points (of type 1 in Definition 13.7) such that if R_i^3 is a primitive relation associated to the transform R^3 of R^2, E is a component of D_{Y_3}, and $\gamma = E \cdot \Omega(R_i^3)$ is not exceptional for $\Omega(R_i^3) \to \Omega(R_i')$, then $\gamma \cap f_3(T(R_i^3)) = \emptyset$. We further have that $f_3(T(R^3)) \cap \gamma = \emptyset$. We see this as follows. Suppose that $p \in f_3(T(R^3)) \cap \gamma$. $\Psi_3(E)$ is a component of D_{Y_2}. Thus p is on the strict transform γ' of $\overline{\Psi_3(E) \cdot \Omega(R_j^2)}$ on Y_3, which is not exceptional for Ψ_2. Since $\gamma' = \overline{E \cdot \Omega(R_j^3)}$, we have a contradiction.

Let

$$W_3 = \left\{ \begin{array}{l} \gamma = \overline{S_{R_i^3}(q) \cdot E} \text{ such that } E \text{ is a component of } D_{Y_3}, \\ R_i^3 \text{ is associated to } R^3 \text{ and } q \in f_3(T(R^3)) \cap U(\overline{R}_i^3) \end{array} \right\}.$$

We have that all $\gamma \in W_3$ contract to a point on Y_1. Let

$$Z_3 = \left\{ \begin{array}{l} q \in U(R^3) - f_3(T(R^3)) \text{ such that there exist } \gamma_i, \gamma_j \in W_3 \\ \text{such that } \gamma_i \neq \gamma_j \text{ and } q \in \gamma_i \cap \gamma_j. \end{array} \right\}$$

Suppose that $q \in Z_3$. Then there exist $\gamma_i = \overline{S_{R_i^3}(p_i) \cdot E_1} \in W_3$ and $\gamma_j = \overline{S_{R_j^3}(p_j) \cdot E_2} \in W_3$ such that $q \in \gamma_i \cap \gamma_j$ and $\gamma_i \neq \gamma_j$.

γ_i and γ_j are exceptional, so they contract to the common point $q_i = q_j \in U(R')$. thus γ_i and γ_j are contained in $\Omega(R_i^3)$ and $\Omega(R_j^3)$ respectively.

The points of Z_3 are prepared 2-points for R^3 (of type 1 of Definition 13.7) by Lemmas 14.1, 14.7. 14.8. Let $\Psi_4 : Y_4 \to Y_3$ be the blow up of Z_3. By Lemma 14.7, there exists a τ-well prepared diagram

$$\begin{array}{ccc} X_4 & \overset{f_4}{\to} & Y_4 \\ \Phi_4 \downarrow & & \downarrow \Psi_4 \\ X_3 & \overset{f_3}{\to} & Y_3 \end{array}$$

of R^3 and Ψ_4. Let R^4 be the transform of R^3 on X_4.

Define

$$W_4 = \left\{ \begin{array}{l} \gamma = \overline{S_{\overline{R}_i^4}(q) \cdot E} \text{ such that } E \text{ is a component of } D_{Y_4}, \\ R_i^4 \text{ is associated to } R^4 \text{ and } q \in \Psi_4^{-1}(f_3(T(R^3))) \cap U(\overline{R}_i^4) \end{array} \right\},$$

$$Z_4 = \left\{ \begin{array}{l} q \in U(R^4) - \Psi_4^{-1}(f_3(T(R^3))) \text{ such that there exist } \gamma_i, \gamma_j \in W_4 \\ \text{such that } \gamma_i \neq \gamma_j \text{ and } q \in \gamma_i \cap \gamma_j. \end{array} \right\}$$

We necessarily have that the curves in W_4 are strict transforms of curves in W_3. We can iterate, blowing up Z_4, and constructing a τ-well prepared diagram, and repeating until we eventually construct a τ-well prepared diagram of R^4

$$\begin{array}{ccc} X_5 & \overset{f_5}{\to} & Y_5 \\ \Phi_5 \downarrow & & \downarrow \Psi_5 \\ X_4 & \overset{f_4}{\to} & Y_4 \end{array}$$

such that Ψ_5 is a sequence of blow ups of prepared 2-points (of type 1 of Definition 13.7) and if $\gamma_1 = \overline{S_{\overline{R}_i^5}(q_i) \cdot E_i}$, $\gamma_2 = \overline{S_{\overline{R}_j^5}(q_j) \cdot E_j}$, for $q_i \in U(\overline{R}_i^5) \cap (\Psi_4 \circ$

$\Psi_5)^{-1}(f_3(T(R^3)))$ (where R^5 is the transform of R^4) and E_1, E_2 components of D_{Y_5}, are such that $\gamma_1 \neq \gamma_2$, then $\gamma_1 \cap \gamma_2 \subset U(R^5) \cap (\Psi_4 \circ \Psi_5)^{-1}(f_3(T(R^3))$.

We now construct pre-relations $\overline{R}_i^*$ on Y_5 with associated primitive relations R_i^* for f_5.

Let $T(R_i^*) = T(R_i^5)$ and let

$$U(R_i^*) = U(R_i^5) \cap (\Psi_4 \circ \Psi_5)^{-1}(f_3(T(R^3))). \tag{15.99}$$

For $q' \in U(\overline{R}_i^*)$, define $\overline{R}_i^*(q') = \overline{R}_i^5(q')$. For $p \in T(R_i^*)$ define $R_i^*(p) = \overline{R}_i^5(f_5(p))$. Let R^* be the relation for f_5 defined by the R_i^*. Let $\Omega(\overline{R}_i^*) = \Omega(\overline{R}_i^5)$.

For all $\overline{R}_i^*$, let

$$V_i(Y_5) = \left\{\gamma = \overline{E_\alpha \cdot S_{\overline{R}_i^*}(q)} \text{ such that } q \in U(\overline{R}_i^*),\ E_\alpha \text{ is a component of } D_{Y_5}\right\}.$$

Recall that these curves are all exceptional for $\Psi_2 \circ \Psi_3 \circ \Psi_4 \circ \Psi_5$.

By our construction, Lemmas 14.5, 14.1, 14.7, 14.9 and 14.8, and Remark 14.2 and 2 of Remark 14.12, every curve $\gamma \in V_i(Y_5)$ is prepared for R^5 of type 6. By (15.99) and our construction of f_4 from f_2, we now conclude that every curve $\gamma \in V_i(Y_5)$ is prepared for R^* of type 6. 1 and 2 of Definition 13.9 thus hold for f_5 and R^*. 3 of Definition 13.9 holds for f_5 and R^* since for all $\overline{R}_i^*$, $V_i(Y_5)$ consists of exceptional curves of $\Omega(\overline{R}_i^*)$ contracting to a nonsingular point $q_i \in \Omega(\overline{R}_i')$. Thus f_5 is τ-very-well prepared with relation R^*.

□

THEOREM 15.12. *Suppose that $f : X \to Y$ is prepared, and $\tau = \tau_f(X) \geq 1$. Then there exists a commutative diagram*

$$\begin{array}{ccc} X_1 & \xrightarrow{f_1} & Y_1 \\ \Phi_1 \downarrow & & \downarrow \Psi_1 \\ X & \xrightarrow{f_1} & Y \end{array}$$

such that Φ_1 and Ψ_1 are products of blow ups of possible centers and f_1 is τ-very-well prepared with a relation R^1.

PROOF. By Theorem 14.4, there exists a commutative diagram

$$\begin{array}{ccc} X_1 & \xrightarrow{f_1} & Y_1 \\ \Phi \downarrow & & \downarrow \Psi \\ X & \xrightarrow{f} & Y \end{array}$$

such that Φ and Ψ are products of 2-curves, and f_1 is τ-prepared. Now by Theorems 15.4, 15.8, 15.10 and 15.11, there exists a commutative diagram

$$\begin{array}{ccc} X_2 & \xrightarrow{f_2} & Y_2 \\ \downarrow & & \downarrow \\ X_1 & \xrightarrow{f_1} & Y_1 \end{array}$$

where the vertical arrows are products of blow ups of possible centers such that f_2 is τ-very-well prepared.

□

CHAPTER 16

Toroidalization

Suppose that $f : X \to Y$ is a proper, dominant morphism of nonsingular 3-folds with toroidal structures D_Y and $D_X = f^{-1}(D_Y)$, such that D_X contains the locus where f is not smooth.

THEOREM 16.1. *Suppose that $\tau \geq 1$ and $f : X \to Y$ is τ-very-well prepared with relation R. Then there exists a τ-very-well prepared diagram*

$$\begin{array}{ccc} X_1 & \overset{f_1}{\to} & Y_1 \\ \downarrow & & \downarrow \\ X & \overset{f}{\to} & Y \end{array}$$

such that the transform R^1 of R is resolved ($T(R^1) = \emptyset$). In particular, f_1 is prepared and $\tau_{f_1}(X_1) < \tau$.

PROOF. Fix a pre-relation $\overline{R}_t$ associated to R on Y, with associated primitive relation R_t. By induction on the number of pre-relations associated to R, it suffices to resolve R_t by a τ-very-well prepared diagram (of R).

Recall (Definition 13.9)

$$V_t(Y) = \left\{ \begin{array}{l} \overline{E \cdot S} \text{ such that } E \text{ is a component of } D_Y, \\ S = S_{\overline{R}_t}(q) \text{ for some } q \in U(\overline{R}_t) \end{array} \right\}.$$

$F_t = \sum_{\gamma \in V_t(Y)} \gamma$ is a SNC divisor on $\Omega(\overline{R}_t)$ whose intersection graph is a forest.

If $\gamma_1 = \overline{E_1 \cdot S_{\overline{R}_t}(q_1)} \in V_t(Y)$ and $q \in \gamma_1$, we will say that γ_1 is good at q if whenever $q \in U(\overline{R}_i)$ for some i, then $S_{\overline{R}_i}(q)$ contains the germ of γ_1 at q (so that $\gamma_1 = \overline{E_1 \cdot S_{\overline{R}_i}(q)} \subset \Omega(\overline{R}_i)$). Otherwise, say that γ_1 is bad at q. Say that γ_1 is good if γ_1 is good at q for all $q \in \gamma_1$.

Let $Y_0 = Y$, $X_0 = X$, $f_0 = f$. We will show that there exists a sequence of τ-very-well prepared diagrams of the transform of R,

$$\begin{array}{ccc} X_{i+1} & \overset{f_{i+1}}{\to} & Y_{i+1} \\ \Phi_{i+1} \downarrow & & \Psi_{i+1} \downarrow \\ X_i & \overset{f_i}{\to} & Y_i \end{array} \tag{16.1}$$

for $0 \leq i \leq m-1$ such that the transform R_t^m of R_t on X_m is resolved.

Suppose that $\gamma_1 \in V_t(Y)$ and $q \in \gamma_1$ is a bad point. By Remark 13.10, we have that $q \in U(\overline{R}_t)$. By (13.8) of Definition 13.5, we have that q is a 2-point. Suppose that E_1, E_2 are the two components of D_Y containing q, ordered so that $\gamma_1 = \overline{E_1 \cdot S_{\overline{R}_t}(q)}$. Let $\gamma_2 = \overline{E_2 \cdot S_{\overline{R}_t}(q)}$.

We will show that q is a good point of γ_2.

γ_1 not good at q implies there exists $j \neq t$ such that $q \in U(\overline{R}_j)$ and the germ of γ_1 at q is not contained in $S_{\overline{R}_j}(q)$. Let

$$u = u_{\overline{R}_t}(q), v = v_{\overline{R}_t}(q), w_t = w_{\overline{R}_t}(q).$$

After possibly interchanging u and v we have that $u = w_t = 0$ are local equations of γ_1, $v = w_t = 0$ are local equations of γ_2 at q. Let $w_j = w_{\overline{R}_j}(q)$. In the equation

$$w_j = w_t + u^{a_{tj}} v^{b_{tj}} \phi_{tj}$$

of (13.7) of Definition 13.5 we thus have $a_{tj} = 0$.

If q is not a good point for γ_2 then there exists $k \neq t$ such that $q \in U(\overline{R}_k)$ and the germ of γ_2 at q is not contained in $S_{\overline{R}_k}(q)$. Let $w_k = w_{\overline{R}_k}(q)$. In the equation

$$w_k = w_t + u^{a_{tk}} v^{b_{tk}} \phi_{tk}$$

of (13.7) we thus have $b_{tk} = 0$. But we must have

$$(0, b_{tj}) \leq (a_{tk}, 0) \text{ or } (a_{tk}, 0) \leq (0, b_{tk})$$

by 4 of Definition 13.5, which is impossible. Thus q is a good point for γ_2.

Suppose that all $\gamma \in V_t(Y)$ are bad. Pick $\gamma_1 \in V_t(Y)$. Since γ_1 is bad there exists $\gamma_2 \in V_t(Y) - \{\gamma_1\}$ such that γ_2 is good at $q_1 = \gamma_1 \cap \gamma_2$ (as shown above). $\gamma_1 \cap \gamma_2$ is a single point since $V_t(Y)$ is a forest. Since γ_2 is bad and $V_t(Y)$ is a forest, there exists $\gamma_3 \in V_t(Y)$ which intersects γ_2 at a single point q_2 and is disjoint from γ_1 such that γ_3 is good at q_2. Since $V_t(Y)$ is a finite set, and the intersection graph of $V_t(Y)$ is a forest, we must eventually find a curve which is good, a contradiction.

Let $\gamma \in V_t(Y)$ be a good curve, so that it is prepared for R of type 6, and is a *-permissible center (Lemma 14.11) and let $\Psi_1' : Y_1' \rightarrow Y$ be the blow up of γ.

By Lemma 14.11 we can construct a τ-very-well prepared diagram of the form of (13.11) of Definition 13.12

$$\begin{array}{ccc} X_1 & \overset{f_1}{\rightarrow} & Y_1 \\ \downarrow & & \downarrow \\ \downarrow & & Y_1' \\ \downarrow & & \downarrow \Psi_1' \\ X & \rightarrow & Y. \end{array} \tag{16.2}$$

where $Y_1 \rightarrow Y_1'$ is a sequence of blow ups of 2-points which are prepared for the transform of R of type 2 of Definition 13.7. Observe that if $\gamma_1 \in V_t(Y)$ is a good curve, with $\gamma_1 \neq \gamma$, then the strict transform of γ_1 in Y_1 is a good curve in $V_t(Y_1)$.

We now iterate this process. We order the curves in $V_t(Y)$, and choose $\gamma = \overline{E \cdot S_{R_t}(q)} \in V_t(Y)$ in the construction of the diagram (16.2) so that it is the minimum good curve in $V_t(Y)$.

We inductively define a sequence of τ-very well prepared diagrams (16.1) by blowing up the good curve in $V_t(Y_i)$ with smallest order, and then constructing a very well prepared diagram (16.1) of the form of (16.2). Then we define the total ordering on $V_t(Y_{i+1})$ so that the ordering of strict transforms in Y_{i+1} of elements of $V_t(Y_i)$ is preserved, and these strict transforms have smaller order than the element of $V_t(Y_{i+1})$ which is not a strict transform of an element of $V_t(Y_i)$. We repeat, as long as R_t^i is not resolved ($T(R_t^i) \neq \emptyset$).

Suppose that the algorithm does not converge in the construction of $f_m : X_m \rightarrow Y_m$ such that the transform R_t^m of R_t is resolved. Then there exists a diagram

$$
\begin{array}{ccc}
\vdots & & \vdots \\
\downarrow & & \downarrow \\
X_n & \stackrel{f_n}{\rightarrow} & Y_n \\
\Phi_n \downarrow & & \downarrow \Psi_n \\
X_{n-1} & \stackrel{f_{n-1}}{\rightarrow} & Y_{n-1} \\
\downarrow & & \downarrow \\
\vdots & & \vdots \\
\Phi_1 \downarrow & & \downarrow \Psi_1 \\
X_0 = X & \stackrel{f_0=f}{\rightarrow} & Y_0 = Y
\end{array}
\tag{16.3}
$$

constructed by infinitely many iterations of the algorithm such that $T(R_t^n) \neq \emptyset$ for all n.

Suppose that $q_n \in U(\overline{R}_t^n)$ is an infinite sequence of points such that $\Psi_n(q_n) = q_{n-1}$ for all n and Ψ_n is not an isomorphism for infinitely many n. q_n is either a 2-point or a 1-point for all n.

First suppose that q_n is a 2-point for all n.

By construction, the restriction of Ψ_n to $S_{\overline{R}_t^n}(q_n)$ is an isomorphism onto $S_{\overline{R}_t^{n-1}}(q_{n-1})$ for all n. Thus the restriction

$$\overline{\Psi}_n = \Psi_1 \circ \cdots \circ \Psi_n : S_{\overline{R}_t^n}(q_n) \rightarrow S_{\overline{R}_t}(q)$$

is an isomorphism, where $q = q_0 = \Psi_1 \circ \cdots \circ \Psi_n(q_n)$. Without loss of generality, we may assume that no Ψ_n is an isomorphism (on Y_n) at q_n. We have permissible parameters $u_i = u_{\overline{R}_t^i}(q_i), v_i = v_{\overline{R}_t^i}(q_i), w_{t,i} = w_{\overline{R}_t^i}(q_i)$ at q_i for all i such that either

$$u_i = u_{i+1}, v_i = v_{i+1}, w_{t,i} = u_{i+1}w_{t,i+1} \tag{16.4}$$

or

$$u_i = u_{i+1}, v_i = v_{i+1}, w_{t,i} = v_{i+1}w_{t,i+1}. \tag{16.5}$$

Suppose there exists $k \neq t$ such that $q_n \in U(\overline{R}_k^n)$ for all n.

Let $w_{k,i} = w_{\overline{R}_k^i}(q_i)$ for $i \geq 0$.

The relations

$$w_{k,i} - w_{t,i} = u_i^{a_{tk}} v_i^{b_{tk}} \phi_{t,k}$$

of (13.7) of Definition 13.5 transform to

$$w_{k,i+1} - w_{t,i+1} = u_{i+1}^{a_{tk}-1} v_{i+1}^{b_{tk}} \phi_{t,k}$$

under (16.4), and transform to

$$w_{k,i+1} - w_{t,i+1} = u_{i+1}^{a_{tk}} v_{i+1}^{b_{tk}-1} \phi_{t,k}$$

under (16.5). But we see that after a finite number of iterations $q_n \notin U(\overline{R}_k^n)$, unless $a_{tk} = b_{tk} = -\infty$. Thus there exists n_0, such that whenever $n \geq n_0$, $q_n \notin U(\overline{R}_k^n)$ if $k \neq t$ and $a_{kt}, b_{kt} \neq -\infty$.

Now suppose that q_n is a 1-point for all n. We have permissible parameters

$$u_i = u_{\overline{R}_t^i}(q_i), v_i = v_{\overline{R}_t^i}(q_i), w_{t,i} = w_{\overline{R}_t^i}(q_i)$$

at q_i for all i such that

$$u_i = u_{i+1}, v_i = v_{i+1}, w_{t,i} = u_{i+1}w_{t,i+1}$$

for all i.

Suppose that there exists $k \neq t$ such that $q_n \in U(\overline{R}_k^n)$ for all n. Let

$$w_{k,i} = w_{\overline{R}_k^i}(q_i)$$

for $i \geq 0$. The relation

$$w_{k,i} - w_{t,i} = u_i^{c_{t,k}}\phi_{t,k}$$

of (13.8) of Definition 13.5 transforms to

$$w_{k,i+1} - w_{t,i+1} = u_{i+1}^{c_{t,k}-1}\phi_{t,k}.$$

Thus after a finite number of iterations, $q_n \notin U(\overline{R}_k^n)$ unless $c_{t,k} = -\infty$. Thus there exists n_0 such that when $n \geq n_0$, $q_n \notin U(\overline{R}_k^n)$ if $k \neq t$ and $c_{t,k} \neq -\infty$.

By our ordering, we have that there exists an n_0 such that if $n \geq n_0$, $\gamma \in V_t(Y_n)$ is good and if k is such that $\gamma \cap U(\overline{R}_k^n) \neq \emptyset$ then $a_{tk}, b_{tk} = -\infty$ (or $c_{tk} = -\infty$), so that the Zariski closures of $\Omega(\overline{R}_k^n)$ and $\Omega(\overline{R}_t^n)$ are the same. Thus all elements of $V_t(Y_n)$ are good for $n \geq n_0$, since otherwise, there would be a bad curve $\gamma_1 \in V(Y_n)$ which intersects a good curve γ_2 at a point q' at which γ_1 is not good. But then we must have that there exists $k \neq t$ such that the Zariski closure of $\Omega(\overline{R}_k^n)$ is not equal to the Zariski closure of $\Omega(\overline{R}_t^n)$, and $q' \in U(\overline{R}_k^n)$, so that $\gamma_2 \cap U(\overline{R}_k^n) \neq \emptyset$, a contradiction.

Our birational morphism of $\Omega(\overline{R}_t^n)$ to $\Omega(\overline{R}_t)$ is an isomorphism in a neighborhood of $U(\overline{R}_t^n)$. Thus we have a natural identification of $V_t(Y_n)$ and $V_t(Y)$, and we see that for $n \geq n_0$, the Ψ_n cyclically blow up the different curves of $V_t(Y)$.

There are points $p_n \in T(R_t^n) \subset X_n$ such that $\Phi_n(p_n) = p_{n-1}$, and $f_n(p_n) = q_n \in U(\overline{R}_t^n)$ for all n. Without loss of generality, we may assume that no Ψ_n is an isomorphism at q_n.

We have that all q_n are 1-points or all q_n are 2-points.

First suppose that all q_n are 2-points.

With the above notation at $q_n = f_n(p_n)$, we have that (16.4) and (16.5) must alternate in the diagram (16.3) for $n \geq n_0$, by our ordering of $V_t(Y_n)$. Let $p = p_0 \in X = X_0$, $q = q_0 = f(p)$.

We have

$$u = u_n, v = v_n, w_t = u_n^{a_n}v_n^{b_n}w_{t,n} \tag{16.6}$$

where

$$u = u_{\overline{R}_t}(q), v = v_{\overline{R}_t}(q), w_t = w_{\overline{R}_t}(q)$$

and a_n, b_n are positive integers which both go to infinity as n goes to infinity.

There exists (by Theorem 4 of Section 4, Chapter VI [**ZS**]) a valuation ν of $\mathbf{k}(X)$ which dominates the (non-Noetherian) local ring $\cup_{n\geq 0}\mathcal{O}_{X_n,p_n}$, and thus dominates the local rings $\mathcal{O}_{X_n,p_n}$ for all n. Without loss of generality, we may identify ν with an extension of ν to the quotient field of $\hat{\mathcal{O}}_{X_n,p_n}$ which dominates $\hat{\mathcal{O}}_{X_n,p_n}$ for all n.

Let x, y, z be permissible parameters for u, v, w_t at p. Suppose that p is a 3-point. Write (in $\hat{\mathcal{O}}_{X,p}$)

$$\begin{aligned} u &= x^a y^b z^c \\ v &= x^d y^e z^f \\ w_t &= x^g y^h z^i \gamma \end{aligned} \tag{16.7}$$

where $xyz = 0$ is a local equation of D_X at p and γ is a unit series.

We may permute x, y, z so that $0 < \nu(x) \le \nu(y) \le \nu(z)$. We have (from (16.6))

$$\nu(w_t) - n\nu(u) - n\nu(v) > 0$$

for all $n \in \mathbf{N}$. Thus

$$0 < (g-na-nd)\nu(x)+(h-ne-nb)\nu(y)+(i-nf-nc)\nu(z) \le ((g+h+i)-nf-nc)\nu(z)$$

for all n. Thus $f = c = 0$, but this is impossible, since $uv = 0$ is a local equation of D_X at p.

There is a similar but simpler algorithm if p is a 1-point or a 2-point and $\tau > 1$, since $w_t = 0$ is supported on D_X at p.

Suppose that $\tau = 1$ (and all q_n are 2-points). We have that $w_t = 0$ is a divisor supported on D_X at p, so that the argument for $\tau > 1$ works in this case also, or we have one of the following two special forms:

p a 1-point

$$u = x^a, v = x^b(\alpha + y), w_t = x^c z \tag{16.8}$$

or p a 2-point

$$u = x^a y^b, v = x^c y^d, w_t = x^e y^f z \tag{16.9}$$

with $ad - bc \neq 0$.

In the diagram (16.2) we have that $q_1 \in U(R_i^1)$ implies $Y_1 \to Y_1'$ is an isomorphism near q_1. $X_1 \to X$ factors above a suitable neighborhood of q_1 as a diagram

$$X_1 = W_m \xrightarrow{\Lambda_m} W_{m-1} \to \cdots \to W_1 \xrightarrow{\Lambda_1} X$$

where $\Lambda_1 : W_1 \to X$ is a sequence of blow ups of 2-curves and 3-points, and each Λ_{j+1} is the blow up of a possible curve Σ_j containing a 1-point and the center $\overline{p}_j$ of ν on W_j, such that $\mathcal{I}_\gamma \mathcal{O}_W$ is not invertible.

Without loss of generality, we may assume that $u = w_t = 0$ are local equations of γ at q.

We have a form (16.8) or (16.9) at the center $\overline{p}_1$ of ν on W_1, where $(a, b) \le (e, f)$ or $(a, b) > (e, f)$ if (16.9) holds.

Suppose that $a \le c$ in (16.8), or $(a, b) \le (e, f)$ in (16.9). Then $\mathcal{I}_\gamma \mathcal{O}_{W_1, \overline{p}_1}$ is invertible. Thus $W_m = W_1$, and above a suitable neighborhood of q_1, $X_1 \to X$ is a sequence of blow ups of 2-curves. Further, we have an expression of the form (16.8) or (16.9) at p_1.

Suppose that $a > c$ in (16.8). Then Λ_2 is the blow up of a curve with local equations $x = z = 0$ at $\overline{p}_1$. Since $\nu(z) > n\nu(x)$ for all $n \in \mathbf{N}$, at $\overline{p}_2$ we have regular parameters x_2, y_2, z_2 defined by

$$x = x_2, y = y_2, z = x_2 z_2$$

and we thus have

$$u = x_2^a, v = x_2^b(\alpha + y_2), w_t = x_2^{c+1} z_2.$$

Iterating, we see that $\overline{p}_m$ has permissible parameters x_m, y_m, z_m such that

$$u = x_m^a, v = x_m^b(\alpha + y_m), w_t = x_m^a z_m.$$

The permissible parameters $u_1, v_1, w_{t,1}$ at q_1 are defined by

$$u = u_1, v = v_1, w_t = u_1 w_{t1}.$$

Thus

$$u_1 = x_m^a, v_1 = x_m^b(\alpha + y_m), w_{t1} = z_m$$

and we have $\tau_{f_1}(p_1) = -\infty$, a contradiction.

We have a similar analysis if $(a, b) > (e, f)$ in (16.9), leading to the conclusions that $\tau_{f_1}(p_1) = -\infty$, a contradiction.

We thus see that for all n in (16.3), $X_n \rightarrow X_{n-1}$ factors as a sequence of blow ups of 2-curves at q_n.

Suppose that there exists n_0 such that p_{n_0} is a 1-point, and thus p_n is a 1-point for all $n \geq n_0$. Then we see that $X_{n+1} \rightarrow X_n$ is an isomorphism at p_n for all $n \geq n_0$. At p_{n_0} there are regular parameters x, y, z such that

$$u_{n_0} = x^a, v_{n_0} = x^b(\alpha + y), w_{tn_0} = x^c z, \tag{16.10}$$

and q_n has regular parameters

$$u_{n_0} = u_n, v_{n_0} = v_n, w_{tn_0} = u_n^{n-n_0} w_{tn}.$$

Substituting into (16.10), we have a contradiction as soon as

$$n > \frac{c}{a} + n_0.$$

Now suppose that p_n is a 2-point for all n. Then a form (16.9) holds at p_1, and at p_n, we have regular parameters x_n, y_n, z_n defined by

$$x = x_n^{r_{11}^n} y_n^{r_{12}^n}, y = x_n^{r_{21}^n} y_n^{r_{22}^n}, z = z_n \tag{16.11}$$

such that $r_{11}^n r_{22}^n - r_{12}^n r_{21}^n = \pm 1$.

Now from (16.6), (16.9) and (16.11), we see that

$$\nu(x^e y^f) - n\nu(x^a y^b) - n\nu(x^c y^d) > 0$$

for all $n \in \mathbf{N}$, a contradiction, since $ad - bc \neq 0$.

The argument is simpler in the case when q_n is a 1-point for all n. (16.6) becomes

$$u = u_n, v = v_n, w_t = u_n^{a_n} w_{t,n} \tag{16.12}$$

where a_n goes to infinity as n goes to infinity.

Suppose that $\tau > 1$. If $p \in f^{-1}(q)$, $u = 0$ is a local equation of D_X at p, and $w_t = 0$ is supported on D_X at p. Thus (16.12) leads to a contradiction.

If $\tau = 1$, there is a similar argument to the above case of q_n a 2-point for all n and $\tau = 1$.

Thus the algorithm converges in a τ-very-well prepared morphism $f_m : X_m \rightarrow Y_m$ such that $T(R_t^m) = \emptyset$, and after iterating for each primitive relation associated to R, we obtain the construction of $f_1 : X_1 \rightarrow Y_1$, as in the conclusions of the theorem, such that f_1 is prepared and $\tau_{f_1}(X_1) < \tau$.

$\square$

CHAPTER 17

Proofs of the main results

Proof of Theorem 1.2

First suppose that X and Y are projective over $\mathbf{k}$. By resolution of singularities and resolution of indeterminacy [**H**] (cf. Section 6.8 [**C6**]), there exists a commutative diagram

$$\begin{array}{ccc} X_1 & \overset{f_1}{\rightarrow} & Y_1 \\ \Phi_1 \downarrow & & \downarrow \Psi_1 \\ X & \overset{f}{\rightarrow} & Y \end{array}$$

where Φ_1, Ψ_1 are products of possible blow ups of points and nonsingular curves supported above D_X and D_Y, such that X_1 and Y_1 are nonsingular and projective. Further, $D_{Y_1} = \Psi_1^{-1}(D_Y)$ and $D_{X_1} = \Phi_1^{-1}(D_X) = f_1^{-1}(D_{Y_1})$ are SNC divisors, and D_{X_1} contains the locus where f_1 is not smooth.

If Y is a curve, the proof of Theorem 1.2 now follows from embedded resolution of singularities (c.f. Introduction to [**C3**]). If Y is a surface, the proof of Theorem 1.2 follows from Theorem 19.11 [**C3**].

Assume that Y is a 3-fold. By Theorem 1.3, we can construct a commutative diagram

$$\begin{array}{ccc} X_2 & \overset{f_2}{\rightarrow} & Y_2 \\ \Phi_2 \downarrow & & \downarrow \Psi_2 \\ X_1 & \overset{f_1}{\rightarrow} & Y_1 \end{array}$$

such that Φ_2 and Ψ_2 are products of possible blow ups of points and nonsingular curves, such that f_2 is prepared for $D_{Y_2} = \Psi_2^{-1}(D_{Y_1})$ and $D_{X_2} = \Phi_2^{-1}(D_{X_1})$.

Now by descending induction on $\tau = \tau_{f_2}(X_2)$ and Theorems 15.12 and 16.1, there exists a commutative diagram

$$\begin{array}{ccc} X_3 & \overset{f_3}{\rightarrow} & Y_3 \\ \Phi_3 \downarrow & & \downarrow \Psi_3 \\ X_2 & \overset{f_2}{\rightarrow} & Y_2 \end{array}$$

such that Φ_2 and Ψ_3 are products of blow ups of possible centers, f_3 is prepared, and $\tau_{f_3}(X_3) = -\infty$.

Thus f_3 is toroidal, and the conclusions of the theorem follow.

Now suppose that X and Y are quasi-projective varieties. There exist projective $\mathbf{k}$-varieties $\overline{X}$ and $\overline{Y}$ such that X is an open subset of $\overline{X}$, and Y is an open subset of $\overline{Y}$. After possibly modifying $\overline{X}$ and $\overline{Y}$ by blowing up $\overline{X} - X$ and $\overline{Y} - Y$, we may assume that $F_1 = \overline{X} - X$ and $F_2 = \overline{Y} - Y$ are closed subsets of pure codimension 1 in $\overline{X}$, $\overline{Y}$ respectively.

Let $\overline{D}_X$ be the Zariski closure of D_X in $\overline{X}$, $\overline{D}_Y$ be the Zariski closure of D_Y in $\overline{Y}$.

Let $D_{\overline{X}} = \overline{D}_X + F_1$, $D_{\overline{Y}} = \overline{D}_Y + F_2$. By resolution of indeterminancy, after possibly modifying $\overline{X}$ by blowing up subvarieties of $\overline{X}$ supported above $D_{\overline{X}}$, we have that the rational map $\overline{f} : \overline{X} \to \overline{Y}$ which extends $f : X \to Y$ is a morphism.

The hypotheses of Theorem 1.2 are satisfied for $\overline{f} : \overline{X} \to \overline{Y}$, so by the first part of this proof, there exists a commutative diagram

$$\begin{array}{ccc} \overline{X}_1 & \xrightarrow{\overline{f}_1} & \overline{Y}_1 \\ \overline{\Phi} \downarrow & & \downarrow \overline{\Psi} \\ \overline{X} & \xrightarrow{\overline{f}} & \overline{Y} \end{array}$$

satisfying the conclusions of Theorem 1.2.

Let $X_1 = \overline{\Phi}^{-1}(X)$, $Y_1 = \overline{\Psi}^{-1}(Y)$, $f_1 = \overline{f}_1 \mid X_1$, $\Phi = \overline{\Phi} \mid D_1$, $\Psi = \overline{\Psi} \mid Y_1$. Then

$$\begin{array}{ccc} X_1 & \xrightarrow{f_1} & Y_1 \\ \Phi \downarrow & & \downarrow \Psi \\ X & \xrightarrow{f} & Y \end{array}$$

satisfies the conclusions of Theorem 1.2.

The proof of Theorem 1.1 is a simplification of the proof of Theorem 1.2.

CHAPTER 18

List of technical terms

(See also Chapter 3, Notation)

admissible center: Definition 12.2
F_g: Chapter 3, Notation
fundamental locus: Chapter 3, Notation
$G_X(f,\tau)$: Definition 9.4
$G_Y(f,\tau)$: Definition 9.4
good at p for f: Definition 11.1
monomial form: Definition 4.4
NF_f: Chapter 3, Notation
nonfinite locus: Chapter 3, Notation
perfect for f: Definition 11.9.
permissible center: Definition 13.11
*-permissible center: Definition 13.12
permissible parameters: before and after Definition 4.1
possible center: Chapter 3, Notation
prepared point or curve of type 1-5: Definition 13.7
prepared curve of type 6: Definition 13.8
prepared morphism
  Prepared morphism from 3-fold to surface: before Remark 4.5
  prepared morphism of 3-folds: Definition 4.6
  τ-prepared morphism: Definition 9.4
  pre-τ-quasi-well prepared: Definition 13.1
  τ-quasi-well prepared: Definition 13.1
  τ-well prepared: Definition 13.5
  τ-very-well prepared: Definition 13.9
relation
  quasi-pre-relation: Definition 12.1
  pre-relation: Definition 12.3
  algebraic pre-relation: Definition 12.4
  primitive relation: Definition 12.5
  relation: Definition 12.5
  algebraic relation: Definition 12.5
resolved quasi-pre-relation: after Definition 12.1
resolved relation: after Definition 12.5
resolving curve: Definition 13.6
$\Sigma(Y)$: Chapter 3, Notation
super parameters: Definition 10.1
$\tau_f(p)$: Definition 9.1

$\tau_f(X)$: after Definition 9.1
τ-quasi-well prepared diagram: after Definitions 13.11 and 13.13
$\Theta(f, Y)$: Definition 11.11
toroidal forms for u, v, w: after Definition 4.3
toroidal forms for u, v: Definition 4.1
toroidal ideal: Chapter 3, Notation
torodial morphism: Definition 4.3
transform of a pre-relation: after Definition 12.2
transform of a relation: after Definition 12.6
uniformizing parameters: Chapter 3, Notation
weakly good at p for f: Definition 11.2
Zariski-Riemann manifold: Chapter 3, Notation

Bibliography

[Ab1] Abhyankar, S., *On the valuations centered in a local domain*, Amer. J. Math. 78 (1956), 321 – 348.

[Ab2] Abhyankar, S., *Algebraic Geometry for Scientists and Engineers*, Amer. Math. Soc., 1990.

[Ab3] Abhyankar, S., *Resolution of Singularities of Embedded Surfaces*, Academic Press, New York, 1966.

[Ab4] Abhyankar, S., *On the ramification of algebraic functions, I.*, Amer. J. Math. 77 (1955).

[AK] Abramovich D., Karu K., *Weak semistable reduction in characteristic 0*, Invent. Math. 139 (2000), 241 – 273.

[AKMW] Abramovich, D., Karu, K., Matsuki, K. and Wlodarczyk, J., *Torification and factorization of birational maps*, JAMS 15 (2002), 531 – 572.

[AMR] Abramovich, D., Matsuki, K., Rashid, S., *A note on the factorization theorem of toric birational maps after Morelli and its toroidal extension*, Tohoku Math J. 51 (1999), 489 – 537, *Correction:* Tohoku Math J. 52 (2000), 629 – 631.

[AkK] Akbulut, S. and King, H., *Topology of algebraic sets*, MSRI publications 25, Springer-Verlag, Berlin.

[BaM] Bartlet, D., Maire, H.M., *Asymptotique des integrales-fibres*, Ann. Inst. Fourier 43, 1267 - 1299 (1993).

[BrM] Bierstone, E. and Millman, P., *Canonical desingularization in characteristic zero by blowing up the maximal strata of a local invariant*, Inv. Math 128 (1997), 207 – 302.

[BEV] Bravo, A., Encinas, S., Villamayor, O.,*A simplified proof of desingularization and applications*, to appear in Revista Matematica Iberamericana.

[Ch] Christensen, C., *Strong domination/weak factorization of three dimensional regular local rings*, Journal of the Indian Math. Soc., 45 (1981), 21 – 47.

[Cr] Crauder, B., *Two reduction theorems for threefold birational morphisms*, Math Ann. 260 (1984), 13 – 26.

[C1] Cutkosky, S.D., *Local factorization of birational maps*, Advances in Mathematics 132 (1997), 167 – 315.

[C2] Cutkosky, S.D., *Local monomialization and factorization of morphisms*, Astérisque 260, 1999.

[C3] Cutkosky, S.D., *Monomialization of Morphisms from 3-folds to surfaces*, Lecture Notes in Mathematics 1786, Springer-Verlag, Berlin, Heidelberg, New York, 2002.

[C4] Cutkosky, S.D., *Local monomialization of trancendental extensions*, Annales de L'Institut Fourier 55 (2005), 1517 – 1586.

[C5] Cutkosky, S.D., *Toroidalization of birational morphisms of 3-folds*, AG/0407258.

[C6] Cutkosky, S.D. *Resolution of Singularities*, American Mathematical Society, 2004.

[C7] Cutkosky, S.D., *Strong Toroidalization of birational morphisms of 3-folds*, preprint, AG/0412497.

[C8] Cutkosky, S.D., *Strong Toroidalization of dominant morphisms of 3-folds*, preprint, AG/0601037.

[CK] Cutkosky, S.D., Kascheyeva, O., *Monomialization of strongly prepared morphisms from nonsingular n-folds to surfaces* J. Algebra (2004), 275-320.

[CP] Cutkosky, S.D. and Piltant, O., *Monomial resolutions of morphisms of algebraic surfaces*, Comm. in Alg. 28 (2000), 5935 – 5959.

[CS] Cutkosky, S.D. and Srinivasan, H. *Factorizations of birational extensions of local rings*, preprint, AC/0601393.

[D1] Danilov, V., *Birational geometry of toric 3-folds*, Math USSR Izv. 21 (83), 269 – 280.

[D2] V. Danilov. *Decomposition of some birational morphisms*, Izv. Akad. Nauk SSSR 44 (1980), 465-477.

[EH] Encinas, S., Hauser, H., *Strong resolution of singularities in characteristic zero*, Comment Math. Helv. 77 (2002), 821 – 845.

[E] Ewald, E., *Blow ups of smooth toric 3-varieties*, Abh. math. Sem. Univ. Hamburg 57 (1987).

[F] Fulton, W. *Introduction to Toric Varieties*, Annals of Math. Studies, Princeton Univ. Press, Princeton, 1993.

[Go] R. Goward. *A simple algorithm for the principalization of monomial ideals*, Transactions of the AMS 357 (2005), 4805-4812.

[G] Grothendieck, A., *Revêtements etales et groupe fondamental (SGA1)*, Lecture notes in mathematics 176, Springer-Verlag, Berlin, Heidelberg, New York, 1970.

[H] Hironaka, H., *Resolution of singularities of an algebraic variety over a field of characteristic zero*, Annals of Math, 79 (1964), 109 – 326.

[K] Karu, K., *Local strong factorization of birational maps*, J. Alg. Geom 14 (2005), 165 – 175.

[KKMS] Kempf, G., Knudsen, F., Mumford, D., Saint-Donat, B., *toroidal embeddings I*, LNM 339, Springer Verlag (1973).

[L] Lichtin, B., *On a question of Igusa, II, uniform asymptotic bounds for Fourier transforms in seceral variables*, Compositio Math. 141 (2005), 192 -2006.

[Li] Lipman, J., *Appendix to Chapter II* in Zariski, O., *Algebraic Surfaces*, second supplemented edition, Springer-Verlag, New York, Heidelberg, Berlin, 1971.

[Mo] Morelli, R., *The birational geometry of toric varieties*, J. Algebraic Geometry 5 (1996), 751 – 782.

[N] Nagata, M., *Local Rings*, John Wiley and Sons, Inc. (Interscience Publishers), New York, 1962.

[O] Oda, T., *Torus embeddings and applications*, TIFR, Bombay, 1978.

[S] Sally, J., *Regular overrings of regular local rings*, Trans. Amer. Math. Soc. 171 (1972) 291 – 300.

[Sh] Shannon, D.L., *Monoidal transforms*, Amer. J. Math 45 (1973), 284 – 320.

[W1] Wlodarcyzk, J., *Decomposition of birational toric maps in blowups and blowdowns*, Trans. Amer. Math. Soc. 349 (1997), 373-411.

[Z] Zariski, O., *The compactness of the Riemann manifold of an abstract field of algebraic functions*, Bull. Amer. Math. Soc., 45 (1044), 683 – 691.

[Z1] Zariski, O., *Introduction to the problem of minimal models in the theory of algebraic surfaces*, Publications of the Math. Soc. of Japan, 1958.

[Z2] Zariski, O., *Algebraic Surfaces*, second supplemented edition, Springer-Verlag, New York, Heidelberg, Berlin, 1971.

[ZS] Zariski, O. and Samuel P., *Commutative Algebra Volume II*, Van Nostrand, Princeton, 1960.

Editorial Information

To be published in the *Memoirs*, a paper must be correct, new, nontrivial, and significant. Further, it must be well written and of interest to a substantial number of mathematicians. Piecemeal results, such as an inconclusive step toward an unproved major theorem or a minor variation on a known result, are in general not acceptable for publication.

Papers appearing in *Memoirs* are generally at least 80 and not more than 200 published pages in length. Papers less than 80 or more than 200 published pages require the approval of the Managing Editor of the Transactions/Memoirs Editorial Board.

As of July 31, 2007, the backlog for this journal was approximately 15 volumes. This estimate is the result of dividing the number of manuscripts for this journal in the Providence office that have not yet gone to the printer on the above date by the average number of monographs per volume over the previous twelve months, reduced by the number of volumes published in four months (the time necessary for preparing a volume for the printer). (There are 6 volumes per year, each usually containing at least 4 numbers.)

A Consent to Publish and Copyright Agreement is required before a paper will be published in the *Memoirs*. After a paper is accepted for publication, the Providence office will send a Consent to Publish and Copyright Agreement to all authors of the paper. By submitting a paper to the *Memoirs*, authors certify that the results have not been submitted to nor are they under consideration for publication by another journal, conference proceedings, or similar publication.

Information for Authors

Memoirs are printed from camera copy fully prepared by the author. This means that the finished book will look exactly like the copy submitted.

Initial submission. The AMS uses Centralized Manuscript Processing for initial submissions. Authors should submit a PDF file using the Initial Manuscript Submission form found at `www.ams.org/cgi-bin/peertrack/submission.pl`, or send one copy of the manuscript to the following address: Centralized Manuscript Processing, MEMOIRS OF THE AMS, 201 Charles Street, Providence, RI 02904-2294 USA. If a paper copy is being forwarded to the AMS, indicate that it is for it Memoirs and include the name of the corresponding author, contact information such as email address or mailing address, and the name of an appropriate Editor to review the paper (see the list of Editors below).

The paper must contain a *descriptive title* and an *abstract* that summarizes the article in language suitable for workers in the general field (algebra, analysis, etc.). The *descriptive title* should be short, but informative; useless or vague phrases such as "some remarks about" or "concerning" should be avoided. The *abstract* should be at least one complete sentence, and at most 300 words. Included with the footnotes to the paper should be the 2000 *Mathematics Subject Classification* representing the primary and secondary subjects of the article. The classifications are accessible from `www.ams.org/msc/`. The list of classifications is also available in print starting with the 1999 annual index of *Mathematical Reviews*. The Mathematics Subject Classification footnote may be followed by a list of *key words and phrases* describing the subject matter of the article and taken from it. Journal abbreviations used in bibliographies are listed in the latest *Mathematical Reviews* annual index. The series abbreviations are also accessible from `www.ams.org/publications/`. To help in preparing and verifying references, the AMS offers MR Lookup, a Reference Tool for Linking, at `www.ams.org/mrlookup/`.

Electronically prepared manuscripts. The AMS encourages electronically prepared manuscripts, with a strong preference for $\mathcal{A}_{\mathcal{M}}\mathcal{S}$-LaTeX. To this end, the Society has prepared $\mathcal{A}_{\mathcal{M}}\mathcal{S}$-LaTeX author packages for each AMS publication. Author packages include instructions for preparing electronic manuscripts, samples, and a style file that generates the particular design specifications of that publication series. Though $\mathcal{A}_\mathcal{M}\mathcal{S}$-LaTeX is the highly preferred format of TeX, author packages are also available in $\mathcal{A}_\mathcal{M}\mathcal{S}$-TeX.

Authors may retrieve an author package from the AMS website starting from `www.ams.org/tex/` or via FTP to `ftp.ams.org` (login as `anonymous`, enter username as password, and type `cd pub/author-info`). The *AMS Author Handbook* and the *Instruction Manual* are available in PDF format following the author packages link from `www.ams.org/tex/`. The author package can also be obtained free of charge by sending email to `tech-support@ams.org` (Internet) or from the Publication Division, American Mathematical Society, 201 Charles St., Providence, RI 02904-2294, USA. When requesting an author package, please specify $\mathcal{A}_\mathcal{M}\mathcal{S}$-LaTeX or $\mathcal{A}_\mathcal{M}\mathcal{S}$-TeX and the publication in which your paper will appear. Please be sure to include your complete mailing address.

After acceptance. The final version of the electronic file should be sent to the Providence office (this includes any TeX source file, any graphics files, and the DVI or PostScript file) immediately after the paper has been accepted for publication.

Before sending the source file, be sure you have proofread your paper carefully. The files you send must be the EXACT files used to generate the proof copy that was accepted for publication. For all publications, authors are required to send a printed copy of their paper, which exactly matches the copy approved for publication, along with any graphics that will appear in the paper.

Accepted electronically prepared files can be submitted via the web at `www.ams.org/submit-book-journal/`, sent via FTP, or sent on CD-Rom or diskette to the Electronic Prepress Department, American Mathematical Society, 201 Charles Street, Providence, RI 02904-2294 USA. TeX source files, DVI files, and PostScript files can be transferred over the Internet by FTP to the Internet node `ftp.ams.org` (130.44.1.100). When sending a manuscript electronically via CD-Rom or diskette, please be sure to include a message identifying the paper as a Memoir.

Electronically prepared manuscripts can also be sent via email to `pub-submit@ams.org` (Internet). In order to send files via email, they must be encoded properly. (DVI files are binary and PostScript files tend to be very large.)

Electronic graphics. Comprehensive instructions on preparing graphics are available at `www.ams.org/jourhtml/`. A few of the major requirements are given here.

Submit files for graphics as EPS (Encapsulated PostScript) files. This includes graphics originated via a graphics application as well as scanned photographs or other computer-generated images. If this is not possible, TIFF files are acceptable as long as they can be opened in Adobe Photoshop or Illustrator. No matter what method was used to produce the graphic, it is necessary to provide a paper copy to the AMS.

Authors using graphics packages for the creation of electronic art should also avoid the use of any lines thinner than 0.5 points in width. Many graphics packages allow the user to specify a "hairline" for a very thin line. Hairlines often look acceptable when proofed on a typical laser printer. However, when produced on a high-resolution laser imagesetter, hairlines become nearly invisible and will be lost entirely in the final printing process.

Screens should be set to values between 15% and 85%. Screens which fall outside of this range are too light or too dark to print correctly. Variations of screens within a graphic should be no less than 10%.

Inquiries. Any inquiries concerning a paper that has been accepted for publication should be sent to `memo-query@ams.org` or directly to the Electronic Prepress Department, American Mathematical Society, 201 Charles St., Providence, RI 02904-2294 USA.

Editors

This journal is designed particularly for long research papers, normally at least 80 pages in length, and groups of cognate papers in pure and applied mathematics. Papers intended for publication in the *Memoirs* should be addressed to one of the following editors. The AMS uses Centralized Manuscript Processing for initial submissions to AMS journals. Authors should follow instructions listed on the Initial Submission page found at `www.ams.org/memo/memosubmit.html`.

Algebra to ALEXANDER KLESHCHEV, Department of Mathematics, University of Oregon, Eugene, OR 97403-1222; email: `ams@noether.uoregon.edu`

Algebraic geometry and its application to MINA TEICHER, Emmy Noether Research Institute for Mathematics, Bar-Ilan University, Ramat-Gan 52900, Israel; email: `teicher@macs.biu.ac.il`

Algebraic geometry to DAN ABRAMOVICH, Department of Mathematics, Brown University, Box 1917, Providence, RI 02912; email: `amsedit@math.brown.edu`

Algebraic number theory to V. KUMAR MURTY, Department of Mathematics, University of Toronto, 100 St. George Street, Toronto, ON M5S 1A1, Canada; email: `murty@math.toronto.edu`

Algebraic topology to ALEJANDRO ADEM, Department of Mathematics, University of British Columbia, Room 121, 1984 Mathematics Road, Vancouver, British Columbia, Canada V6T 1Z2; email: `adem@math.ubc.ca`

Combinatorics to JOHN R. STEMBRIDGE, Department of Mathematics, University of Michigan, Ann Arbor, Michigan 48109-1109; email: `FRS@umich.edu`

Complex analysis and harmonic analysis to ALEXANDER NAGEL, Department of Mathematics, University of Wisconsin, 480 Lincoln Drive, Madison, WI 53706-1313; email: `nagel@math.wisc.edu`

Differential geometry and global analysis to LISA C. JEFFREY, Department of Mathematics, University of Toronto, 100 St. George St., Toronto, ON Canada M5S 3G3; email: `jeffrey@math.toronto.edu`

Dynamical systems and ergodic theory to AMIE WILKINSON, Department of Mathematics, Northwestern University, 2033 Sheridan Road, Evanston, IL 60208-2730; email: `transactions@math.northwestern.edu`

Functional analysis and operator algebras to DIMITRI SHLYAKHTENKO, Department of Mathematics, University of California, Los Angeles, CA 90095; email: `shlyakht@math.ucla.edu`

Geometric analysis to WILLIAM P. MINICOZZI II, Department of Mathematics, Johns Hopkins University, 3400 N. Charles St., Baltimore, MD 21218; email: `trans@math.jhu.edu`

Geometric analysis to MLADEN BESTVINA, Department of Mathematics, University of Utah, 155 South 1400 East, JWB 233, Salt Lake City, Utah 84112-0090; email: `bestvina@math.utah.edu`

Harmonic analysis, representation theory, and Lie theory to ROBERT J. STANTON, Department of Mathematics, The Ohio State University, 231 West 18th Avenue, Columbus, OH 43210-1174; email: `stanton@math.ohio-state.edu`

Logic to STEFFEN LEMPP, Department of Mathematics, University of Wisconsin, 480 Lincoln Drive, Madison, Wisconsin 53706-1388; email: `lempp@math.wisc.edu`

Partial differential equations to GUSTAVO PONCE, Department of Mathematics, South Hall, Room 6607, University of California, Santa Barbara, CA 93106; email: `ponce@math.ucsb.edu`

Partial differential equations and dynamical systems to PETER POLACIK, School of Mathematics, University of Minnesota, Minneapolis, MN 55455; email: `polacik@math.umn.edu`

Probability and statistics to KRZYSZTOF BURDZY, Department of Mathematics, University of Washington, Box 354350, Seattle, Washington 98195-4350; email: `burdzy@math.washington.edu`

Real analysis and partial differential equations to DANIEL TATARU, Department of Mathematics, University of California, Berkeley, Berkeley, CA 94720; email: `tataru@math.berkeley.edu`

All other communications to the editors should be addressed to the Managing Editor, ROBERT GURALNICK, Department of Mathematics, University of Southern California, Los Angeles, CA 90089-1113; email: `guralnic@math.usc.edu`.

Titles in This Series

890 **Steven Dale Cutkosky,** Toroidalization of dominant morphisms of 3-folds, 2007

889 **Michael Sever,** Distribution solutions of nonlinear systems of conservation laws, 2007

888 **Roger Chalkley,** Basic global relative invariants for nonlinear differential equations, 2007

887 **Charlotte Wahl,** Noncommutative Maslov index and eta-forms, 2007

886 **Robert M. Guralnick and John Shareshian,** Symmetric and alternating groups as monodromy groups of Riemann surfaces I: Generic covers and covers with many branch points, 2007

885 **Jae Choon Cha,** The structure of the rational concordance group of knots, 2007

884 **Dan Haran, Moshe Jarden, and Florian Pop,** Projective group structures as absolute Galois structures with block approximation, 2007

883 **Apostolos Beligiannis and Idun Reiten,** Homological and homotopical aspects of torsion theories, 2007

882 **Lars Inge Hedberg and Yuri Netrusov,** An axiomatic approach to function spaces, spec tral synthesis and Luzin approximation, 2007

881 **Tao Mei,** Operator valued Hardy spaces, 2007

880 **Bruce C. Berndt, Geumlan Choi, Youn-Seo Choi, Heekyoung Hahn, Boon Pin Yeap, Ae Ja Yee, Hamza Yesilyurt, and Jinhee Yi,** Ramanujan's forty identities for Rogers-Ramanujan functions, 2007

879 **O. García-Prada, P. B. Gothen, and V. Muñoz,** Betti numbers of the moduli space of rank 3 parabolic Higgs bundles, 2007

878 **Alessandra Celletti and Luigi Chierchia,** KAM stability and celestial mechanics, 2007

877 **María J. Carro, José A. Raposo, and Javier Soria,** Recent developments in the theory of Lorentz spaces and weighted inequalities, 2007

876 **Gabriel Debs and Jean Saint Raymond,** Borel liftings of Borel sets: Some decidable and undecidable statements, 2007

875 **C. Krattenthaler and T. Rivoal,** Hypergéométrie et fonction zêta de Riemann, 2007

874 **Sonia Natale,** Semisolvability of semisimple Hopf algebras of low dimension, 2007

873 **A. J. Duncan,** Exponential genus problems in one-relator products of groups, 2007

872 **Anthony V. Geramita, Tadahito Harima, Juan C. Migliore, and Yong Su Shin,** The Hilbert function of a level algebra, 2007

871 **Pascal Auscher,** On necessary and sufficient conditions for L^p-estimates of Riesz transforms associated to elliptic operators on $\mathbb{R}^n$ and related estimates, 2007

870 **Takuro Mochizuki,** Asymptotic behaviour of tame harmonic bundles and an application to pure twistor D-modules, Part 2, 2007

869 **Takuro Mochizuki,** Asymptotic behaviour of tame harmonic bundles and an application to pure twistor D-modules, Part 1, 2007

868 **Gelu Popescu,** Entropy and multivariable interpolation, 2006

867 **Vilmos Totik,** Metric properties of harmonic measures, 2006

866 **William Craig,** Semigroups underlying first-order logic, 2006

865 **Nathanial P. Brown,** Invariant means and finite representation theory of $C*$-algebras, 2006

864 **John M. Lee,** Fredholm operators and Einstein metrics on conformally compact manifolds, 2006